MATERIALS UNDER EXTREME CONDITIONS

Molecular Crystals at High Pressure

MATERIALS UNDER EXTREME CONDITIONS

Molecular Crystals at High Pressure

Roberto Bini • Vincenzo Schettino

University of Florence, Italy

ICP

Imperial College Press

Published by

Imperial College Press
57 Shelton Street
Covent Garden
London WC2H 9HE

Distributed by

World Scientific Publishing Co. Pte. Ltd.

5 Toh Tuck Link, Singapore 596224

USA office: 27 Warren Street, Suite 401-402, Hackensack, NJ 07601

UK office: 57 Shelton Street, Covent Garden, London WC2H 9HE

Library of Congress Cataloging-in-Publication Data
Schettino, Vincenzo.
Materials under extreme conditions : molecular crystals at high pressure / Vincenzo Schettino, Roberto Bini, University of Florence, Italy.
pages cm
Includes bibliographical references and index.
ISBN 978-1-84816-305-8 (hardcover : alk. paper)
1. High pressure chemistry. 2. Materials at high pressures. 3. Molecular crystals. I. Bini, Roberto (Chemist) II. Title.
QD538.S34 2014
548'.842--dc23

2013038985

British Library Cataloguing-in-Publication Data
A catalogue record for this book is available from the British Library

In-house Editor: Darilyn Yap

Printed in Singapore by B & Jo Enterprise Pte Ltd

Dedication

to Margherita, Martina and Stefania,
to Giulia and Anna Margherita

Preface

The pressure variable and its effects on the structure and properties of molecules and materials have played a role of primary importance in the development of chemistry and to implement important chemical processes. For example, one could mention the high-pressure synthesis of ammonia from the elements as a starting point for the production of nitrogen-based fertilizers. In more recent times we could mention the use of supercritical fluids, like water and carbon dioxide, as friendly solvents in green chemistry processes. In these and in many other extensive applications of high-pressure techniques in chemistry, interest has been generally confined to gas phase, liquids and solutions with pressures not exceeding 1 GPa. Higher pressure regimes, above 1 GPa, where most materials are in the solid crystalline state, have prevalently been the province of solid state physicists, material scientists, geochemists and geophysicists, and planetary scientists. This can be surprising considering the large variety of chemical phenomena observed in this pressure range. As a matter of fact, the study of the behaviour of materials at very high pressures significantly overlaps with findings and principles of fields like solid state chemistry, crystal engineering and supramolecular chemistry. From a fundamental point of view, application of high pressures to a crystalline array of molecules produces a fine-tuning of the intermolecular distances, orientations and conformations, parameters eligible as potential reaction coordinates. Therefore, high-pressure experiments offer unique possibilities to study chemical reaction mechanisms in elemental materials in a greatly simplified chemical environment. The molecular confinement realized at high pressures produces a substantial overlap of the electron distributions, disclosing new information on fundamental atomic and molecular properties. Indeed, the behaviour of molecules at high pressures can be quite different from what we are used to in normal laboratory practice, to such an extent that it has been suggested that the periodic table of the elements, the basic tool of a chemist, should be revisited as if the high pressure was adding a third dimension to the table.

The purpose of this book has been to summarize and collect the nevertheless large body of chemical reactions that have been reported at very high pressures on solid materials. The attention has been focused on the physical chemistry of molecular crystals that are particularly sensitive to an increase of pressure since the weakness of the intermolecular interactions makes them highly compressible. Attempts have consequently been made to outline similarities of chemical behaviour at high pressures depending on similarities of the molecular structure and of the intermolecular interactions. It actually turns out that high-pressure chemical reactions in the solid state are highly collective or aggregate events. It is the ambition of the authors that this tentative comprehensive presentation of ultrahigh-pressure chemical phenomena can stimulate some novel interest in high-pressure science in the chemical community.

Foreword

In less than 20 years the study of the properties of molecular crystals under extreme conditions of pressure and temperature, has made formidable progress, thanks to the use of several diffraction and spectroscopic methods. These techniques, in combination with the diamond anvil cells, allowed us to deal with samples of very small dimensions and to exploit their particular sensitivity to variations of interatomic distances and of bond lengths and angles, the main molecular factors affected by variations of the unit cell dimensions in molecular crystals.

Changes of temperature and pressure give rise to important physical effects, inducing phase transitions and even opening the road to new unknown crystal structures stable only under extreme conditions. These changes, however, affect not only the physical, but also the chemical properties of molecular crystals or liquids, since they directly act on the conformational structure of the molecular units, giving rise to a redistribution of their internal energy. This is conveniently analysed considering that external intensive parameters, such as temperature and pressure, shape the form of the inter- and intramolecular potentials reducing both the interatomic and intermolecular distances. Volume contraction thus forces the molecules to explore higher regions of the repulsive parts of the potential to reach new energy minima. This also increases the relative weight of higher terms in the parametric expansion of the intermolecular potential, inducing important changes in its anharmonic part.

Chemical reactivity changes due to the coupled effect of pressure and temperature are actually the main concern of this book, to my knowledge the first book dealing in-depth with 'details of all aspects of chemical reactivity under extreme conditions. This book, which specifically addresses the study of the position and intensity of vibrational bands in molecular crystals under very high pressures and low temperatures, originates from the merging of the theoretical competencies of Prof. Vincenzo Schettino, a well-recognized expert of the theoretical treatment of the lattice dynamics of molecular crystals, with the experimental skills and competencies of his former student Roberto Bini, today one of the world's leading experts in high-pressure experiments.

The number of molecular crystals discussed in this book, mostly studied in the Florence laboratory where the authors work, either to individuate the full range of their phase transitions or to follow in time their reactivity, is impressive. Some highlights can give the reader a sufficiently detailed image of the approach that characterizes the research reported in the book and can underline the high level of sophistication reached in the experimental collection of data, and in the theoretical interpretation of the results, either in structural or in reaction rate experiments. These include detailed investigations of the crystal structure and phase transitions from simple to rather complex molecules, as well as a documented analysis of their reaction rates in different types of reactions. As a simple but impressive example of the deep interest of the authors in the application of high-pressure measurements to basic theoretical and experimental problems in the field, we consider worth reporting the collection of data concerning the structural changes occurring in crystals of the simplest molecular systems, diatomic molecules, that configure the basic experimental and theoretical procedures when extended to larger molecular systems.

In the case of solid nitrogen, the phase diagram shows a large variety of different structural arrangements in the unit cell and the increase of pressure, starting from the well-known Pa3 structure stable at normal pressure, leads progressively to the formation of new phases in which the quadrupole-quadrupole interactions dominant at low pressure are progressively substituted by a full reorganization of the triple bond, ultimately leading to polymeric non-molecular structures and eventually to a conductive polymorphic material. Of particular interest is the case of solid oxygen. Molecular oxygen has an open shell configuration with two unpaired electrons and possesses a magnetic moment that adds magnetic interactions to the intermolecular potential, making it basically different from that of nitrogen. At high pressure a beautiful strongly red-coloured phase has been obtained in which the oxygen molecules are coupled by π-π^* interactions in prismatic $(O_2)_4$ in an orthorhombic unit cell, a structure consistent with the vibrational spectra and confirmed by *ab initio* density functional simulations. Even more impressive is the occurrence of as many as five different structural phases occurring in the range up to 40 GPa for solid carbon dioxide, all retaining a centrosymmetric structure. Above 2000 K solid carbon dioxide decomposes into amorphous carbon including single diamond-like crystallites. At pressures higher than 40 GPa carbon dioxide, even at temperatures lower than 2000 K, gives rise to a number of polymeric forms with structures similar to those found in cristobalite, stishovite, coesite and amorphous silica analogues.

At extremely high pressures inducing an increase of the electronic kinetic energy, molecular crystals respond either by the formation of metallic states, by the decomposition into single atoms or into ionic structures, more adequate to the extremely high packing. Metallization of solid hydrogen, predicted as early as 1935, has been experimentally claimed only very recently and a number of metallic systems including oxygen, sulphur, iodine, bromine, hydrogen sulphide and even some four or five atomic molecules, have been realized in several laboratories.

More important for the theoretical understanding of the reaction rate dependence on variations of the structural conformation is the extension of the research to irreversible reactions induced by pressure such as the polymerization reactions of unsaturated hydrocarbons as well as of aromatic systems.

At normal pressure irreversible polymerization processes can be activated either by photons or by suitable catalysts. Pressure represents, however, an alternative activation technique to allow the molecules to overcome the energy barrier that separates the reactants from the final polymeric products. High pressure therefore offers a variety of possible reaction pathways not feasible under normal conditions since, in the multidimensional potential surface, it can select specific channels not occurring at normal pressure. The number of these new reaction channels is further amplified by the synergy of the simultaneous application of high pressure and photoactivation procedures, that allows the system to explore additional reaction pathways not accessible without jumping over potential barriers in the multidimensional energy surface. The collective nature of lattice phonon propagation in crystal can further assist in propagating energy in the crystals in order to reach reaction active sites such as defects or dislocations that can help in overcoming energy potential barriers.

One of the most interesting polymerization reactions is that of acetylene. In the crystal, the acetylene molecules are lined up with an optimal orientation for a topochemical addition to *trans* polyacetylene and in fact at high pressure and in the absence of irradiation, pure *trans* polyacetylene is obtained with complete delocalization of the π electron system. Irradiation with laser light gives origin to the production of gauche kinks whose relative abundance increases by increasing the pressure.

Very important for practical applications is the polymerization of ethylene to produce both *cis* and *trans* polyethylene polymers in which the ratio of the two forms depends sensibly on the pressure.

More complex is the behaviour under pressure of butadiene that, depending upon the reaction conditions, can produce either polybutadiene through an ionic mechanism which favours the growth of a linear polymer or even vinylcyclohexene through a multistep dimerization procedure.

Polymerizations of aromatic hydrocarbons build a series of very complex reactions, owing to the weakly intermolecular forces binding the molecules in the crystal structures, giving rise to a large number of different crystalline phases stable under high pressure. The phase diagrams of several aromatic and hetero-aromatic hydrocarbons including even furane, thiophene, indole and similar molecules have been studied either to follow their polymerization or their ring opening reactions. Here we shall only consider the most studied compound of the series, crystalline benzene, just to give an idea of the broad range of interests of the authors of the book. In the case of benzene at extreme high pressure, amorphous hydrogenated carbon material is obtained.

The large variety of interests that characterizes the present book makes it an invaluable instrument for young researchers interested in the behaviour of matter under extreme conditions and an indispensable support to those, chemists, physicists, as well as geologists and mineralogists, interested in the study of crystalline phases and in the comprehension of the structural parameters and intermolecular potentials that make molecular crystals a unique form of solid matter.

Salvatore Califano
Professor Emeritus
University of Florence

Acknowledgments

We would like to thank Professor Salvatore Califano for carefully reading the manuscript and for his encouragement and suggestions.

We would like to express our special thanks to the researchers, postdocs and students that during the last ten years followed, with an enthusiastic daily dedication, the research topics at the basis of the present book. A special acknowledgement goes to M. Ceppatelli, M. Citroni, F.A. Gorelli and M. Santoro who have been protagonists of the scientific and technical achievements which constitute the backbone of this book and were also able to create a unique human environment.

Contents

Preface vii

Foreword ix

Acknowledgments xiii

1. Introduction 1

2. Historical Survey 15

3. Elasticity and Equation of State 23

3.1 Stress and strain 24
3.2 Elasticity and anisotropy in molecular crystals 31
3.2.1 Elastic constants 31
3.2.2 Temperature and pressure dependence of the elastic constants 47
3.2.3 Single crystals and polycrystals 53
3.2.4 Variation of crystal parameters with pressure . . . 55
3.3 Parametric equations of state 70

4. High-pressure Technical Survey 79

4.1 The piston-cylinder device 80
4.1.1 Large volume piston-cylinder apparatus 85
4.2 The opposed anvil devices 86
4.3 Multi-anvil devices 89
4.4 The diamond anvil cell 93
4.4.1 Diamonds 102

4.4.2 Gasket 109
4.4.3 Loading techniques 113
4.4.4 Compression media 116
4.5 High- and low-temperature techniques 118
4.5.1 Low-temperature devices 119
4.5.2 Resistive heating 121
4.5.3 Laser heating 123
4.6 Pressure measurement 128
4.7 Probing techniques based on electromagnetic radiation . . 140
4.7.1 Optical spectroscopy 141
4.7.2 X-ray diffraction 149
4.7.3 Neutron diffraction 155
4.7.4 Nuclear magnetic resonance 157

5. Principles of Chemical Reactivity Under Pressure 159

5.1 Pressure effects on chemical equilibria 160
5.2 Pressure effects on reaction rates 165
5.3 Environmental effects at high pressure 177
5.4 Effects of high pressure on the electronic structure 185

6. Chemical Reactions in Molecular Crystals 193

6.1 Reversible reactions 196
6.1.1 Solid nitrogen at high pressure: the archetypal energetic material 196
6.1.2 Red oxygen 201
6.1.3 Carbon dioxide: a multiform solid 203
6.1.4 Formic acid 208
6.1.5 Sulphur. Polymeric and molecular phases 209
6.1.6 Symmetry breaking and ionization at high pressures 211
6.1.7 Metallization at high pressures 214
6.2 Irreversible reactions 217
6.2.1 Unsaturated hydrocarbons 218
6.2.2 Aromatics and heteroaromatics 225
6.2.3 Miscellanea 232
6.2.4 Energetic materials 241
6.2.5 Photochemistry at high pressures 245

Bibliography 261

Appendix A 327

Index 349

Chapter 1

Introduction

All physical and chemical transformations are controlled by the thermodynamical variables of temperature, pressure and chemical potential [1]. Acting on these external or internal parameters it is possible to affect the equilibrium composition of the systems and the kinetics of the transformation. The practical and theoretical implications of controlling the physical and chemical processes are evident. In the interest of the present book we shall confine the discussion to the effects of temperature and pressure and, without much loss of generality, we shall refer mainly to chemical processes.

Changing the temperature of a system by exchange of heat with the environment has a double effect. On one side there will be a change of the internal energy of the system, with the excess energy redistributed among the internal vibrational degrees of freedom and the translational and rotational motions of the molecules, the latter two corresponding to acoustic and librational phonons in crystalline systems. When the system is composed of complex or flexible molecules a conformational reorganization can also be induced by a temperature change. A redistribution of the excess internal energy in the electronic energy levels is not of much interest for the systems and the conditions considered in the present book. A change of temperature also produces a change of volume of the system by the effect of the anharmonicity of the interaction potential. On the contrary, the exchange of mechanical energy by an isothermal change of the external pressure applied to the system will result in a mere change of volume. It is therefore seen that changes of temperature or pressure both have the common primary effect of a volume dilatation or compression of the system.

Decreasing the volume of the system by lowering the temperature or by increasing the pressure produces a shortening of the interatomic and intermolecular distances. The effect of a volume contraction on the structure of the system can be qualitatively viewed in different converging ways. Our attention will be focused on molecular systems which, even at moderate pressures, are found in the crystalline state. As usually, the intermolecular interaction potential can be decomposed in an attractive and in a repulsive component [2–6]. In molecular crystals and in the absence of, or neglecting, electrostatic interactions that have a slower decay with separation, the attractive part of the interaction potential mainly arises from London dispersion forces and typically depends on the inverse sixth power of the intermolecular distance r. The repulsive part of the interaction potential substantially arises from electron overlap and from electron-electron correlation and typically depends on the inverse 12th power of the intermolecular separation. This is the form of the Lennard-Jones potential [5–10],

$$E = 4\epsilon \left[-\left(\frac{r}{\sigma}\right)^{-6} + \left(\frac{r}{\sigma}\right)^{-12} \right] \tag{1.1}$$

where ϵ is the value of the energy minimum at equilibrium and σ is the finite intermolecular distance at which the repulsive and attractive terms are equal (and opposite in sign).

In Figure 1.1 the Lennard-Jones potential for argon is shown with the following values of the parameters $\frac{\epsilon}{k} = 119.8$ K and $\sigma = 3.405$ Å [2]. It can easily be seen that decreasing the intermolecular distance below the equilibrium configuration, the repulsive energy rapidly increases and otherwise inaccessible regions of the potential energy surface are explored. Therefore, the system enters a region of possible electronic instability and a structural reorganization can be expected. We must expand our considerations beyond the description of the system as governed by purely central forces, as it is actually the case for the molecular crystals which are the main object of our treatment and, as it will be discussed in a following chapter, can be highly anisotropic as a consequence of both the intrinsic molecular shape and the type of intermolecular interactions (van der Waals, electrostatic, hydrogen bonding). The search for a new energy minimum can be pursued by a re-orientation of the molecules in the crystal structure, by phase transitions of different types [10] and, at higher reduced volumes, by a reorganization of the general bonding structure through chemical reactions. The effects of a volume contraction on the electronic structure can best be considered from the point of view of the density functional theory [11]. In the limit of the electron gas behaviour the repulsive electron-electron interaction varies with density ρ as $\rho^{2/3}$ (or in the isotropic case as r^{-2}) while the attractive electron-nuclei interaction varies as $\rho^{1/3}$ (or as r^{-1} in the isotropic case).

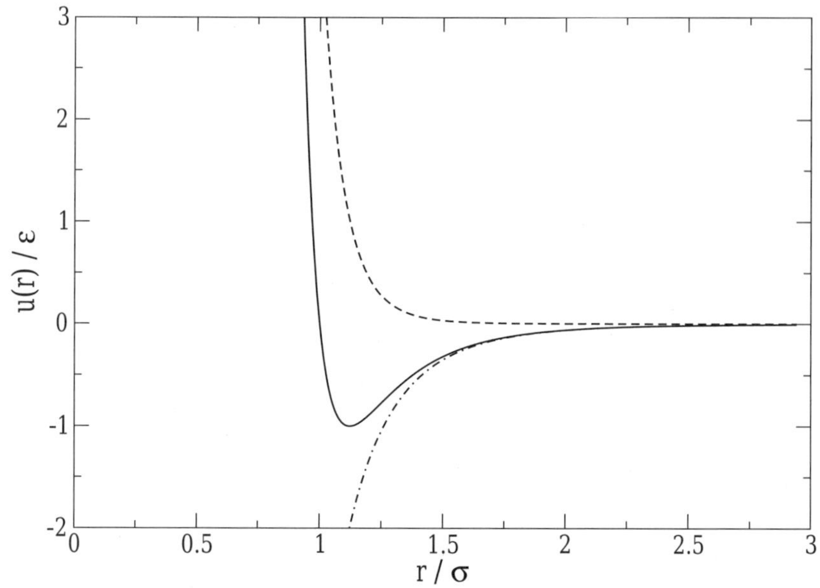

Fig. 1.1 Lennard-Jones potential of argon showing separately the attractive (dash dot) and repulsive (dash) contributions.

Also within this approach it is seen that at reduced volumes the repulsive electron-electron energy increases more rapidly and finally becomes equal to the attractive terms, thus leading to possible transformations and chemical reactions allowing the system to reach a new energy minimum.

The types of transformations that can be induced in a molecular system by a decrease of volume, and specifically by increasing the pressure, have been discussed by Drickamer and Frank [12, 13] who consider the four types of transformation listed in Table 1.1 occurring at increasing volume contraction. The boundaries between the different types of transitions considered in Table 1.1 are rather loosely defined. To some extent a change of the electronic configuration may occur in all cases of volume reduction, but the extent of reorganization can be substantial only at higher compressions when the electron overlap can lead to changes of configuration interaction resulting in new ground and excited states. This situation is of specific interest in the present work which has the purpose of discussing molecular crystals and other molecular aggregates that, when subjected to high pressures, exhibit a chemical behaviour associated with structural changes that can be drastically different from the known or expected behaviour under normal conditions.

Table 1.1 Transitions occurring in molecular systems at increasing volume contraction

	Type
I	Atomic rearrangements (e.g. reorientation of molecules)
II	Polymorphism, Phase transitions
III	Discontinuous electronic transitions, Structural changes
IV	Continuous electronic transitions, New reactive ground state

Following Drickamer [12, 13], the changes of the electronic structure occurring at reduced volumes can be schematically described considering a two-level system where the ground and excited states can be considered as representative of the HOMO (Highest Occupied Molecular Orbital) and LUMO (Lowest Unoccupied Molecular Orbital) molecular orbitals in a molecular system. The changes are represented in Figure 1.2 where, on the left side, the potential energy curves at the initial volume V_0 are shown while, on the right, the energy curves at the reduced volume V are depicted. For the present purposes the electronic structure can be characterized by the vertical (Franck–Condon) excitation energies E_0 and E_0', by the thermal excitation energies E_{th} and E_{th}' and by the differences between the ground and excited state geometries $\Delta = R_e - R_0$ and $\Delta' = R_e' - R_0'$. The basic structural change $R_0' - R_0$ is also of great relevance. The diagram of Figure 1.2 is oversimplified. In general, we shall be dealing with multidimensional energy surfaces. In addition, it has been assumed that the energy surfaces are purely quadratic thus neglecting the important role of

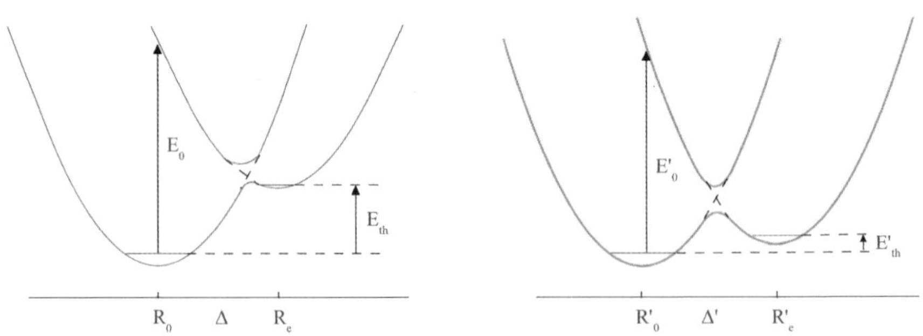

Fig. 1.2 Evolution of the electronic structure as a function of the volume of the system. Left: electronic structure at the initial volume V_0; right: electronic structure at the reduced volume V.

anharmonicity. More elaborate models of the pressure effects on electronic states have been discussed [14–18]. Nevertheless, the experimentally observed essential features of the problem can be captured from the figure. It has been reported that the frequency of the electronic transitions shows a small red shift on pressurization (i.e., $E_0' < E_0$) and, in addition, there is a general broadening of the absorption (and emission) lines. Nevertheless, as it can be appreciated from the figure, the thermal excitation energy can considerably lower and a significant population of the excited state can occur even at room temperature. This gives rise at high pressures to a reduction (or even a cancellation) of the HOMO-LUMO gap as it is actually represented in the figure and can have important consequences on the high-pressure mechanisms of chemical reactions as will be discussed in the following chapters. Another important feature emerging from the figure is that at highly reduced volumes the structures of the ground and excited states could become very similar, if not coincident, and that some energy barriers can reduce to a significant extent.

All the above considerations apply to any volume reduction of the system independent of its origin. However, there are in practice significant differences in the effects that can be induced by a temperature decrease or by an increase of pressure (or by a change in composition). To illustrate the point, let us consider a typical molecular crystal composed of rigid molecular units held together by weak van der Waals interactions like the C_{60} fullerene crystal [19]. Figure 1.3 shows the evolution of the cubic unit cell parameter as a function of temperature from 0 to 360 K. This is a pretty large temperature interval for most molecular crystals before melting or thermal decomposition. It can be seen from the figure that in this ample temperature interval, a volume variation of only 3–4% occurs despite the fact that molecular crystals are rather soft in view of the weakness of the intermolecular interactions. The same systems (molecular crystals) can be subjected, before decomposition, to a much larger volume variation by application of an external pressure. As an example in Figure 1.4 the isothermal equation of state of solid methane at 300 K is reported for a compression up to 210 GPa [20]. In the first instance, the figure demonstrates the extraordinary stability of methane to compression, a fact that is of relevance to the physics of the extraterrestrial giant planets Neptune and Uranus [21]. In the ample pressure range shown in the figure, methane is stable and experiences a unit cell volume variation from 590 to 250 A^3. At such a large volume contraction ($V/V_0 = 0.42$) the shortening of interatomic distances is substantial, the electronic structure is expected to

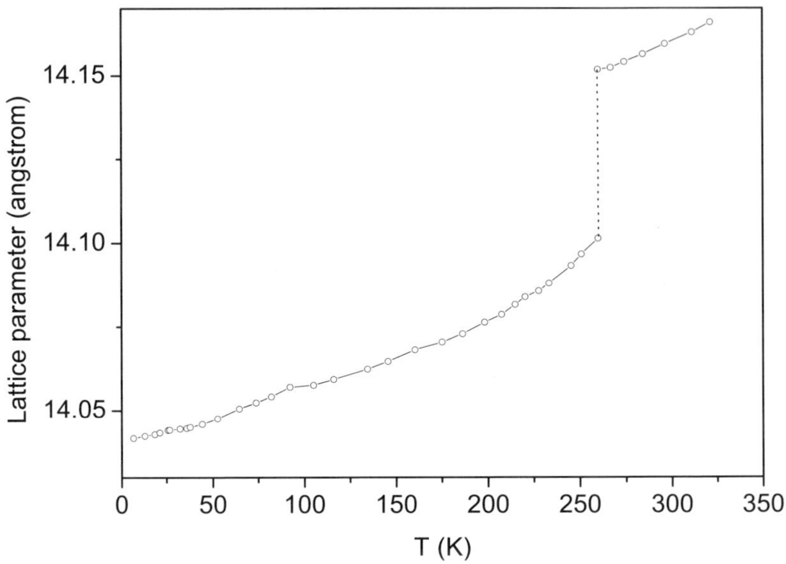

Fig. 1.3 Variation with temperature of the cubic unit cell parameter of fullerene (re-adapted from I. F. David, et al. *Europhys. Lett.* **18** (1992) 219, Copyright (1992), with permission from IOP).

change significantly and the chemical behaviour can be completely different than under normal conditions. It is therefore seen that acting on the pressure variable, volume variations can be explored in much greater detail.

The same point can be stated in a different way considering, for instance, the variation of the free energy induced in the CsI crystal by feasible variations of the temperature and external pressure [22]. This is shown in Figure 1.5. It can be seen that by tuning the pressure, a much more extended region of the free energy surface can actually be explored than by changing the temperature.

The Gibbs free energy increase occurring at high pressures is mainly due to the PV term of the free energy,

$$G = E + PV - TS. \tag{1.2}$$

As an example, in Table 1.2 the variation of the PV term of the free energy of some typical molecular crystals (methane [20], benzene [23], nitromethane [24] and acetaminophen [25]) subjected to pressurization in

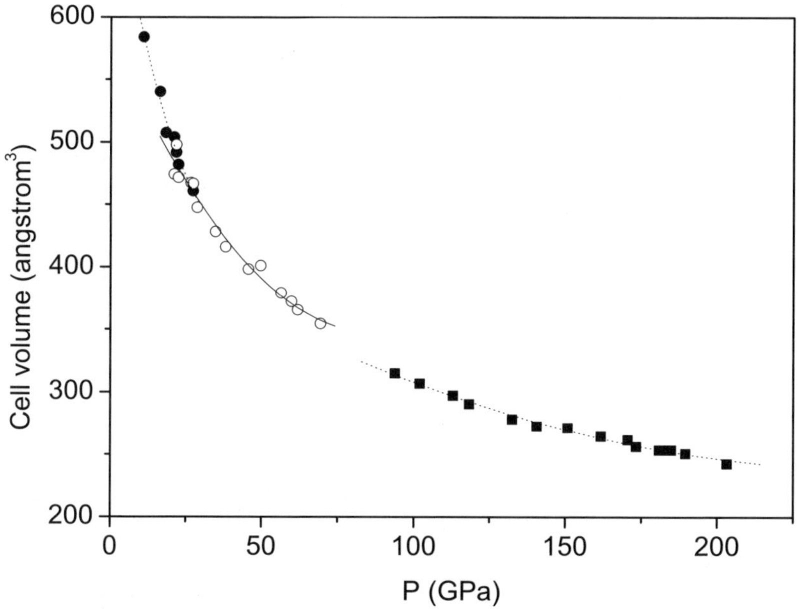

Fig. 1.4 Isothermal (300 K) equation of state of solid methane. The P-T stability range of the rhombohedral (filled dots), simple cubic (empty dots) and high pressure cubic (filled squares) phases are illustrated as reported in Ref. [20]. (Readapted from L. Sun, et al. *Chem. Phys. Lett.* **473** (2009) 7274, Copyright (2009), with permission from Elsevier).

Table 1.2 Free energy increase per unit cell of some molecular crystals at high pressure

Crystal	P_0 (GPa)	P (GPa)	Δ(PV)(eV)	Cell occupancy
Methane	12	202	222	4
Benzene	1	27	22	2
Nitrobenzene	2	27.5	27	2
Acetaminophen	1	4	11	2

different ranges is summarized. The reported variation of the PV term refers to the unit cell of the crystal and the occupancy of the unit cell is also reported in the table for reference. It can be seen that the free energy variation exceeds 10 eV per molecule, the typical dissociation energy of a chemical bond. A free energy increase of this kind necessarily brings about a redefinition of the electronic distribution in the molecules and therefore a different physical and chemical behaviour.

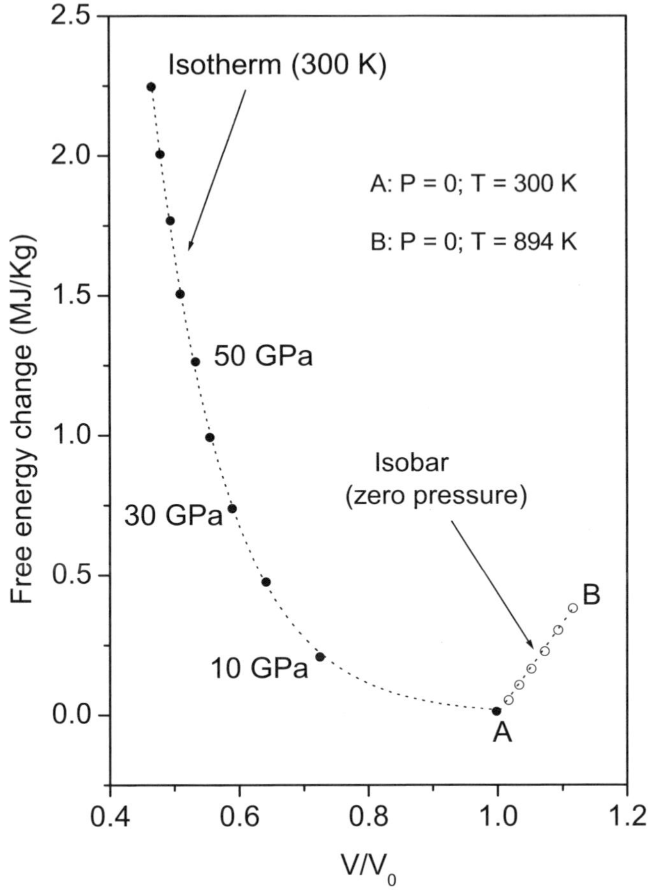

Fig. 1.5 Free energy variation of the CsI crystal changing pressure along the room temperature isotherm, and changing temperature along the zero pressure isobar [22].

The following set of rules of thumb describing the behaviour of elements, compounds and crystals under high pressures has been discussed by Prewitt and Downs [26]:

- On compression, the greatest distortion is observed between more weakly bound atoms
- Short bonds are the strongest and long bonds are the weakest
- On compression, bonds become more covalent
- On compression, the coordination number increases
- The oxygen atom is more compressible than the cations

Table 1.3 Schematic representation of the evolution of temperature and pressure in the Earth's interior

Temperature (K)	Depth (Km)	Pressure (GPa)
298	0	10^{-4}
1900	400	14
3000	2900	140
4500	5100	330
5000	6400	364

- Angle bending depends on the coordination number
- Oxygen-oxygen interactions are very important for compressibility
- On compression, structures rearrange in terms of closest packed atomic arrays
- At high pressure, elements behave like the heavier elements of the same group

Although these rules were specifically intended by the authors for mineral physics and chemistry they are of a more general, albeit qualitative, application. Some of these rules have been rediscussed in a review article by Grochala et al. [27] who also recast some of these rules in a different or more explicit form. For the purposes of the present work it is worth mentioning some of Grochala's rules and particularly rule D (stating that orbital symmetry or energy barriers may affect the reversibility of high-pressure reactions), rule I (stating that electron disproportionation and ionization or polarization can be competitive with straight close packing) and finally, rule J (stating that at high pressure virtual orbitals can be significantly occupied). It is worth remarking that in these rules the cooperativity of the phenomena occurring at high pressures in crystals is not evident. This aspect has, however, been brought to attention by Drickamer [28].

The importance of exploring the behaviour of matter under high pressures derives from the fact that the range of high pressures encountered in the universe is extremely ample, varying from the familiar 1 atm pressure at sea level on our planet to 10^7 atm at the Earth's centre, to 10^{15} atm at the centre of the Sun, to 10^{30} atm at the centre of a neutron star. In Table 1.3 the evolution of pressure and temperature with depth in the Earth is shown. From these data it is easily understood how a large variety of applications of high-pressure science and technology are available and have been exploited in geophysics and geochemistry and, more generally, in

Earth and planetary sciences [29]. In fact, following the pioneering work of Percy W. Bridgman [30], it is nowadays possible to reach, in the laboratory, pressures in the order of hundreds of gigapascals (GPa). An overview of some high-pressure techniques will be presented in a following chapter of this book.

The fields of application of high-pressure science of interest to chemists are quite extended. At the beginning of the last century it was realized that there are minerals which are stable or kinetically stable, although thermodynamically unstable, at normal pressure and temperature, but can only be synthesized at the high pressures and temperatures of the Earth's interior. This is the basis for the extensive research devoted to exploit the modified chemical properties to obtain recoverable novel materials of technological interest by chemical reactions at high pressures [31]. The most spectacular achievement in this field has been the preparation of artificial diamonds and other superhard materials [32–34].

High-pressure experiments are of great relevance in the biosciences and in the study of biomolecules [35–39]. In the first instance, it has been demonstrated that some living organisms can survive at high pressures and others (barophyls) can even develop better at high pressures. This is of importance, for instance, also in connection with the possible origin of life in the deep ocean or in other unfriendly environments [40]. The application of high-pressure methods in food science is a continuously developing field [41, 42]. The capability of high pressure to stabilize some specific polymorphs is a source of interest in the pharmaceutical industry [43, 44] as a tool to obtain species more amenable to processing and with facilitated absorption. From the point of view of basic science, by changing the pressure it is possible to finely tune the intermolecular distances and the intermolecular interactions in biological systems without chemically changing the environment (as by changing the pH or by substituting appropriate chemical groups) and is a source of novel information on the structure-function relationships.

Of paramount importance is the impact of high-pressure experiments on our basic knowledge of atomic behaviour and electronic structure. It has been discussed that at high pressures the atomic properties can change to such an extent that the periodic table of the elements (see Figure 1.6) should be revisited at high pressure or even that a new periodic table is needed in these conditions [31, 45]. There are many experimental observations that support this view. At normal pressure, elements with a metallic behaviour are found in an extended but well-defined section of the periodic table. At

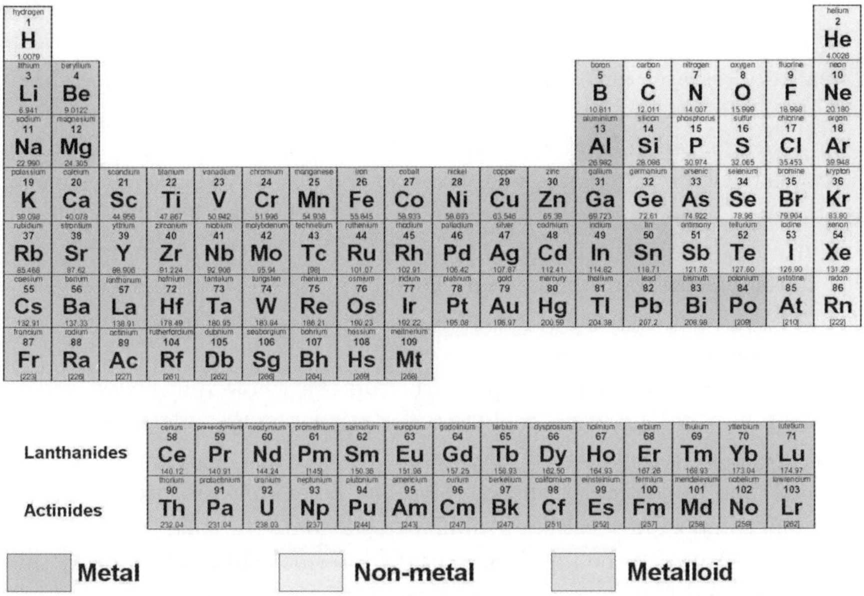

Fig. 1.6 The periodic table of the elements.

very high pressure, however, metallization has been observed for elements like oxygen [46, 47] and the halogens (iodine and bromine) [48–50] or the rare gases [51] and has also been predicted for hydrogen [52], although the experimental observation has been elusive so far in this latter case. Solid solutions of potassium and magnesium with iron have been observed at high pressures [53, 54]. This is a clear indication of the change of the s character of the outer electrons in alkali and alkaline earth elements in general. In several cases, and notably in the IV and V group of the periodic table, it has been found that at high pressures the light elements tend to behave like the heavier elements of the same group. This has been a source of unpredicted chemical behaviour of the low-Z elements leading, for instance, to the formation of extended non-molecular three-dimensional phases of carbon dioxide and of nitrogen, as will be described in a following chapter. A remarkable finding is the behaviour of argon (a rare gas) in comparison with iron [22, 55]. It has been reported that at very high pressure, the melting point of argon becomes higher than that of iron, a quite unexpected finding considering the periodic properties of the elements in their usual meaning.

The purpose of the present book is to discuss the chemical reactions that have been reported on molecular crystals at high pressures. There is a wealth of research activity on high-pressure inorganic and organic chemistry in solution [56–60] that will not be discussed here, although several basic concepts are in common. Chemical reactions at high pressure in liquids or molecular aggregates will, however, be considered whenever useful to elucidate experiments in related molecular crystals. The literature in this area is quite extensive on the one side, but on the other, many further developments can be foreseen. Focus will be on several facets of this topic. First, it is of interest to consider the novelty of the products that can be obtained at high pressures at variance with the chemical behaviour observed in normal laboratory practice. On the whole, high pressure appears a powerful activation tool that can ensure highly selective reaction pathways. This possibility can be further enhanced when other activation methods, and in particular photo-activation, are coupled to pressurization. Of primary importance is the possibility to elucidate the solid state reaction mechanisms through the fine-tuning of the intermolecular interactions that can be realized by changing the pressure. Attention will be particularly focused on the structural aspect of the molecular crystals and on the relation between the crystal (and molecular) structure and the mechanisms of the high-pressure reactions.

The approach of the present work is at the crossroads and in the overlap region of several disciplines. The basic encounter between high-pressure science and technology and molecular crystals is by itself of particular interest first, because molecular crystals are an extremely large ensemble of widespread systems of great relevance in many fields of application and secondly because, as already mentioned in this introduction, the softness of molecular crystals makes the tuning realized by pressurization particularly effective and advantageous. Since the purpose of the present work is to discuss high-pressure chemical reactions in molecular crystals there is a fundamental overlap with the field of solid state chemistry [61]. Reactivity in solids is governed by the fixed distances and relative orientations of the molecules in the crystal and by the reduced mobility in the crystalline array that may require severe thermal activation or activation by other means, such as the addition of impurities or initiators. Application of pressure to the solid allows a fine-tuning of the intermolecular distances and orientations and a reduction of some energy barriers. Therefore, application of high pressure is an important alternative means of activation of solid state reactions and of the selectivity of the reaction pathways. The high-pressure tuning of the crystal structure is also akin to the principles of crystal engineering [9]. Finally, it has been anticipated that at high pressures thermal occupation of the electronic excited states becomes possible. This establishes a bridge with the developing field of photo-crystallography [62] and the more general field of laser chemistry [63]. It will in fact be shown that synergistic activation of chemical reactions by application of high pressure and by selective laser irradiation is a powerful tool to both address the pathway of chemical reactions and to enlighten the mechanisms of solid state reactivity.

The outline of the book is as follows. A brief historical survey is presented in Chapter 2, illustrating early attempts to modify the properties of materials by application of mechanical forces and the slow development of modern high-pressure technology with particular reference to chemistry. In Chapter 3, the elastic properties of molecular crystals will be discussed focusing attention on the anisotropy of these systems which plays a basic role in the high-pressure reaction mechanisms. The experimental methods to produce high static pressures and to monitor *in situ* high-pressure phenomena will be reviewed in Chapter 4 with particular reference to diamond anvil cell technology. The basic physical chemistry concepts of high-pressure science, including the pressure dependence of the chemical equilibrium, of the chemical kinetics and of the electronic structure, will be reviewed in Chapter 5. Finally, Chapter 6 will be devoted to the discussion of the chemical reactions that have so far been studied in molecular crystals with attempts to establish possible correlations, at least in groups of systems with comparable molecular structures.

Chapter 2

Historical Survey

Substantial progress in high-pressure science and in its applications to material science and related fields has only occurred in the last century, and in particular, thanks to the pioneering work of Percy W. Bridgman [30]. The slow but continuous progress in high-pressure technology and in applications for the measurements of a variety of physical properties of materials in the 19th and approximately the first half of the 20th century, have been described in detail by Bridgman [30], but his historical reconstruction only occasionally referred to specific applications in chemistry.

The beginning of the study, in the modern scientific sense, of the effects of pressure on chemical reactions can be traced back to the second half of the 19th century and, if we wish to identify a precise accomplishment as a real starting point, most appropriately to the year 1884, when Henri L. Le Chatelier first formulated his famous principle stating that in response to an increase in pressure, a chemical reaction will proceed in the direction corresponding to a decrease of the volume [64, 65]. Le Chatelier's original statement reads: *Any system in stable chemical equilibrium, subjected to the influence of an external cause which tends to change its temperature or its condensation (pressure, concentration, number of molecules in unit volume), either as a whole or in some of its parts, can only undergo such internal modifications that would, if produced alone, produce a change of temperature or of condensation of opposite sign to that resulting from the external cause.* The formulation of the principle by Le Chatelier was in essence a generalization of a previous observation by van't Hoff [66] on the effect of temperature on chemical equilibria. Le Chatelier also recognized that most of the laws of what he called "*mécanique chimique*" were implicitly contained in the works of Gibbs, even though at that time the latter were not easily accessible to chemists. Actually, as already noted by

Hamman [67], the essence of the Le Chatelier principle was anticipated 20 years before by J. Thomson [68] in a series of papers which, however, were not concerned with chemical reactions but with research on regelation and liquefaction. Thomson's statement reads: *If any substance be in a condition in which it is free to change its state of molecular arrangement and if a mechanical work is applied to it in such a way as that the occurrence of the change of state will make it lose that mechanical work then the substance or system will pass into the changed state.* The observations of the effects of pressure by Le Chatelier were not confined to chemical reactions, but were extended to other properties including the case of transformations in condensed phases, such as the dimorphism of silver iodide, and, in particular, the effect of pressure on the transition temperature or on the solubility of salts.

However, it is evident that the simplest observations of the Le Chatelier principle were concerned with chemical reactions involving gas-phase reagents or products, where the volume effects produced by a change of pressure were larger. In this sense, an earlier perception of the relevance of the pressure parameter in chemical phenomena can be recognized in the pneumatic chemistry developed in England in the 18th century, when methods to collect and manipulate gases in appropriate vessels were developed, together with the necessary analytical procedures [69]. It is to this period, for instance, the formulation of Boyle's law connecting the volume and the pressure of a gas dates back, a law that, together with the other gas laws, played an important role in the establishment of the modern atomic theory.

The progress in studies of high-pressure effects in material science and chemistry, particularly in condensed phases, had been rather slow until the work of Bridgman, mainly because of technological difficulties bound to the unavailability of appropriate materials for vessels to support high hydrostatic pressures [70]. However, it can be said that a naive or at least a practical perception that the application of pressure, or of a mechanical force, could have an effect on the properties of materials, or even on transformations (chemical reactions) of substances, can be found in some written documents or in some practices of antiquity.

Concerning the origin and formation of minerals and rocks, Plato in *Timaeus* [71] says: *...the air, being weighty, when it is thrust and poured around the mass of earth, presses it hard and squeezes it into the space that the new made air quitted. Thus the earth when compressed by the air into a mass that will not dissolve in water, forms stones.* In this sentence, besides the concept that pressurization can have an effect on the formation of minerals, albeit in the framework of the Aristotelic thinking that all substances are made from four elements, we may recognize a first intuition of the concept of packing that is of paramount importance in high-pressure phenomena. The same intuition is expressed in many other places where the thinking of Greek atomistic philosophers are shared and for instance, in the *Pneumatica* of Hero (10–70 AD) [72] : *Every body is composed of minute particles, between which are voids less than these particles of the body...in proportion as any one of these particles recedes, some other follows it and fills the vacant space*, or in the Latin poet Lucretius (96–55 BC) which in Book V of his *De Rerum Natura* [73] writes: *Although bodies are thought to be solid and filled ... they are mixed with voids ... and therefore between bodies of equal volume the one appearing lighter contains a larger amount of empty spaces while the heavier one indicates in itself a larger amount of matter and inside a smaller amount of voids.* These expressions, if the minute particles (atoms) were endowed with all the necessary properties, including not only the volume and shape but also the weight, which is not necessarily the case in ancient Greek scientific thinking, would not be much different from our present day scientific definitions.

The first grasp by mankind that an applied force, or a pressure, according to our understanding, can have an effect on the properties of materials very much precedes ancient Greek philosophy and can be associated with the earliest stages of metallurgy, occurring very likely in the 6th millennium BC or earlier [74–76]. The earliest metals collected by man were native metals that occur as such in nature and were initially collected as special kinds of stones and, like other raw materials, on the basis of man's recognition of properties like the colour or texture. Among the earliest metals used by mankind, copper was certainly the most important and copper production and working, the earliest branch of metallurgy, determined the transition from the Stone to the Metal Age [77]. The breakthrough was the discovery that copper (and other metals) could be reshaped to obtain serviceable tools. The important point was that by cold working (hammering), copper could not only be reshaped but also hardened. Lucretius in Book VI of *De Rerum Natura* [73] writes: *... and you see to reduce them (the metals) by hammering in the shape of cutting sharp and thin blades, you see sharpening arms, cutting tools ... won yielded the strength of those metals unable to stand such strong impacts...* Work hardening characteristics allowed obtaining tools with the required overall strength and edge sharpness [78]. Nowadays, cold working is accomplished by lamination under appropriate presses. According to our present understanding, work hardening of metals depends on the number and distribution of dislocations and defects in the material. The energy released to the material by cold working of some type increases the number of dislocations which interfere with each other in such a way that their mobility is reduced, as is the malleability of the metal, to such an extent that the material becomes fragile and can be cracked on further working. It may appear surprising to find in Theophrastus' (?–287 BC) *De Sensu* [79] the following explanation, attributed to Democritos, of the hardness of iron as compared with lead: *If you compare what is hard or soft with what is heavy or light there is a difference also as to the position and internal distribution of voids, and therefore iron is harder while lead is heavier, in fact iron has an irregular composition and here and there has frequent and also extended voids while in other parts is highly compact, in general however has more voids than lead. Lead, which contains less voids, has a regular and uniform composition in all its parts and for this reason it is certainly heavier, but softer, than iron...* The malleability can be restored by annealing and recrystallization.

Independent from any scientific understanding, practical methods of repeated alternative work hardening and annealing were known to ancient men. We do not know of written documentation with technical details on metal working in ancient times. It is instead well documented that at its beginning, the metallurgic art shared a ritual character. The figure of the blacksmith was surrounded by an atmosphere of magic and sacral character, very much like the shaman [80]. Indeed, a proverb quoted by Eliade [80] says that *blacksmiths and shamans are of the same nest* [81]. Therefore, the technical skills of the blacksmith could only be handed down, generally from father to son, by rigid and secret initiation rituals. However, the thermal and mechanical history of ancient metal artefacts can be revealed by archaeometric methods and, in particular, by the study of its texture. The very first appearance of copper artefacts has been documented by excavations at the site of Cayonu Tepesi, in south eastern Anatolia, where some 50 copper artefacts have been discovered, including awls, pins and hooks [82]. Radiocarbon analysis has dated these artefacts to the period 8400–7500 BC and archaeometric analysis has shown that they are made of native copper, cold hammered by stone tools to double the hardness of the starting material. There is also evidence of some mild annealing and recrystallization of these artefacts [83]. The practice of cold working of metals (by hammering or other, more profitable methods) has continued up to the present day.

The piston-cylinder device has been the most used apparatus in studies of high-pressure phenomena, at least until the opposed anvil system and the diamond anvil cell were developed, and today it is still most important for large volume applications. However, the principle of this device is more than 2000 years old, since it was first exploited in the 3rd century BC by Ctesibius from Alexandria (285–222 BC). Ctesibius was a son of a barber and the story says that as a child he dropped a lead ball in a tube, thus producing a loud sound by the compression of the air. From this Ctesibius understood that air was a substance and that from the compression of the air (and of water likewise) other bodies could be moved. This observation was the basis for the many inventions of Ctesibius who we can consider a mechanical engineer *ante litteram* and the father of pneumatics and hydraulics. He described his many inventions in a book that unfortunately has been lost, but they are mentioned in many places and in particular, in the *Pneumatica* of his pupil, Hero of Byzantium (260–180 BC), in the *Pneumatica* of Philo (10–70 AD) and by Vitruvius [84, 85]. The most important invention by Ctesibius was the water pump, a device used to raise water to a certain height, exerting a pressure with a piston in a bronze cylinder connected to pipes through one-way valves. The principle of the Ctesibius machine (called *ctesibia machina* by Vitruvius) has remained unchanged to the present day despite the many improvements in materials and technology. Piston pumps, derived from the hydraulic pump of Ctesibius, are described at length in Agricola's *De Re Metallica* [86, 87]. The principle of the water pump is the basis of the piston cylinder device for high-pressure generation.

It would be frustrating to search in antiquity for documented practices of chemical transformations under the effect of an applied pressure. However, a fairly clear and explicit description of a chemical reaction, occurring by the effect of mechanical activation (application of pressure and stress), is found for the mercury (quicksilver) extraction from cinnabar according to the reaction,

$$HgS + Cu \rightarrow Hg + CuS.$$

The description is reported by Theophrastus (372–287 BC) in his *De Lapidibus* [88, 89] in the following form:*...it is clear that art imitates nature, and yet produces its own peculiar substances, some for their utility, some merely for their appearance, like wallpaint, and some for both purposes, like quicksilver; for even this has its uses. It is made by pounding cinnabar with vinegar in a copper mortar with a copper pestle. And perhaps one could* *find several things of this kind.* The same method of preparation is mentioned in other ancient sources, and in particular in Vitruvius' *The Ten Books of Architecture* [84] and in Pliny's *Natural History* [90], but is not mentioned any later and is absent in Agricola's *De Natura Fossilium* [91], the first treatise of mineralogy in the modern sense or in the *De Re Metallica* where six methods of preparation of quicksilver, but not this one, are described. However, the quicksilver preparation according to Theophrastus has been correctly defined as the first example of a mechanochemical reaction [92–94].

In the interest of the present work, which is essentially focused on transformations at very high pressure and therefore, on materials in condensed phases, a major historical concern has been the attempts to ascertain whether a liquid or a solid is compressible, a difficult topic to investigate in the absence of technological advances to generate sufficiently high pressure and to measure small volume variations. Even though the already mentioned concept, of the Greek atomistic philosophers, that substances are composed of a mixture of small particles of matter (atoms) and voids, was not against the idea that bodies could be compressed by the reduction of the void's content, the very first experiments on liquid compressibility were made by the scientists of the Accademia del Cimento, the first European scientific society founded in Florence in 1657 by the initiative of pupils of Galileo under the sponsorship of Leopoldo of Tuscany. The experiments performed in the academy were described in a book published in Florence in 1967 [95] and among them there is an attempt to determine whether water is compressible. The book and the description of the *Esperienze intorno alla compressione dell'acqua* (Experiments on the compression of water) can be accessed via the online library of the Museo e Istituto di Storia della Scienza in Florence. Three experiments are described in which water is compressed either by a gas, by a column of mercury or by a stress on a silver container. The conclusion was that water is not compressible. However, it was not excluded that if a higher pressure than feasible at the time (a few tens of atmospheres) could be attained, a different conclusion could be reached. This became actually possible more than 100 years later when, in 1761, Canton definitely demonstrated that water is compressible [96]. The seminal work of Canton opened the way to a large number of measurements of the compressibility of liquids in the 18th and 19th centuries, as summarized by Bridgman [30]. In this context, it is particularly noteworthy to mention the work of Tait [97] on the compressibility of fresh water, marine water, mercury and glass, following the great scientific expedition of the HMS Challenger in the years 1872–1876 . The results of this work led to the formulation of the Tait equation of state,

$$\frac{V_0 - V}{P} = \frac{A}{\Pi + P}, \tag{2.1}$$

where V_0 is the reference volume and A and Π are parameters to be determined and are characteristic of the substance (the latter being temperature dependent). The Tait equation of state goes beyond the simple power expansion in the variable, and in the original or in its linearized form, as well as in various modified forms [98, 99], is still in use to the present day for liquids and solids up to pressures of thousands of atmospheres [100]. The empirical Tait parameters have been correlated with the parameters of the intermolecular potentials [98]. In the early studies of the compressibility of solutions, the pressure effects on the volume of solutions of ionic salts attracted particular interest and were interpreted in terms of electrostriction [67], i.e. the contraction of the solvent around dissolved ions and charged groups, a phenomenon which accounts both for changes in the molecular structure and for solvent-solute interactions.

One of the most important outcomes of the Challenger voyage and of other contemporary oceanographic expeditions [101], was the discovery that living specimens could be found at depths of about five miles and thus at pressures of about 800 atm. This finding catalysed more accurate studies of life under pressure [102] and a few years later, Regnard started a more systematic study of the effect of high pressures on living organisms [103]. This was the starting point of an intensive research activity on pressure effects in bioscience [35] that was further stimulated by a paper of Bridgman [104] who reported on *...a fact of possible biological interest ... If the white of an egg is subjected to hydrostatic pressure at room temperature, it becomes coagulated, presenting an appearance much like that of a hard-boiled egg... The effect of temperature which is not large seem to be such that the ease of coagulation increases at low temperatures, contrary to what one may expect.* The observation of Bridgman is important, since it gave rise to the investigation of the effect of high pressures on structural biology.

According to a quotation from a non-specified source reported by Balny [35], the interest in understanding the effects of high pressures on biological molecules stimulated Hite [105], the first to explore the pressure treatment of foods, to investigate chemical reactions under high pressures and to describe an apparatus to perform chemical reactions under these conditions.

At the beginning of the 20th century, progress on high-pressure reactions in condensed phases started and it is worth mentioning the work of Conant [106, 107] on polymerization reactions, since polymerization is a typical reaction occurring at sufficiently high pressures. Quite important is also the activity of Cohen [108, 109], who investigated not only the effect of pressure on the chemical equilibria, but also on the reaction rates. Extensive references to early works on chemical reactions at high pressures can be found in Bridgman [30, 110] and in Cohen [108]. A bibliography of high-pressure work in the period 1900–1960 has been published by Merrill [111]. A description of the development of high-pressure research in different countries at the beginning of the last century can be found in Bradley [112].

Chapter 3

Elasticity and Equation of State

The equation of state (EOS), in the usually accepted formulation, gives the interconnection between the thermodynamic parameters of pressure P, volume V, temperature T and number of particles n, or of their conjugate variables, in the form,

$$P = P(V, T, n), \tag{3.1}$$

or in the inverted form,

$$V = V(P, T, n). \tag{3.2}$$

The equation of state of solids, and the underlying mechanical properties, is a topic of particular importance in high-pressure research, since at sufficiently high pressure, most materials are ultimately in the solid state. This topic has been discussed in great detail in an extensive number of books [1, 6, 113–119] and review articles [120–130] that encompass the many facets of the problem, including:

- The basic thermodynamic relations (3.1 and 3.2)
- The extension to a pressure-temperature regime of atomization or metallization, or even transformation to plasmas
- The connection with microscopic and electronic structures, with intermolecular interactions at work and with the microscopic dynamics of the material
- The onset of phase transitions
- The range of applicability
- The experimental methods for its direct determination
- The characterization with parameters
- The comparison of experiments with theoretical derivations at various levels of approximation

- The theoretical foundation of the semi-empirical formulations of common use

Here, it is only appropriate to present a brief overview of the general problem while the focus will be particularly on molecular solids, that are of major interest in the present work, and their peculiarities. As a matter of fact, molecular solids are characterized by strong intramolecular bonds that are essentially (even if not always) stable in the not too high-pressure regime, and by weak intermolecular (van der Waals and hydrogen bonding) interactions. Therefore, molecular solids are highly compressible, without necessarily much deformation of the intramolecular structure. From ambient conditions to a pressure of tens of GPa, they can experience a volume contraction by a factor of up to 2 or more, corresponding to a compression,

$$\eta = \frac{V}{V_0} \geq 0.5. \tag{3.3}$$

This compression corresponds to a $\sim$20% shortening of the intermolecular separation and this is of primary importance when the purpose is to investigate the chemical reactivity at high pressures.

In the usual representation of the profile of a chemical reaction, the free energy is reported as a function of a suitable reaction coordinate. The simplest reaction coordinate can be some kind of interatomic or intermolecular distance. The equation of state (according to the basic thermodynamic definition) monitors, in essence, just the energy of the material, and the underlying structural evolution of the system, as a function of the linear contraction,

$$\frac{l}{l_0} = \left(\frac{V}{V_0}\right)^{\frac{1}{3}}. \tag{3.4}$$

Therefore, knowledge of the equation of state is of basic importance in any study of chemical reactions, and of any other chemical and physical property at high pressure, since it is a consequence of the variation of the free energy as a function of an interatomic or intermolecular separation.

3.1 Stress and strain

A complete knowledge of the equation of state is a very difficult task, notably for the possible anisotropic character of the compression phenomena that, as mentioned, can be particularly prominent in molecular solids.

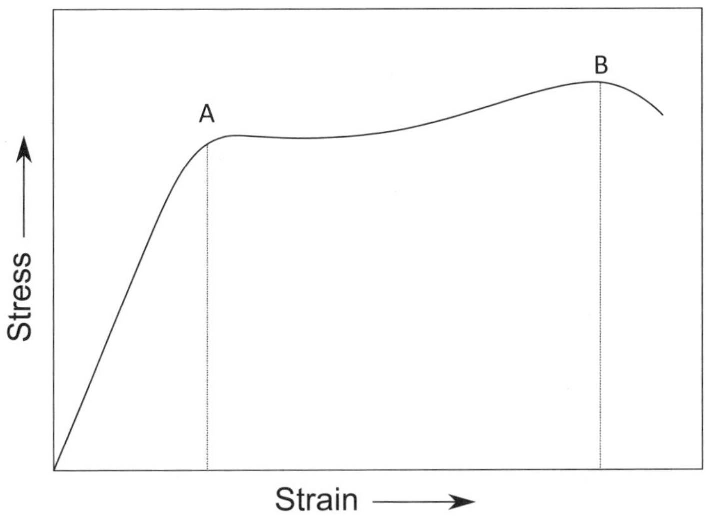

Fig. 3.1 Schematic behaviour of the stress-strain curve of a solid. A and B represent the yield strength and the ultimate tensile points, respectively.

The typical form of the stress-strain curve in a solid material is illustrated in Figure 3.1. In the elastic regime, corresponding to the linear behaviour up the yield point A, the deformation in response to an external stress (application of pressure in our case) is substantially instantaneous and is fully reversible on downloading of the pressure. At larger values of stress, the strain increases non-linearly and this regime corresponds to a plastic deformation that is not completely reversible. At even larger values of stress (i.e. above point B), the solid can deform catastrophically even upon a minor increase of the stress. In the interest of the present book on molecular crystals we shall mostly be confined in the elastic regime. In fact, when the strain is increased within this regime, in many cases the changes of the electronic arrangement of the molecules can lead to some chemical transformation before the onset of plastic deformations, unless the chemical reaction is itself considered as a catastrophic deformation. This is, however, only a convenient or approximate point of view. Actually, it has been reported that after the pressurization of molecular crystals, releasing of the pressure does not lead precisely to the initial situation as revealed by the observation that, e.g., vibrational frequencies [23] do not recover the initial zero-pressure values, as if a residual pressure remained active in the downloaded sample. In this respect, however, other effects could come into play such as the generally polycrystalline character of the samples or the presence of minor impurities of some kind. It has also been reported, for instance in the case of the nitromethane crystal [24], that some hysteresis can be observed on releasing the pressure and this can be interpreted as due to some significant changes of the intermolecular interactions at high pressures, changes that are made difficult to remove by the presence of small energy barriers.

We begin by recalling some basic definitions. When a stress of a general form (e.g., variation of pressure, temperature or composition, electric or magnetic fields, ...) is applied to a solid, each point **r** suffers a displacement (strain) that is a function of the position and can be expressed by a strain function u(**r**). As a consequence, the infinitesimal distance dr between two points (which in a crystal can be connected to an interatomic distance) changes and its value dr' in the stressed sample can, assuming that the strain function is continuous and in the limit of small deformations, be expressed as,

$$dr' = dr + e(\mathbf{r})dr, \tag{3.5}$$

where $e(r)$ is the derived tensor of the strain function,

$$e_{ij} = \frac{\partial u_j}{\partial r_i}. \tag{3.6}$$

The strain is an adimensional tensor field which should not conform to the symmetry of the crystal in the general case. Neglecting translational and rotational displacements, only the symmetric part of the tensor e is of interest and we may define the strain tensor Σ as,

$$\Sigma = \frac{1}{2}(e + \widetilde{e}), \tag{3.7}$$

where $\widetilde{e}$ indicates the transpose. The diagonal elements Σ_{ii} of the strain tensor represent the fractional lengthening of an element aligned along the i axis, while the off-diagonal elements Σ_{ij} are connected to the angular deformation of two vectors aligned along the i and j axes, respectively, in the unstrained solid. The off-diagonal elements thus represent shear deformations. The trace of the strain tensor gives the volume dilation, i.e., the fractional increase of the volume element.

The strain field, acting on the material, must then be associated with an external or internal force F acting on each volume element: here we shall only be concerned with externally applied forces. It is then possible to define, with reference to an external force per unit volume $F(\mathbf{r})$, a stress tensor τ with elements given by,

$$F(r_i) = \sum_j \frac{\partial \tau_{ij}}{\partial r_j}. \tag{3.8}$$

If the system is at equilibrium and no translation or rotation is involved, the stress tensor is symmetric. The diagonal element τ_{ii}, defines the force acting perpendicularly to a surface of the volume element (i.e., perpendicularly to the i-th axis), whereas the off-diagonal elements τ_{ij}, define shear stresses acting tangentially (e.g., τ_{ij} defines the force acting on a face perpendicular to j along the i-th direction). The stress and the strain tensors are related to each other through an operational tensor s_{ijkl}, whose structure depends on the selected coordinates (see, for example, ref. [131]) and, in the linear response theory and for an elastic solid obeying Hooke's law, we have,

$$\Sigma_{ij} = \sum_{kl} s_{ijkl}\tau_{kl}. \tag{3.9}$$

The relation between the strain and the stress tensors can more conveniently be expressed in a matrix notation adopting the so-called Voigt notation. To this purpose we define a six-dimensional strain vector ϵ with components given by the following correspondence with the elements of the strain tensor:

ϵ_1	ϵ_2	ϵ_3	ϵ_4	ϵ_5	ϵ_6
Σ_{11}	Σ_{22}	Σ_{33}	$2\Sigma_{23}$	$2\Sigma_{13}$	$2\Sigma_{12}$

With the same correspondence, but with the exclusion of factor 2 for the components derived from the off-diagonal elements, we define a six-dimensional stress vector τ. The 4-th rank tensor s_{ijkl}, is converted to a 6 x 6 matrix S by associating the first two and the second two indices to a single index as shown above for the strain and stress tensor, respectively. In addition, a factor 2 is considered when m or n equal 4,5,6 ($2s_{ijkl} = S_{mn}$) and a factor 4 when both m and n equal 4,5,6 ($4s_{ijkl} = S_{mn}$). In the Voigt notation, the strain and stress components are then connected by the relation in a matrix form,

$$\epsilon = \mathbf{S}\tau, \tag{3.10}$$

through the matrix $\mathbf{S}$ of the elastic moduli or compliance coefficients expressed in units of GPa^{-1} (or, in general, stress^{-1}). The inverse relation,

$$\tau = \mathbf{C}\epsilon, \tag{3.11}$$

Table 3.1 Number of independent components of the elastic constants tensor in crystalline systems

triclinic	21
monoclinic	13
orthorhombic	9
tetragonal	7 or 6
trigonal	7 or 6
hexagonal	5
cubic	3

involves the matrix $\mathbf{C}$ ($\mathbf{C} = \mathbf{S}^{-1}$) of the stiffness coefficients or elastic constants expressed in units of GPa. The matrices of the elastic constants and of the elastic moduli are symmetric and in the most anisotropic case (i.e., for a non-centrosymmetric triclinic crystal structure) they involve 21 independent components. The number of the independent components reduces with increasing symmetry and in cubic crystal systems, there are only 3 independent non-zero components, e.g.,

$$c_{11} = c_{22} = c_{33} \tag{3.12}$$

$$c_{44} = c_{55} = c_{66} \tag{3.13}$$

$$c_{12} = c_{23} = c_{32}. \tag{3.14}$$

The number of independent components in the various crystalline systems is summarized in Table 3.1. For an isotropic system (this, for example, can be the case for polycrystalline materials) there is the additional relation,

$$c_{44} = \frac{1}{2}(c_{11} - c_{12}), \tag{3.15}$$

that further reduces the number of independent components, thus showing that cubic crystals, as far as their elastic properties are concerned, are not actually isotropic. The structure of the matrices of stiffness and compliance coefficients for an isotropic solid and for solids belonging to the various crystalline systems is shown in Figure 3.2. For an isotropic solid, one can easily obtain,

$$\tau_1 + \tau_2 + \tau_3 = (c_{11} + 2c_{12})(\epsilon_1 + \epsilon_2 + \epsilon_3). \tag{3.16}$$

For a uniform compression, the diagonal terms of τ give the negative of the pressure:

$$\begin{pmatrix} a & b & b & 0 & 0 & 0 \\ b & a & b & 0 & 0 & 0 \\ b & b & a & 0 & 0 & 0 \\ 0 & 0 & 0 & (a-b)/2 & 0 & 0 \\ 0 & 0 & 0 & 0 & (a-b)/2 & 0 \\ 0 & 0 & 0 & 0 & 0 & (a-b)/2 \end{pmatrix}$$

isotropic

$$\begin{pmatrix} a & b & b & 0 & 0 & 0 \\ b & a & b & 0 & 0 & 0 \\ b & b & a & 0 & 0 & 0 \\ 0 & 0 & 0 & c & 0 & 0 \\ 0 & 0 & 0 & 0 & c & 0 \\ 0 & 0 & 0 & 0 & 0 & c \end{pmatrix}$$

cubic

$$\begin{pmatrix} a & b & c & d & -e & 0 \\ b & a & c & -d & e & 0 \\ c & c & f & 0 & 0 & 0 \\ d & -d & 0 & g & 0 & 2e \\ -e & a & 0 & 0 & g & 2d \\ 0 & a & 0 & 2e & 2d & (a-b)/2 \end{pmatrix}$$

trigonal A

$$\begin{pmatrix} a & b & c & d & 0 & 0 \\ b & a & c & -d & 0 & 0 \\ c & c & e & 0 & 0 & 0 \\ d & -d & 0 & f & 0 & 0 \\ 0 & 0 & 0 & 0 & f & 0 \\ 0 & 0 & 0 & 0 & 0 & (a-b)/2 \end{pmatrix}$$

trigonal B

$$\begin{pmatrix} a & b & c & 0 & 0 & d \\ b & a & c & 0 & 0 & -d \\ c & c & e & 0 & 0 & 0 \\ 0 & 0 & 0 & f & 0 & 0 \\ 0 & 0 & 0 & 0 & f & a \\ d & -d & 0 & 0 & 0 & g \end{pmatrix}$$

tetragonal A

$$\begin{pmatrix} a & b & c & 0 & 0 & 0 \\ b & a & c & 0 & 0 & 0 \\ c & c & d & 0 & 0 & 0 \\ 0 & 0 & 0 & e & 0 & 0 \\ 0 & 0 & 0 & 0 & e & 0 \\ 0 & 0 & 0 & 0 & 0 & f \end{pmatrix}$$

tetragonal B

$$\begin{pmatrix} a & b & c & 0 & 0 & 0 \\ b & a & c & 0 & 0 & 0 \\ c & c & d & 0 & 0 & 0 \\ 0 & 0 & 0 & e & 0 & 0 \\ 0 & 0 & 0 & 0 & e & 0 \\ 0 & 0 & 0 & 0 & 0 & (a-b)/2 \end{pmatrix}$$

hexagonal

$$\begin{pmatrix} a & b & c & 0 & 0 & 0 \\ b & d & e & 0 & 0 & 0 \\ c & e & f & 0 & 0 & 0 \\ 0 & 0 & 0 & g & 0 & 0 \\ 0 & 0 & 0 & 0 & h & 0 \\ 0 & 0 & 0 & 0 & 0 & i \end{pmatrix}$$

orthorhombic

$$\begin{pmatrix} a & b & c & 0 & 0 & l \\ b & d & e & 0 & 0 & m \\ c & e & f & 0 & 0 & n \\ 0 & 0 & 0 & g & p & 0 \\ 0 & 0 & 0 & p & h & 0 \\ l & m & n & 0 & 0 & i \end{pmatrix}$$

monoclinic

$$\begin{pmatrix} a & b & c & d & e & f \\ b & g & h & i & l & m \\ c & h & n & o & p & q \\ d & i & o & r & s & t \\ e & l & p & s & u & v \\ f & m & q & t & v & z \end{pmatrix}$$

triclinic

Fig. 3.2 Structure of the matrix of the stiffness or compliance coefficients for the various crystalline symmetries. Trigonal A refers to the 3 and $\bar{3}$ classes, trigonal B to all other trigonal classes. Tetragonal A refers the the classes 4, $\bar{4}$ and 4/m, tetragonal B to all other tetragonal classes.

$$\begin{pmatrix} -p & 0 & 0 \\ 0 & -p & 0 \\ 0 & 0 & -p \end{pmatrix}$$

while the sum of the diagonal elements of ϵ,

$$\begin{pmatrix} \epsilon_1 & 0 & 0 \\ 0 & \epsilon_2 & 0 \\ 0 & 0 & \epsilon_3 \end{pmatrix},$$

gives the fractional volume contraction,

$$\frac{\Delta V}{V} = \epsilon_1 + \epsilon_2 + \epsilon_3. \tag{3.17}$$

The volume bulk modulus K,

$$K = -V\left(\frac{\partial P}{\partial V}\right)_T \tag{3.18}$$

for an isotropic solid can be defined as,

$$K = \frac{c_{11} + 2c_{12}}{3}. \tag{3.19}$$

Considering a uniaxial compression in a cubic crystal (e.g., compression along axis 1) the Young's modulus,

$$E = \frac{p}{\epsilon_1} \tag{3.20}$$

is given by,

$$E = c_{11} - \frac{2c_{12}^2}{c_{11} + c_{22}}. \tag{3.21}$$

Additional restraints are imposed on the tensor of the elastic constants by the fact that the lattice energy must be positive definite. If the interaction forces are central and the lattice points are inversion centres, the Cauchy relations,

$$c_{23} = c_{44} \tag{3.22}$$

$$c_{31} = c_{55} \tag{3.23}$$

$$c_{12} = c_{66} \tag{3.24}$$

$$c_{14} = c_{56} \tag{3.25}$$

$$c_{25} = c_{64} \tag{3.26}$$

$$c_{36} = c_{45}, \tag{3.27}$$

must be obeyed and the number of independent constants in the general case reduces to 15.

3.2 Elasticity and anisotropy in molecular crystals

3.2.1 *Elastic constants*

Molecular crystals are, in most cases, composed by asymmetric molecular units and, apart from cases of orientational disorder, the crystal structure has low symmetry as well. In addition, the weak intermolecular interactions can have a distinctly directional character, such as occurs in the important class of hydrogen-bonded systems, contributing to the anisotropy of the system. Therefore, the number of independent elastic constants to be determined to fully characterize the elastic properties of the system and their anisotropy is large. On the other hand, when the purpose is to study chemical reactions at high pressure, knowledge of the elastic anisotropy is of fundamental importance, since it gives information on the direction of the most favourable shortening upon pressurization of the intermolecular contacts that are eligible as reaction coordinates. Elastic constants are not accessible directly by experiments but are related to the velocity of sound propagation. For a continuum medium, the relation is given by the Christoffel equations which can be written as,

$$DC\widetilde{D}u = \rho v^2 u, \tag{3.28}$$

where C is the tensor of the elastic constants, D is a 3 by 6 matrix containing the components of a unit vector l in the direction of the wave propagation in the form,

$$D = \begin{pmatrix} l_1 & 0 & 0 & 0 & l_3 & l_2 \\ 0 & l_2 & 0 & l_3 & 0 & l_1 \\ 0 & 0 & l_3 & l_2 & l_1 & 0 \end{pmatrix},$$

$\widetilde{D}$ is its transpose, ρ is the density, v is the wave velocity and u is the strain vector. The Christoffel equations can be written more explicitly as,

$$\sum_{jk} (c_{ijkl} e_j e_k - \rho v^2 d_{il}) u_l = 0. \tag{3.29}$$

If the elastic constants are known, the Christoffel equations directly give the wave velocities in any desired direction as eigenvalues. To obtain the elastic constants the solution of Eq. 3.29 must be reverted and the wave velocities must be measured along different directions and, in the most general case, the elastic constants are obtained by appropriate fitting procedures. The acoustic wave velocities are obtained directly by ultrasonic techniques in their alternative variants [132] or from the slope of acoustic modes dispersion curves measured by Brillouin scattering [133–135]. Brillouin spectroscopy can be particularly convenient for fragile molecular crystals as it is a non-contact technique. Alternatively, elastic constants can be obtained from analysis of diffuse X-ray scattering [136–138]. Details of the experimental methods to obtain elastic constants can be found in the references contained in the appendix, reporting their determination in specific crystalline systems. The experimental determination of the elastic constants poses some particularly relevant problems in molecular crystals since in general, single crystals of sufficient quality and dimensions are needed and the wave velocity must be obtained along different direction, in order to efficiently solve the Christoffel equations for the elastic constants. For this reason, the attainment of elastic constants from models of the intermolecular interactions and computational methods is of particular interest in molecular crystals [6, 139]. In the elastic limit, the elastic constants are the second derivatives c_{ij} of the lattice free energy with respect to the elastic strain ϵ,

$$\Phi = \frac{1}{2}\sum_{ij} c_{ij}\epsilon_i\epsilon_j. \tag{3.30}$$

In molecular crystals and at low temperature, the free energy can safely be approximated by the lattice energy. If a model of the intermolecular potential is available, commonly in the form of two-body atom-atom potentials [5, 6], the energy is minimized at the experimental structure and then a homogeneous strain is applied, allowing the strained structure to relax at a new minimum. From the variation of the strain energy, the elastic constants can be derived analytically or numerically [140, 141]. Alternatively, the problem can be approached by lattice dynamical methods [142, 143], calculating the vibrational frequencies and generating phonon dispersion curves whose slopes, in the long wave limit, give the wave propagation velocity in desired directions. Once the wave velocities are available, the elastic constants are obtained in the same way as in the experimental methods mentioned previously. For both these computational approaches, the energy minimization is ideally carried at zero temperature and anharmonic effects are neglected. These effects can approximately be taken into account in a quasi-harmonic approximation, minimizing the structure at the effective temperature, including temperature and anharmonic effects entirely in the volume dilatation. In the lattice dynamics method, advantage is taken of the fact that in molecular crystals, intramolecular forces are generally much stronger than intermolecular interactions and therefore molecular vibrations occur at much higher frequencies than translational and orientational (lattice) vibrations [5]. The approximation of the separation of internal and external vibrations is thus adopted and the molecular crystal is treated as an assembly of rigid molecular units with only six degrees of freedom per molecular unit. It is evident that large flexible molecules may have several low-frequency internal vibrational modes that can couple with external vibrations and the rigid body approximation is not strictly valid. In other words, molecular flexibility can contribute to crystal elasticity. Alternatively, the elastic constants can be calculated through molecular dynamics or Monte Carlo methods [144]. In molecular dynamics, effective harmonic frequencies and sets of dispersion curves can be obtained from the harmonic fluctuation of the normal coordinates and the wave velocities can be obtained from them. In principle, molecular dynamics has the advantage that in the *ab initio* variant, the intermolecular forces are calculated from first principles and this can be particularly advantageous at high pressure. In addition, in molecular dynamics anharmonic and temperature effects are automatically included in the treatment. In molecular dynamics and Monte Carlo simulations, the compliance tensor can be more directly sorted out through the fluctuations of the elastic tensor [145],

$$S_{ijkl} = \langle \epsilon_{ij} \epsilon_{kl} \rangle \frac{\langle V \rangle}{kT}. \tag{3.31}$$

In Appendix A, the available elastic constants for a number of molecular crystals are collected, together with the indication of the crystal structure, the temperature of the determination and the experimental methods used. Whenever more determinations are available, it should be considered that there can be a significant spread among different determinations of the elastic constants. A comparison of the elastic constants obtained with various experimental methods is reported in [146] for crystals of energetic materials and this gives a general idea of the range of the values of available elastic constants sets. In Appendix A, the elastic constants of the rare gas crystals are also reported, since the type of intermolecular interactions is the same (van der Waals) as in molecular crystals. In addition, the elastic constants of few polymeric crystals, which are mixed molecular-covalent systems, are also reported. For comparison, it can be useful to consider the elastic constants of two solids that are prototypes of solids with stronger interatomic interactions (units of GPa), e.g., NaCl: $c_{11} = 49.47$, $c_{12} = 12.88$, $c_{44} = 12.87$ and Cu: $c_{11} = 168$, $c_{12} = 128.8$, $c_{44} = 75.7$.

Previously, a collection of elastic constants of organic molecular crystals has been reported by Haussühl [147], with the purpose of finding possible trends or correlations between the molecular and crystal structure and the elastic constants. A general discussion of the structure-elasticity correlation in molecular crystals has also been attempted by Day et al. [139, 148]. A similar correlation with the molecular structure, albeit confined only to Young's modulus has been proposed by Roberts et al. [149]. In the Day and Haussühl reports [139, 147], the full set of elastic constants is reported for only about 60 molecular crystals. In the collection of Appendix A, the number of systems considered, including elastic constants from computational modelling, is more than twice as large, thus constituting a larger basis for correlations of general character or within classes of similar systems.

Haussühl [147] has made an attempt to find regularities in relation to structural properties, in analogy with similar findings in ionic crystals [150], but the variance in structure of molecular crystals is so pronounced that simple rules of thumb are not apparent, at least at first sight. It can be seen from the appendix that the elastic constants of molecular crystals are generally much smaller (reference is made in particular, to the diagonal elements c_{11}, c_{22} and c_{33}) than for ionic or metallic solids, on account of weaker (mostly dispersive) intermolecular interactions. However, these differences in the elastic constants significantly reduce if the molecular units have strongly dipolar groups or if hydrogen bond networks dominate the crystal arrangement.

Starting from the simpler group of rare gas solids, it is seen from the table reported in Appendix A, that the stiffness significantly increases with the polarizability of the units and the strength of the interactions and, ultimately, with the dimensions of the units. Considering, for example, the diagonal c_{11} constants, it can be seen that in the order Ne:Ar:Kr:Xe at low temperature, the elastic constant increases in the ratio 1:2.5:3:3.1 which can be compared favourably with the ratio of atomic radii of 1:1.9:2.3:2.8. Similar trends have been found in isotypical ionic crystals [150]. Although it may be improper to extend this concept to molecular crystals in general, in view of the already mentioned variability of molecular structure and of crystal packing, it can be noted that this seems to be the tendency in other groups of homologous systems such as benzene-naphthalene-anthracene or in the series benzene-biphenyl-paraterphenyl, with a pronounced monotonous increase of the elastic constants in one particular direction (i.e., c_{33} in these particular cases) corresponding to the long molecular axis. A similar trend is not really apparent in the dyad hydrogen chloride, hydrogen bromide whose elastic constants were, however, obtained in the cubic orientationally disordered phase. In any case, these simple correlations should be taken with some caution, since the stiffness is mostly a manifestation of the intermolecular interactions in the crystal and these could change along a series of similar molecules.

It can be noted that in crystals of isoelectronic molecules (as is the case for the nitrogen, carbon monoxide and the carbon dioxide, nitrous oxide dyads), the elastic constants are very similar as a consequence of nearly identical crystal structure and intermolecular interactions. In some cases, the elastic constants have been measured for crystals of hydrogenated and deuterated molecules as in ice I_h, methane, diphenyl and p-terphenyl (see the appendix). Considering that for the isotopic crystals, the molecular and crystal structure and the intermolecular interactions are identical, the elastic stiffness constants are essentially equal, if allowance is made for the non-negligible errors associated with the experimental determination of these parameters.

Of considerable interest are the cases where the elastic constants have been reported for different polymorphs, because the differences must be ascribed only to the changes in crystal structure and intermolecular interactions. The elastic constants reported in Appendix A, for NH_3-III and NH_3-V (both of cubic symmetry) are not directly comparable since they were obtained at considerably different pressures. However, it can be seen that the c_{11}, c_{44} and c_{12} constants are in the ratio 1:0.504:0.672 in NH_3-III and in the ratio 1:0.225:0.625 in NH_3-V, indicating a significant variation of the elastic anisotropy with the change of crystal packing. A more complex variation of the elastic constants and of the anisotropy is observed in the polymorphs of 4-methyl benzophenone, associated with the change from a trigonal to a monoclinic crystal structure. Other cases of polymorphism will be considered in the following sections with reference to the elasticity of pharmaceuticals and polymers. The most extraordinary case of polymorphism is encountered in ice, with 15 possible polymorphs observed in the explored temperature and pressure range up to the symmetrization of the hydrogen bond [151, 152]. The elastic properties of ice have been reviewed by Gagnon et al. [152]. In the appendix, the elastic constants of seven polymorphs of water are collected, showing a variety of crystal structures (cubic, hexagonal, tetragonal, monoclinic) with the associated range of elastic constants and anisotropy. It can be noted that the elastic constants of the methane clathrate hydrate SI compares quite well with those of hexagonal ice I_h [153], as one would expect considering that the lower density of clathrate is compensated by the repulsive interactions of the methane guest molecule with the water cage. It can be argued in this case that the methane guest molecule does not affect the overall elastic properties of clathrate: this may be reasonable considering that the guest molecule does not deform the host cages significantly beside contributing to their stability [154].

As can be seen from the appendix, in many molecular crystals, and in particular in aromatic crystals (benzene, naphthalene, anthracene, phenantrene, acenaphtene, tolan, stilbene, durene), when the intermolecular interactions are mostly of the non-directional van der Waals type, the diagonal elements c_{11}, c_{22} and c_{33} of the elasticity tensor, corresponding to the linear components, are rather similar, indicating a substantial elastic isotropy. In these cases, most of the anisotropy appears in the diagonal shear components c_{44}, c_{55} and c_{66}. This can be associated with the possibility of reorientation of the molecular units as a response to the applied stress. Larger anisotropy can be observed in crystals of molecular units with large electrical dipoles, as a consequence of the directionality of dipole-dipole interactions. This, for instance, is the case in the crystal of m-dinitrobenzene $C_6H_4(NO_2)_2$ where it is found that c_{33} is twice as large as c_{11} or c_{22}. This can be attributed to the interaction of the local NO_2 dipolar groups.

An important source of elastic anisotropy in molecular crystals is hydrogen bonding. In the orthorhombic β polymorph of resorcinol (m-dihydroxybenzene), the structure is dominated by a strong hydrogen-bonding network in the *bc* crystal plane, and as a consequence the c_{33} and c_{22} elastic constants are definitely larger than c_{11}. In the tetragonal crystal of pentaerythritol, the hydrogen-bonding network develops in the *ab* crystal plane, whereas the interactions between the hydrogen-bonded molecular sheets are of the van der Waals type. Consequently, the stiffness constant in the *ab* plane represented by c_{11}=c_{22} is much larger than c_{33}. Also in urea, hydrogen bonding is a relevant intermolecular interaction, but in this case it gives rise to stronger interactions along the *c* axis and as a consequence the c_{33} elastic constant is the largest. However, it can be noted that hydrogen bonding is not necessarily a source of strong anisotropy, as far as the linear diagonal stiffness constants are concerned. This, for instance is the case for the monoclinic polymorph of carbamazepine. In fact, the molecular units are hydrogen bonded in pairs that are further packed in the crystal by adirectional van der Waals interactions, which ultimately determine the stiffness of the crystal. Therefore, the elastic constants of carbamazepine resemble those of the aromatic compounds mentioned above. The same considerations apply for malonic acid and for citric acid. The case of dianin clathrand and of its clathrates is of some interest [155, 156]. Dianin, whose molecular structure is shown in Figure 3.3, has hydroxyl groups that allow the formation of hexamers with trigonal

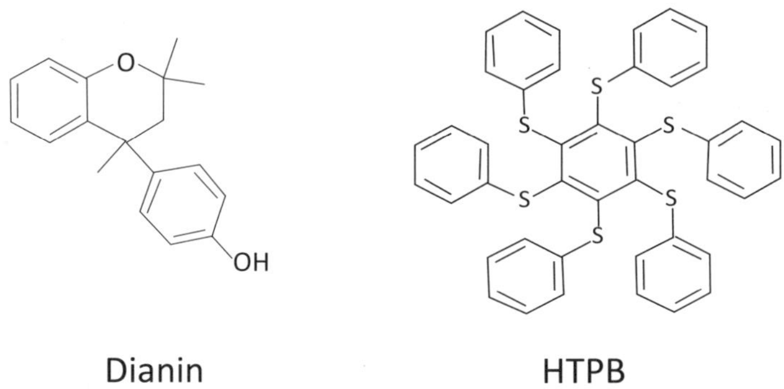

Fig. 3.3 Molecular structure of two clathrand precursors.

symmetry, which associate in couples by van der Waals interactions to form host cages for guest molecules. As seen from the appendix, the clathrand by itself has a moderate but distinct anisotropy, as in usual van der Waals crystals, because the hydrogen bonds are confined within the cage. With inclusion in the cage of ethanol (clathrate I) or heptanol (clathrate II), the guest molecules deform the clathrand cage to some extent, with the former becoming distinctly more isotropic. In the appendix, the elastic constant of HTPB, another clathrand precursor whose structure is shown in Figure 3.3, is also reported; for the interpretation of the observed elastic constants, anharmonicity of the potential has been invoked.

An example of extreme anisotropy is reported in the succinic acid monoclinic crystal [157], where hydrogen-bonded chains are formed parallel to the c crystal axis: different chains are held together by weak van der Waals interactions. It has been reported that in the succinic acid crystal, the c_{33} elastic constant is the largest and almost 13 times larger than the elastic constants in the other two perpendicular directions. This degree of anisotropy is almost comparable with that found in linear polymers (see the following sections).

As already noted, information on elasticity anisotropy may be contained in the shear elastic constants and the nature or extent of the anisotropy can be hard to grasp at first sight. A clearer and graphical view of the elastic anisotropy can be grasped from the linear compressibility, β giving the decrease in length of a line with direction cosines l_i. Under a unit hydrostatic pressure, β can be obtained from the compliance tensor s according to,

$$\beta = \sum_{ijk} s_{ijkk} l_i l_j, \tag{3.32}$$

where the l's are the direction cosines of the direction of interest. Examples of the linear compressibility in molecular crystals will be given in the following sections.

Crystalline polymers are mixed covalent-molecular crystals bound by covalent bonds along the polymeric chains which, in turn, are held together by van der Waals forces. It is useful in the present context to briefly consider some peculiarities of the elastic properties of polymers [158], since they allow the introduction of the important problems regarding the intramolecular flexibility contribution to the elastic constants, an issue that is also of general relevance in strictly molecular crystals. Polymeric materials for technological use are mostly composite or amorphous materials, as a consequence of a variety of factors including: the distribution of molecular weights, variability of conformation and stereoregularity, distribution of crystalline regions embedded in disordered regions and blending. Consequently, the deformation mechanism of polymers is highly complicated [159] and basic information is most useful. Slow cooling of the melt or stretching of the polymers favours the orientation of the polymeric chains, thus increasing the crystallinity, and using X-ray diffraction or micro Brillouin scattering , crystalline regions can be studied. In the appendix, the calculated elastic constants of a few polymeric crystals, polyethylene (PE), poly-vinyl-alcohol (PVA), poly-vinylidene-fluoride (PVF), Nylon 6 α and γ and polypropylene, are reported. These refer to ideal, perfectly crystalline polymers and depict a limiting elasticity of actual polymeric materials. These parameters are an important reference to interpret the elasticity of polymeric materials in terms of their molecular and crystal structure [160, 161].

In agreement with the fact that the polymeric crystals are bound by covalent bonds in one direction and by van der Waals interaction in the others, the elastic constant c_{33} along the chain is expected to be one order of magnitude, or more, larger than c_{11} and c_{22}. It can be seen from the table reported in the appendix, that this is actually found for PE and also for PVA, PVF and Nylon 6 α. These are polymers with chains in planar zig-zag conformation and the compressional energy distributes in the bond stretching and bond bending coordinates. It can be assumed that the different ratio between c_{33}, c_{11} and c_{22} in the four polymers considered here, is due to a different distribution of the compressional energy between bond stretching and bending. Considering the directions perpendicular to the polymeric chains, it can be seen that PE is rather isotropic, since the chains are packed by van der Waals forces, while the other polymers are significantly stiffer than PE and more anisotropic. This arises because the chains are also held by hydrogen bonds. The anisotropy in this plane is more relevant in nylon, since hydrogen bonding gives rise to sheets which are stacked by van der Waals forces and the stiffness is larger in the plane of the sheets than in the perpendicular direction. The lower anisotropy of Nylon 6 γ compared with Nylon 6 α, is due to the fact that the polymeric chain of the former is less elongated because of the different conformation around the peptide bond.

Polypropylene is the prototype of polymers with helical chain conformation. In these systems, the compressional energy distributes prevalently in the softer bond and torsional angles, and therefore the elastic constant in the chain direction is considerably smaller (42.44 against 320 GPa in polyethylene) than in the linear chain polymers considered before. In the plane perpendicular to the chain, where the interaction between the side methyl groups predominates, the stiffness is very low. To interpret experimental anisotropy in this plane Tashiro et al. suggested that anharmonicity of the methyl group rotation plays an important role [162].

It is remarkable that the degree of anisotropy encountered in polymers is completely absent in the crystals of long-chain hydrocarbons like $C_{33}H_{68}$ and $C_{36}H_{74}$ [163, 164]. In composite or amorphous polymeric materials, the anisotropy is reduced if not almost completely lost [165, 166].

The interplay of intermolecular strain and intramolecular flexibility has a particular importance in determining the mechanical properties of biomolecules and biopolymers (proteins, DNA, polysaccharides) [167, 168]. Some attention has been devoted to determine the elastic constants of protein crystals [169–173]. The relevance of the problem derives from the circumstance that the response of biomolecules to external stresses regulates, to a significant extent, their biological function and activity. On the other hand, biomolecules are generally hydrated and bound in a complex way to their environment, but knowledge of the mechanical properties of biomolecular crystals is basic starting information.

Within this perspective the elastic constants of tetragonal lysozyme crystals, whose crystal arrangement [174] is shown in Figure 3.4, have been measured and discussed in some considerable detail as a case study. The

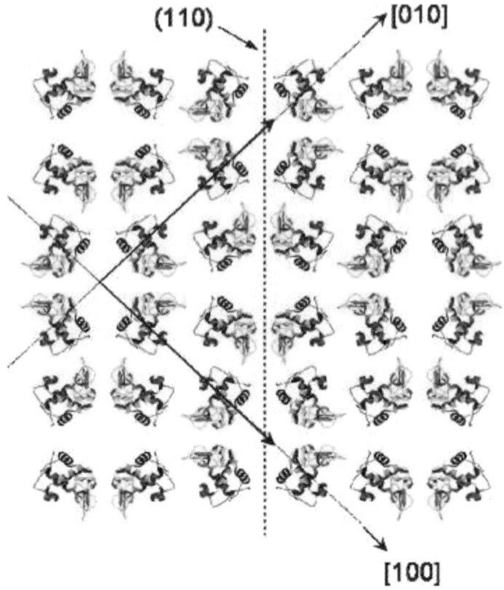

Fig. 3.4 Packing of the lysozyme protein in the tetragonal crystal viewed along the unique axis (reprinted with permission from A. Zamiri and S. De *Langmuir* **26** (2010) 4251, Copyright (2010), American Chemical Society).

elastic constants of lysozyme, crystal dehydrated at 42%, and of hydrated lysozyme are reported in the appendix. It can be seen that both hardness and anisotropy increase on dehydration. This emphasizes the role of water content and mobility on the general mechanical properties of biomolecules. A point of great relevance is the relative contribution of the intrinsic flexibility of a protein macromolecule and of intermolecular interactions determining the crystal packing. The issue has also been discussed on the basis of molecular dynamics simulations [175]. On the whole, it can be noted that the elastic constants are not dissimilar from those of "normal" molecular crystals.

There are various classes of molecular crystals where the mechanical properties and the behaviour under pressure are important, not only for the correlation with the structural properties but also for relevant practical purposes. This is the case for pharmaceuticals as has been discussed by Fabbiani and Pulham [176]. Mechanical properties are important for processing, such as compaction, solubility and the bioavailability of pharmaceuticals. In addition, pharmaceuticals frequently occur in more polymorphs that differ as far as bioavailability, through their solubility, processing and storage are concerned [177]. Examples reported in the appendix include aspirin, paracetamol (acetaminophen), carbamazepine, primidone and BRL61063, whose molecular structures are shown in Figure 3.5.

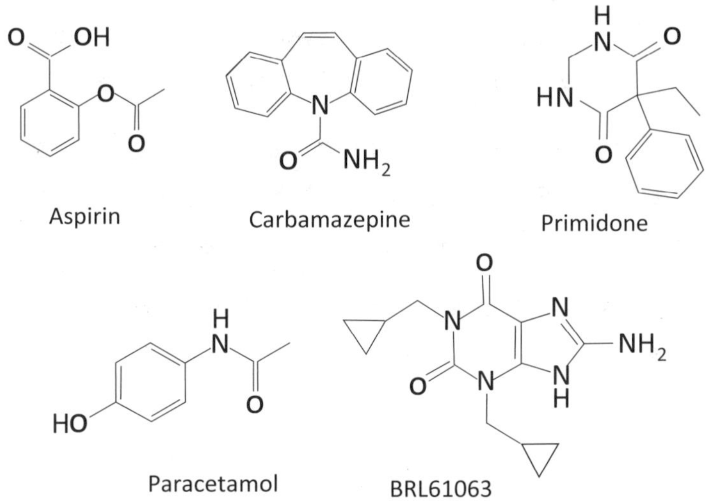

Fig. 3.5 Molecular structure of representative pharmaceuticals.

Aspirin, acetylsalicylic acid, is found in two monoclinic polymorphs and the elastic constants have been measured for the stable form I and calculated for form II. The molecules are bound in hydrogen-bonded dimers which are packed in the crystal by van der Waals interactions. Aspirin is an overall soft material and this accounts for the good processability. In Figure 3.6, the packing of aspirin I is shown together with a section of the linear elastic compressibility. The figure also shows that the largest linear stiffness is actually in the direction of the hydrogen bonds.

Carbamazepine , 5H-dibenz[b,f]azepine-5-carboximide, can exist in four different polymorphs [178] and the elastic constants have only been determined for the monoclinic form III. Molecules in the crystals are again bound in dimers by hydrogen bonding. It can be seen from the table that the diagonal tensile constants indicate a rather isotropic system. However, as can be seen from Figure 3.7, the linear compressibility is rather anisotropic, particularly in the *ac* and *ab* planes. The shear deformability is of particular importance for the processing of pharmaceuticals. It can noted that the Young's modulus of form III (7.3 GPa) differs considerably from that of the trigonal polymorph II (13.2 GPa) and this shows how changes in the crystal structure can significantly alter the mechanical properties [179]. A

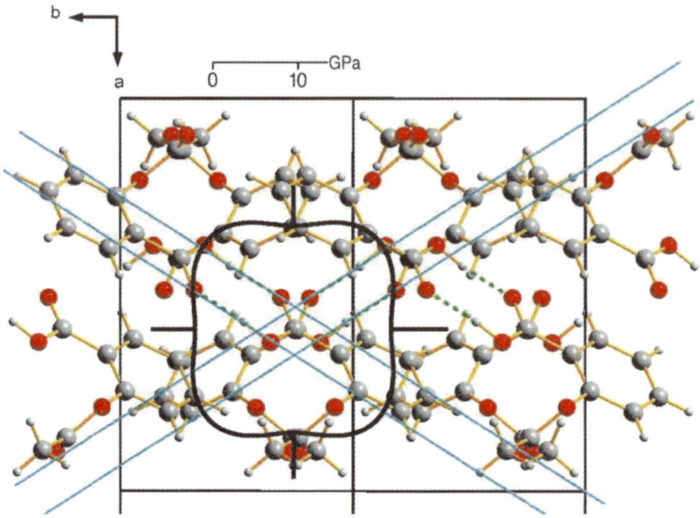

Fig. 3.6 Projection in the *ab* plane of the structure of aspirin I showing the directions of the hydrogen bonds and the section of the linear compressibility (reprinted with permission from J. D. Bauer, et al. *Cryst. Growth Des.* **10** (2010) 3132, Copyright (2010), American Chemical Society).

similar difference has been found for the two polymorphs of sulphathiazole (with values of 10.5 and 14.6 GPa) [180].

What is also quite significant is the difference in anisotropy of the linear diagonal constants of the three crystal polymorphs of pharmaceutical BRL61063, a system of complex molecular structure 1,3-di(ciclopropylmethyl)-8-aminoxantine [177], as shown in Figure 3.5.

Paracetamol (parahydroxyacetanilide), an analgesic and antipyretic drug of common use whose structure is reported in Figure 3.5, is a case study of interest to show the correlation between crystal structure and elasticity. Paracetamol crystallizes in the monoclinic form I (which is the stable form), in a metastable orthorhombic form II (orthorhombic) [181, 182] and in an additional non-characterized form. The calculated elastic constants for forms I and II are reported in the appendix. In the monoclinic form, the elastic constants for elongations along the principal axes are quite similar, while c_{55} is definitely smaller than c_{44} or c_{66} in the metastable orthorhombic form II. It can be seen that the values of the constants and the anisotropy change considerably in the two polymorphs and the changes can be intuitively understood with reference to the crystal packing shown in Figure 3.8. In both forms the molecules are hydrogen bonded to form sheets par-

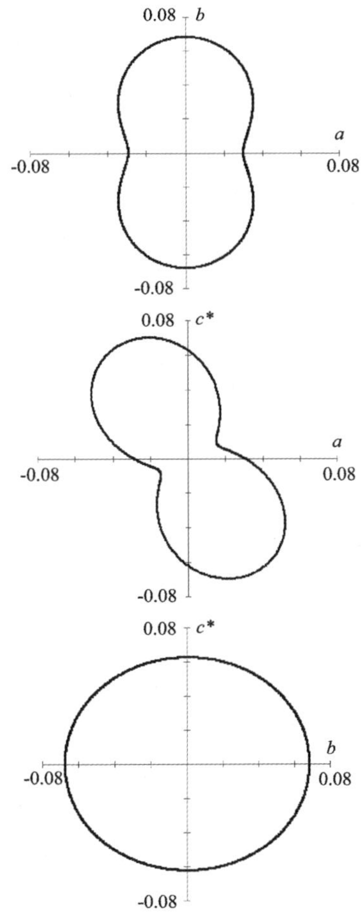

Fig. 3.7 Linear compressibility of carbamazepine form II in the main crystal planes (reprinted with permission from H. Mohapatra and C. J. Eckhardt *J. Phys. Chem. B* **112** (2008) 2293, Copyright (2008), American Chemical Society).

allel to the *ab* plane, but the pleated sheets configuration of form I moves to a planar configuration in Form II. This explains the large value of the c_{11} constant, since the crystal axis a coincides with the direction of the hydrogen bonds. The crystal packing also accounts for the low value of the c_{33} constants, since the intermolecular interactions between the sheets along the c axis are only of the van der Waals type. The resistance to slippage of the molecular sheet, one with respect to the other, is expected to be low as is argued from the small value of the shear constant c_{55}. The mechanical

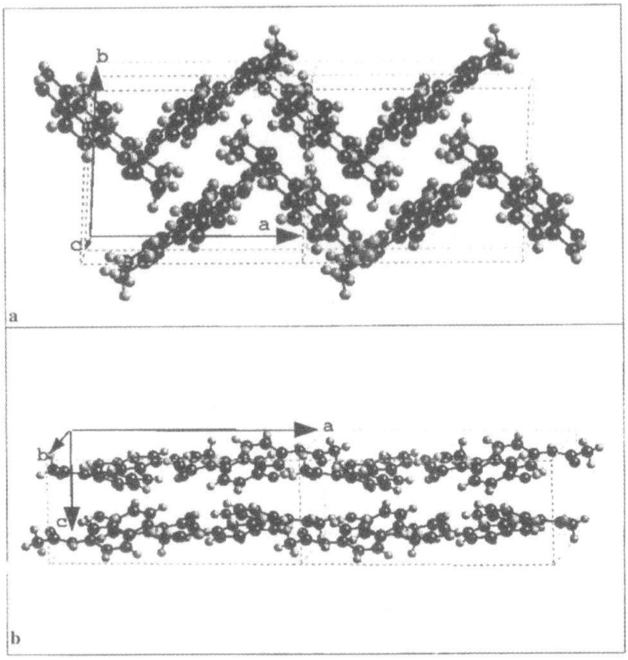

Fig. 3.8 Packing of the paracetamol molecules in polymorph I (a) and in polymorph II (b) (reprinted with permission from T. Beyer, G. M. Day, S. L. Price *J. Am. Chem. Soc.* **123** (2001) 5086, Copyright (2001), American Chemical Society).

properties of paracetamol are of particular relevance for practical purposes. In fact, the stable form I has too high a shear constant for processing and unfavourable morphology. Therefore, several methods have been discussed to improve the processing, such as exploiting the effect of dislocations on the mechanical properties, the formation of cocrystals, or the direct use of appropriately stabilized polymorph II [183, 184].

For primidone, an anticonvulsant that exhibits an orthorhombic and a monoclinic polymorph, the elastic constants have been reported only partially and the Young's moduli of the two forms are rather close [149, 185].

The response of energetic materials to external stimuli is of primary importance for their applications, since during their use they experience extreme high pressure and temperature. Generally, energetic materials are employed as composite materials with binders and the deformation processes must be considered at the microscopic, mesoscopic and macroscopic level [146, 176]. Nevertheless, performance depends on the morphology, orientation and energy dissipation in the primary stages of detonation, which are basic properties of the energetic components and therefore, their mechanical properties are a necessary prerequisite. In the appendix, the elastic constants of a few representative secondary explosives (HMX, RDX, PETN, CL-20 and TNAD), whose molecular structures are shown in Figure 3.9, are reported. As soft molecular crystals, all these materials can

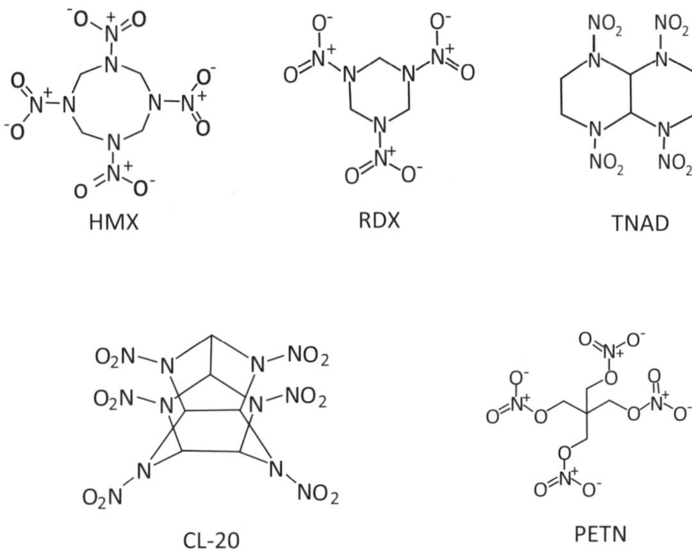

Fig. 3.9 Molecular structure of representative secondary explosives.

exist in different polymorphs and the question arises, which is the most appropriate for use on the basis of their mechanical properties? Inspection of appendix shows that for these systems, the elastic constants have values close to the typical values of molecular crystals and the anisotropy of the elastic constants is evident but not very prominent. However, this is an important issue, since there are clear indications of the importance of shear deformation in the detonation processes. A better, or rather a more immediate, view of the anisotropy can be gained from the linear compressibility which is reported in Figure 3.10 for the β polymorph of HMX [186]. It can be seen that the a crystal axis is the least compressible and this has been associated with weaker intermolecular interactions in the bc crystal plane, in agreement with the observed ease of cleavage. The secondary explosives considered here are rather complex and flexible molecules, and the

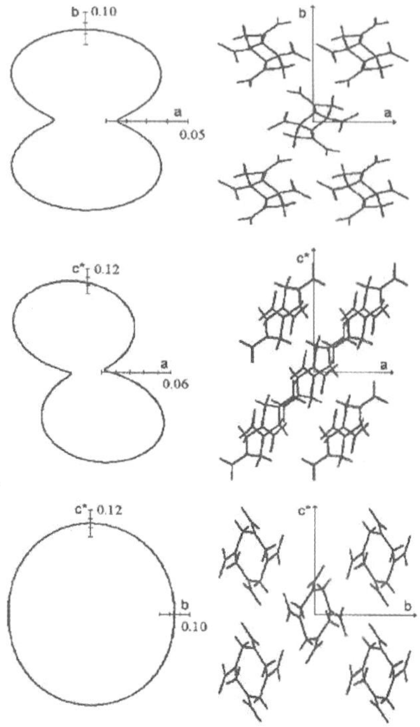

Fig. 3.10 Linear compressibility of the β polymorph of secondary explosive HMX in the main crystallographic planes (reprinted with permission from L. L. Stevens and C. J. Eckhardt, *J. Chem. Phys.* **122** (2005) 174701, Copyright (2005), American Institute of Physics).

possible role of phonon-phonon and phonon-vibrational exciton couplings in determining the overall elastic properties has been discussed.

The Christoffel equations, from which the elastic constants can be obtained, are strictly devised for an elastic continuum and, therefore, the coupling of acoustic modes with librations or internal modes of the molecular units in molecular crystals is neglected. However, when these latter modes occur at low frequency, the decoupling is not effective [5] with consequences on the values of the elastic constants. Examples of contributions of translation-rotation coupling and internal vibrations to the crystal elastic constants have already been mentioned above. The most extreme case is represented by the fullerene C_{60} crystal. In fact, it has been estimated that the intrinsic stiffness of the C_{60} units is higher than 700 GPa [187], but the linear elastic constants of the simple cubic crystals (sc) at room temperature range from 6.6 to 14.9 GPa, as can be seen from the appendix, and are comparable with those typical of aromatic crystals. In this particular case, the molecular flexibility is of no relevance for the elastic properties of the crystal and only some rotational-translational coupling could come into play. The situation can be significantly different in highly flexible molecules, like biomolecules, with many low-frequency vibrational modes. For crystalline C_{60}, an abrupt increase of the elastic constants is observed at 275 K on lowering the temperature [188, 189], associated with the occurrence of the phase transition to the face-centred cubic (fcc) low-temperature phase.

3.2.2 *Temperature and pressure dependence of the elastic constants*

Neglecting the electronic contribution, the Helmholtz free energy for a molecular crystal can be written as,

$$F(V,T) = F_0 + \frac{1}{2}\sum_{jk} h\nu_{jk}(V) + k_B T \sum_{jk} \exp - \left[\frac{h\nu_{jk}(V)}{k_B T}\right], \tag{3.33}$$

where the first term is the static lattice energy at zero temperature, the second term is the zero-point vibrational energy with j and k numbering the phonon mode and its wavevector, respectively, and the last term is the vibrational contribution. The temperature and pressure dependence of the elastic constants, as second derivatives of the free energy with respect to strain, will be mediated by the volume dependence of the vibrational frequencies of the lattice, volume dependence that is expressed by Grüneisen parameters,

$$\gamma_{jk} = \frac{d\log \nu_{jk}}{d\log V}. \tag{3.34}$$

Therefore, the temperature and pressure dependence of the elastic constants involves the anharmonicity of the interatomic interaction potential, since in a purely harmonic crystal no volume variation would occur. In the already mentioned quasi-harmonic approximation, the effect of temperature or pressure is mastered, minimizing the structure of the crystal at the effective volume, thus accounting for an effective anharmonicity.

The temperature effect on the elastic constants in molecular crystals has been studied by experiments and computations. Haussühl [147] reported the thermoelastic constants,

$$Tij = \frac{dc_{ij}}{dT}, \tag{3.35}$$

of a number of molecular crystals. The thermoelastic constants are negative, indicating the expected stiffening of the crystals at low temperature. In Figure 3.11, the elastic constants of the benzene orthorhombic crystal as a function of temperature are reported [190], showing mostly a linear variation. However, the variation at the lowest temperatures should be more complex, approaching a zero slope at zero temperature [191]. It can be

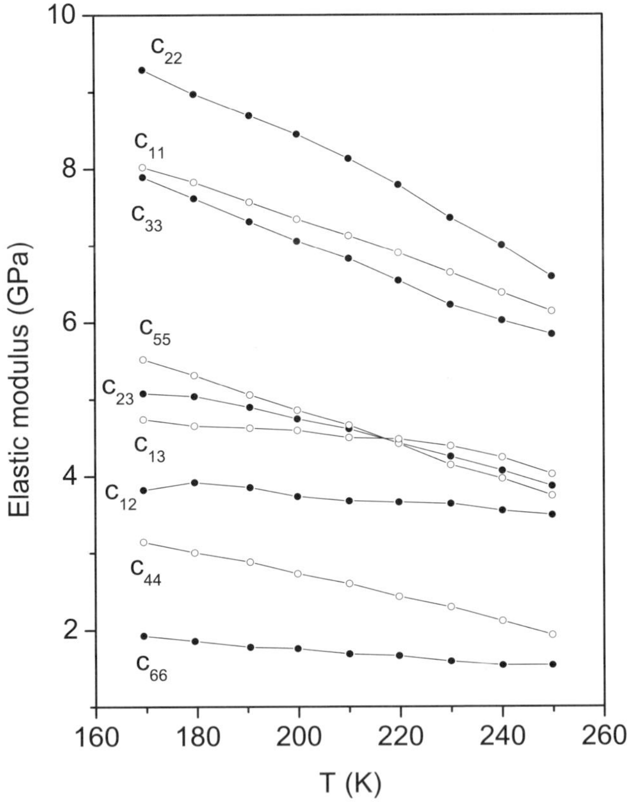

Fig. 3.11 Elastic constants of the orthorhombic benzene crystal as a function of temperature (readapted with permission from J. C. W. Heseltine, et al. *J. Chem. Phys.* **40** (1964) 2584, Copyright (1964), American Institute of Physics).

noted from Figure 3.11, that the various thermoelastic constants differ appreciably among them and, therefore, the anisotropy of the crystal elasticity changes with temperature. The linear behaviour, observed in benzene and in many other cases, obviously applies in the absence of phase transitions when discontinuities or instabilities of some of the elastic constants can be observed. As an example, in Figure 3.12, a plot of the Young's modulus of C_{60} as a function of temperature is reported, showing the discontinuity at the transition temperature from the simple cubic to the face-centred cubic structure [192].

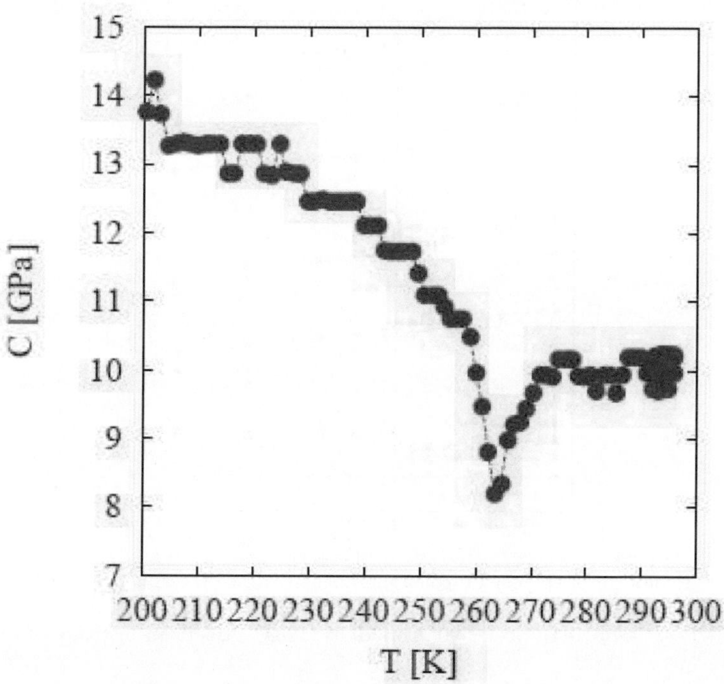

Fig. 3.12 Evolution of the Young's modulus of fullerene C_{60} as a function of temperature through the sc-fcc phase transition (reprinted with permission from Acoustic phonon dispersion in single-crystal C_{60} by M. Haluska, et al, *J. Phys.: Condens. Matter* **11** (1999) 1009, Copyright (1999), IOP Publishing Ltd).

Another example of abrupt changes of the elastic constants through a phase transition is shown in Figure 3.13 for the deuterated p-terphenyl crystal, which undergoes a phase transition from an orthorhombic to a monoclinic structure [193].

The volume contraction on increasing the pressure is generally much larger than on lowering the temperature. Consequently, the stiffening of

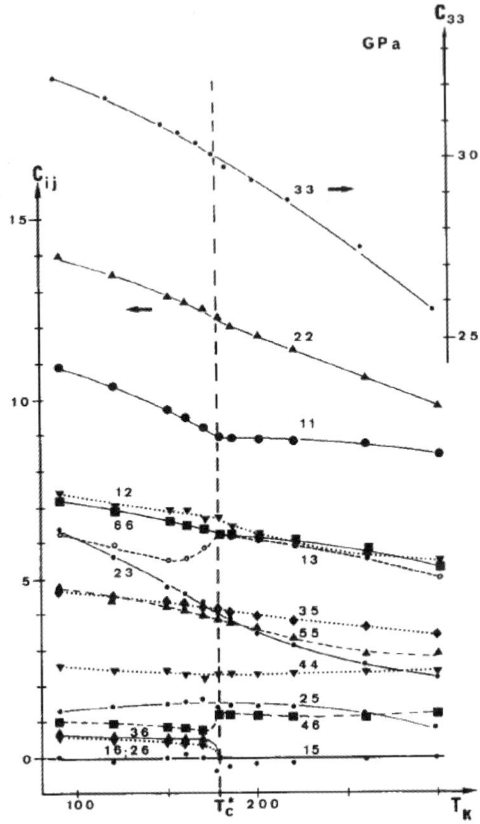

Fig. 3.13 Evolution of the elastic constants of perdeutero p-terphenyl as a function of temperature through the phase transition (reprinted with permission from C. Ecolivet, et al. *J. Chem. Phys.* **81** (1984) 599, Copyright (1984), American Institute of Physics).

the crystals at high pressures is more substantial than on lowering the temperature. As an example, in Figure 3.14, the elastic constants as a function of pressure are reported for methane. Again, an overall linear trend is observed in the range of the pressure considered. However, in several instances the experimental trends have been fitted by a second order polynomial.

In the appendix, the enormous difference of the reported elastic constants of bromine from those of iodine can be noted. The difference is, however, only apparent and simply means that the elastic constants of bromine have been determined at a much higher pressure [194, 195]. When

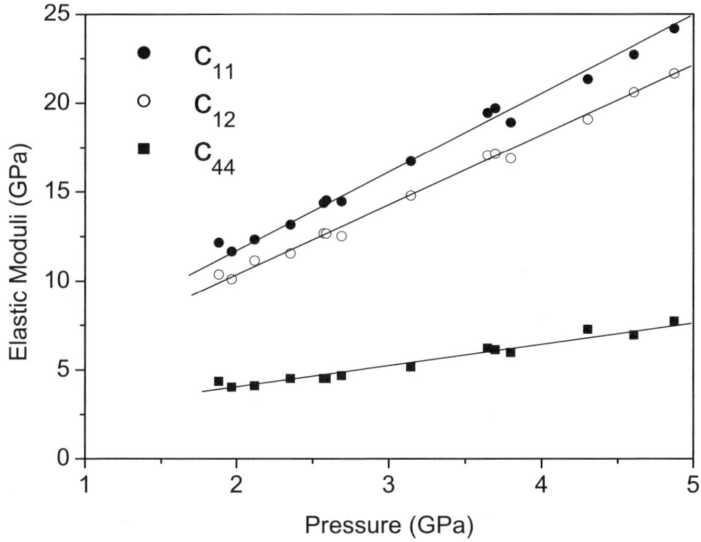

Fig. 3.14 Evolution of the elastic constants of the methane crystal as a function of pressure (re-adapted with permission from S. Sasaki, et al. *Physica B: Condensed Matter* **219-220** (1996) 380, Copyright (1996), Elsevier).

comparing elastic constants of different homologous systems, care should be taken to perform the comparison at corresponding temperatures and pressures.

An interesting correlation has been discussed for the elastic constants, as a function of pressure, of the orientationally disordered (cubic) crystals of simple molecules including krypton, methane, ice VII, NH_3-III, H_2S, HCl, considering the interplay of the translational-rotational coupling and the hydrogen bonding, whenever they are present [196]. It is found that the elastic constants of NH_3-III, H_2S and of HCl are quite similar over the pressure range up to 4.5 GPa. The anisotropy of all these crystals, defined as,

$$A = 2\frac{c_{44}}{c_{11} - c_{12}}, \tag{3.36}$$

is rather different and has a different profile as a function of pressure as can be seen from Figure 3.15. The anisotropy increases with pressure because of the increase in the translational-rotational coupling which, in turn, is prevented by the strengthening of the hydrogen bonding.

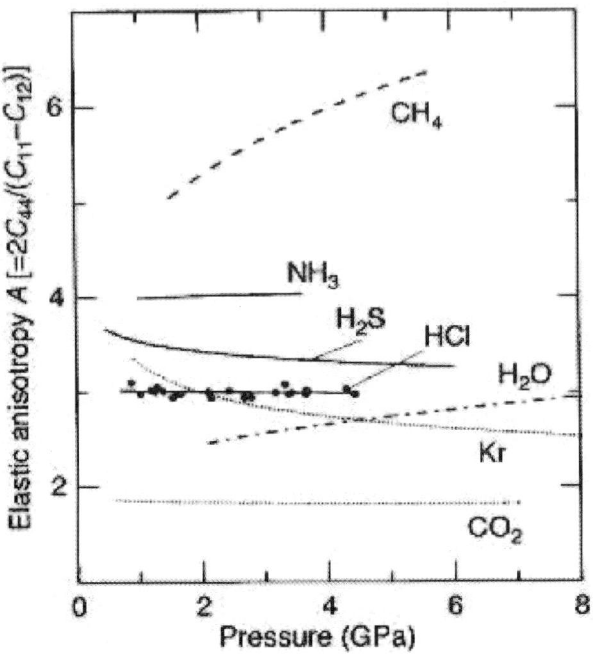

Fig. 3.15 Evolution with pressure of the anisotropy for a number of cubic crystals (reprinted with permission from H. Shimizu, et al. *Phys. Rev. B* **59** (1999) 11727, Copyright (1999), by the American Physical Society).

Both temperature and pressure produce a primary volume variation. Therefore, as argued from the expression of free energy [3.33], their effect is mediated by the volume dependence of the vibrational frequencies. Similarities are therefore expected between temperature and pressure effects on the crystal structure and on the mechanical properties of solids. This issue has been discussed by Hazen et al. [197–199] in terms of the so-called inverse relationship, illustrated in the idealized form in Figure 3.16, showing that the effect on the structural parameters of lowering (or raising) the temperature can be compensated by lowering (or raising) the pressure.

To illustrate the point, for example, it is worth mentioning the variation with pressure of the unit cell volume of paracetamol and of the linear strain along the principal directions of the strain ellipsoid (see the following section), reported in Figure 3.17, which is very similar to the behaviour observed when the temperature is reduced from 300 to a few kelvin at ambient pressure [200].

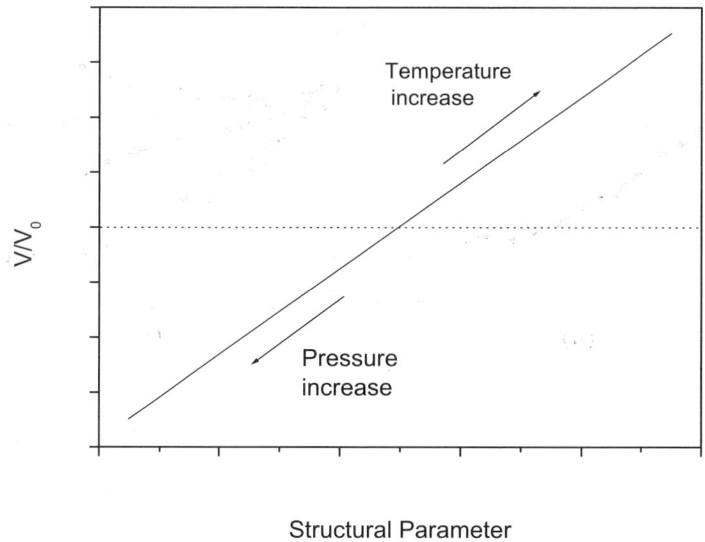

Fig. 3.16 Inverse relationship: idealized comparative representation of the temperature and pressure effects on crystallographic parameters.

3.2.3 *Single crystals and polycrystals*

In many practical applications, solid molecular materials (e.g., polymers, pharmaceuticals, biomaterials, energetic materials) are used as polycrystals or composites and are therefore, more or less, macroscopically isotropic. For this kind of material Voigt [201] considered the orientational average of the elasticity tensor c, assuming conditions of constant strain and continuity across the grain boundaries, and obtained for the bulk modulus B_V and for the shear modulus G_V the expressions,

$$B_V = \frac{1}{9}[c_{11} + c_{22} + c_{33} + 2(c_{12} + c_{13} + c_{23})], \tag{3.37}$$

$$G_V = \frac{1}{15}[c_{11} + c_{22} + c_{33} - c_{12} - c_{13} - c_{23} + 3(c_{44} + c_{55} + c_{66})]. \tag{3.38}$$

Alternatively, Reuss [202] considered a random array of crystallites under uniform stress with the condition of continuity and obtained the bulk modulus B_R, and the shear modulus G_R, from the orientational average of the compliance tensor s,

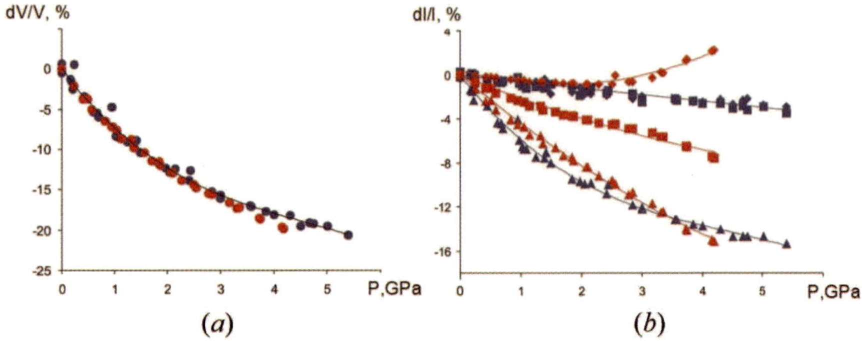

Fig. 3.17 Variation of the unit cell volume (a) and linear strain in the principal directions (b) as a function of pressure of paracetamol. Red and blue symbols refer to the monoclinic and orthorhombic polymorph, respectively. (Reprinted with permission from High-pressure diffraction studies of molecular organic solids. A personal view, by Elena V. Boldyreva *Acta Cryst. A* **64** (2008) 218, Copyright (2008), by the International Union of Crystallography).

$$\frac{1}{B_R} = [s_{11} + s_{22} + s_{33} + 2(s_{12} + s_{13} + s_{23})], \tag{3.39}$$

$$G_R = \frac{15}{4(s_{11} + s_{22} + s_{33} - 4(s_{12} - s_{13} - s_{23}) + 3(C_{44} + C_{55} + C_{66})}. \tag{3.40}$$

Later, Hill [203] showed that the Voigt and Reuss averages actually represent an upper and a lower boundary, respectively, of the effective elastic moduli of a polycrystal and it is usual practice to take as the effective moduli, the arithmetic (or geometric) mean of B_V and B_R, and of G_V and G_R. Several other alternative averaging procedures for polycrystals have been proposed and discussed [6, 139, 204, 205]. Apart from this, a major problem arises in molecular crystals as the large anisotropy of these systems leads to preferential crystal growth along specific directions and it may be difficult to obtain perfectly random orientation of the crystallites. This can manifest in deviations from the expected average moduli. The effect of impurities and defects, particularly at the grain boundaries, can also lead to deviations from the expected averages. Treatment (e.g., compaction) of polycrystalline powders by uniaxial pressurization can, in turn, lead to partially anisotropic macroscopic materials by induced reorientation of the crystallites.

For all the molecular van der Waals crystals reported in the appendix, the bulk moduli have also been obtained and in several cases compared with the averages. The bulk moduli, which are obviously more amenable to experimental determination, are available for many other molecular crystals.

3.2.4 *Variation of crystal parameters with pressure*

An intuitive appreciation of anisotropy can be gained from the pressure evolution of the lattice parameters. In the ethylene crystal (which belongs to the monoclinic crystal system), it has been found experimentally from X-ray diffraction on the isotopomer C_2D_4 crystal [206], that when the pressure is increased from 20 to 1930 MPa at 90°C, the unit cell contraction is highly anisotropic, as can be seen from Figure 3.18. More specifically, the linear contraction $\frac{\ell}{\ell_0}$ along the a, b and c crystal axes is 0.976, 0.84 and 0.985, respectively. This anisotropic compressibility has been qualitatively reproduced in an *ab initio* molecular dynamics simulation in the Car–Parrinello approach [207], showing that the compressibility along the b axis is larger, although the difference is not as large as found experimentally. This anisotropy of the compressibility has a clear influence on the high-pressure behaviour of the ethylene crystal and in particular, on the polymerization reaction, as will be discussed later in this book. The knowledge of the evolution with temperature of the unit cell parameters is particularly useful in systems where the elastic constants are not available.

It is of interest to mention the variation with pressure, determined by X-ray diffraction, of the unit cell volume and unit cell parameters of the monoclinic phase of the benzene crystal [208]. The results for the unit cell parameters are shown in Figure 3.19. The anisotropy is considerably smaller than in the ethylene crystal. However, it should be noted that the importance of the (electronic) structural rearrangement induced by pressure and the anisotropy of the intermolecular interactions are not fully evident from the simple evolution of the crystal parameters. However, they play an important role in the high-pressure amorphization reaction of the benzene crystal to be discussed in a later chapter.

The anisotropy of crystal distortion in molecular crystals under pressure can, however, be much more pronounced than in the cases discussed above. The case of the monoclinic polymorph of acetaminophen (paracetamol), whose molecular structure is shown in Figure 3.5, is of particular interest to illustrate the point. Acetaminophen is an important analgesic and antipyretic which can be crystallized in the polymorph I (monoclinic

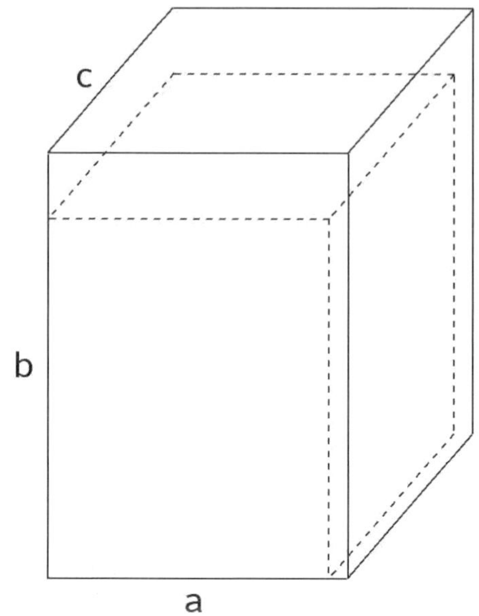

Fig. 3.18 Comparison of the unit cell of crystalline perdeuteroethylene at low (full line) and high (dotted line) pressure.

$P2_{1/n}$) [25] and in the polymorph II (orthorhombic Pbac) [209]. The monoclinic phase has been studied by X-ray diffraction under pressures up to 4 GPa [25]. Although the pressure-volume diagram is rather regular, as can be seen from Figure 3.20, the variation of the unit cell parameters shown in the figure is highly anisotropic. In particular, it is found that the c crystal axis first decreases with increasing pressure up to 2 GPa and then increases. The compressibility behaviour must be interpreted in terms of the response of intermolecular interactions to compression. In the present case, the intermolecular interactions are dominated by stronger (O-H$\cdots$O) and weaker (N-H$\cdots$O) hydrogen bonds. The crystal arrangement is characterized by a network of hydrogen bonds in the ac crystal plane, while the interaction between the hydrogen-bonded layers is controlled by van der Waals interactions between the aromatic rings. It turns out that the larger compressibility is found to occur along the b axis which is perpendicular to the hydrogen-bonded layers. Generalization of the concept that a larger compressibility occurs along the directions of weaker intermolecular interactions should be taken with some caution. In fact, in acetaminophen the refinement of the structure shows that in the hydrogen-bonded layer,

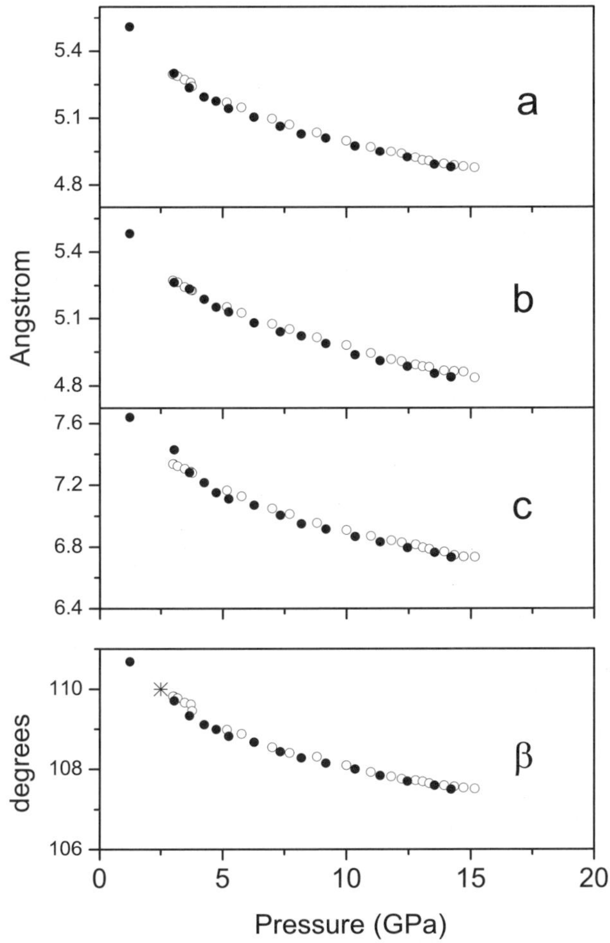

Fig. 3.19 Evolution of the crystal parameters in crystalline form II of benzene with pressure during compression (filled dots) and decompression (empty dots) experiments.

the increase of pressure leads to a slight shortening of both the (O-H$\cdots$O) and the weaker (N-H$\cdots$O) hydrogen bonds. However, on the whole, the hydrogen-bonded layer contracts in one direction and expands in another as the pressure increases. This is due to the occurrence of a continuous conformational change of the molecular framework, induced by the pressure, since a planar structure can itself affect the stability of the hydrogen bond network. The rationale is that in the crystal, it is not the minimum of a single intermolecular interaction but the cooperative effect of all the inter-

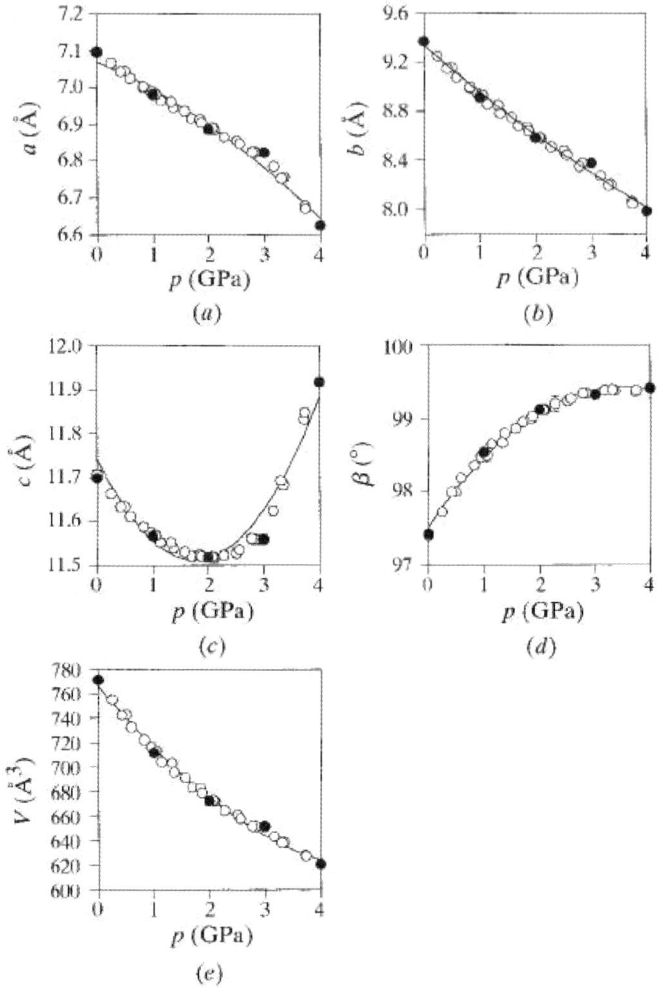

Fig. 3.20 The variation of the unit cell parameters and the unit cell volume with pressure in form I of acetaminophen (reprinted with permission from E. V. Boldyreva et al. *Acta Cryst.* *B* **56** (2000) 299, Copyright (2000), by the International Union of Crystallography).

actions at work that really matters. Phenacetin differs from acetaminophen in the substitution of the hydrogen in the O-H group by an ethyl group. As a consequence, only weaker N-H· · · O hydrogen bonds are operative. Although in phenacetin a distinct anisotropy is also observed, anomalies in the lattice parameters upon varying the pressure are not observed [210].

There is an overall similarity in the packing of acetaminophen and of chlorpropamide, a drug of far more complex molecular structure (4-chlor-N-((propylaamino-carbonyl)-benzene sulphonamide), used as an antidiabetic agent. In this system, the highest compressibility is observed along the b axis of the orthorhombic crystal [211]. This finding correlates with a structure, formed of pleated sheets parallel to the ac plane, dominated by strong hydrogen bonds.

Molecular crystals with hydrogen bond chains or layers or three-dimensional arrays are particularly amenable to this kind of analysis to disentangle the interplay of hydrogen bonds, van der Waals interactions and conformational rearrangements of flexible molecular units. The matter has been reviewed by Boldyreva [200, 212–218].

The pentaerithritol crystal has been studied by X-ray diffraction [219]. The low-temperature phase is ordered and tetragonal, and is stable at high pressure up to 1.2 GPa. Increasing the pressure, it is seen that the compressibilities along and perpendicular to the unique axis are very different. The results are shown in Figure 3.21. Actually, between 0.1 and 1155 MPa, the linear average compressibilities are $k_a = 4.6410^{-2}$ GPa and $k_c = 0.8310^{-2}$ GPa, differing by a factor of six. The explanation can again be found by considering the molecular arrangement in the crystal, consisting of sheets of strongly hydrogen-bonded arrays of molecules in a plane perpendicular to the unique axis, with weaker van der Waals interactions between different sheets.

The direct connection of the anisotropic linear compression along the crystal axes with the structural arrangement and the type of intermolecular interactions, has been clearly evidenced in an X-ray diffraction study of the high-pressure polymorph (monoclinic $P2_{1/c}$) of 1,3-cyclohexanedione [220]. The structural arrangement in the present case consists of chains of hydrogen-bonded molecules in the enolic form. The chains form planar sheets where the hydrogen-bonded chains are held together by strong electrostatic dipole-dipole interactions (the dipole moment of the enolic form of cyclohexanedione is larger than 5.0 D). Weaker van der Waals interactions are active between different sheets. Consequently, the linear compressibility along the b axis, perpendicular to the sheets, is larger than the compressibility along the other two axes. The same considerations apply to the similar crystals of cyclopentanedione [221] and 2-methylcyclopentanedione [222]. In all these cases the molecular framework is flexible and this, as already noted, contributes to determine the overall molecular arrangement. However, the correlation between the direc-

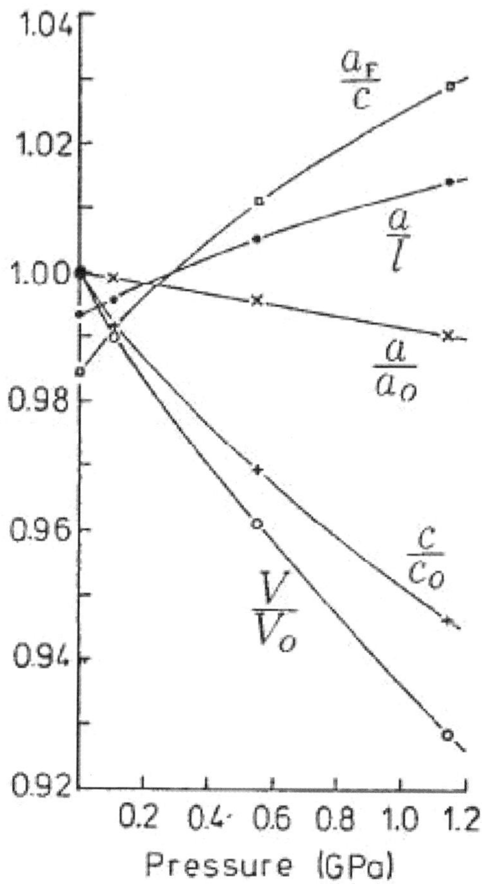

Fig. 3.21 Pressure variation of the unit cell parameters and unit cell volume of pentaerythritol (reprinted with permission from A. Katrusiak, *Acta Cryst. B* **51** (1995) 873, Copyright (1995), by the International Union of Crystallography).

tion of the larger compressibility and the strength of the intermolecular interactions is not always as simple. For instance, in the dimedone (5,5-dimethyl-1,3-dimethylcyclopentanedione) crystal, it has been found [223] that the crystal is actually less compressible in the direction perpendicular to the molecular planes of hydrogen-bonded units. The same is the case for p-benzoquinone [215], where the molecular framework is rigid and the molecules are arranged in flat layers held together by weak C-H$\cdots$O hydrogen bonds. The compressibility is found to be larger in the plane of the layers.

Differences in linear compressibilities can arise from a variety of different arrangements in the crystal. The crystals of simple amino acids are of considerable interest as a model for the packing and properties of more complex peptides and biopolymers in general. Glycine, the simplest amino acid, has been found to crystallize in three different polymorphs at ambient pressure. Two polymorphs, $\alpha-$ and $\beta-$glycine, are monoclinic ($P2_1/n$ [224] and $P2_1$ [225] space groups, respectively) while $\gamma-$glycine is trigonal ($P3_1$) [224, 226]. For the latter polymorph, a phase transition to a $\delta-$form has been observed at high pressure [226]. The characteristic structural feature of all the glycine polymorphs is the linking of zwitterions in a network of N-H$\cdots$O hydrogen bonds. In $\alpha-$glycine, antiparallel hydrogen-bonded layers are formed and the interaction between the layers is of the van der Waals type. In the β polymorph, the layers are bound to each other by other hydrogen bonds to form a three-dimensional network, whereas in $\gamma-$glycine they are linked to form polar helices that are further hydrogen bonded in a three-dimensional array. The head-to-tail hydrogen bond of the zwitterions are rather strong and not very compressible. As a consequence, in $\alpha-$glycine, the highest compressibility is observed in the direction perpendicular to the plane of the hydrogen bonds, i.e., in the direction of the weakest intermolecular interactions. Similar considerations can be extended to the anisotropy of the linear strain of L-serine [227]. The most rigid direction in the crystal coincides with the a direction, which is parallel to the direction of the head-to-tail N-H$\cdots$O hydrogen bonds of the zwitterion units.

Peculiarities of the structural arrangements in connection with anisotropy of linear compressions have also been identified in molecular systems without hydrogen bonds. This is the case, for instance, of sodium oxalate in the pressure range up to 3.8 GPa [228]. The conformationally flexible oxalate ions are arranged in close-packed layers nearly parallel to the ac crystal plane of the monoclinic $P2_1/c$ crystal. It is found that the largest compressibility is in the direction perpendicular to the oxalate layers. A remarkable case of linear strain anisotropy has been reported for a mixed molecular-covalent system, the three-dimensional polymer obtained from C_{60} fullerene. At room temperature, the C_{60} lattice is held together by weak van der Waals interactions with a corresponding small value of the bulk modulus. At high pressures and temperatures, a three-dimensional polymer has been obtained [229]; its equation of state has been determined from X-ray diffraction [230] and fitted to the Vinet equation of state that will be discussed in the next section. A volume bulk modulus of 280 GPa has thus been derived, to be compared with the value of 443 GPa of diamond. As a consequence of different bonding characters, the bulk moduli along macroscopic axial and radial directions of 230 and 320 GPa, respectively, have been found.

These effects have been investigated in considerable detail for the nitromethane crystal to find hints for the mechanism of the pressure induced reaction. Nitromethane is stable at room temperature up to 25 GPa always remaining in the same monoclinic space group [24]. At low pressure, the methyl group freely rotates around the C-N bond as is shown in Figure 3.22. As the pressure is increased, the free rotation of the methyl group is gradually hindered. There is no evidence of this change in the pressure volume equation of state nor in the linear compression along the three crystal axes of the monoclinic cell as can be seen from Figure 3.23, which shows that the compression is rather smooth and isotropic in the present case. However, by accurate refinement of the crystal structure it has been found that while at low pressure, the staggered and eclipsed conformations of the methyl versus the nitro group are equally populated, corresponding to the free rotation of the methyl group, upon increasing the pressure a gradual locking into the eclipsed conformation occurs, a locking that is finally complete at $\sim$10 GPa. The evolution with pressure of the relevant rotation angle and of the population of the two limiting conformations is shown in Figure 3.24. The conformational rearrangement induced by the pressure can be explained in terms of the intermolecular interactions. In fact, at high pressure, the eclipsed molecules can become involved in an extended network of C-H$\cdots$O hydrogen bonds of different strengths. The high-pressure structural arrangement is shown in Figure 3.25. The behaviour of nitromethane nicely illustrates the interplay of the intramolecular flexibility of the molecular units and the intermolecular interactions. As the network of hydrogen bonds is formed, the infrared and Raman spectra of the nitromethane crystal change significantly to such an extent that the (low pressure) structure, with freely rotating methyl groups, and the (high pressure) structure, with molecules in the eclipsed conformation, definitely have different spectra. The evolution with pressure of the infrared spectrum is shown in Figure 3.26. Looking only at the vibrational spectra, one could argue that in nitromethane there are actually two different crystalline phases. Indeed, it has been found that the high-pressure phase is stable on releasing the pressure down to 4 GPa and that at this pressure, the ambient pressure structure is recovered only on heating at 50°C. On the whole, it appears that nitromethane can be taken as an example of a crystal exhibiting an

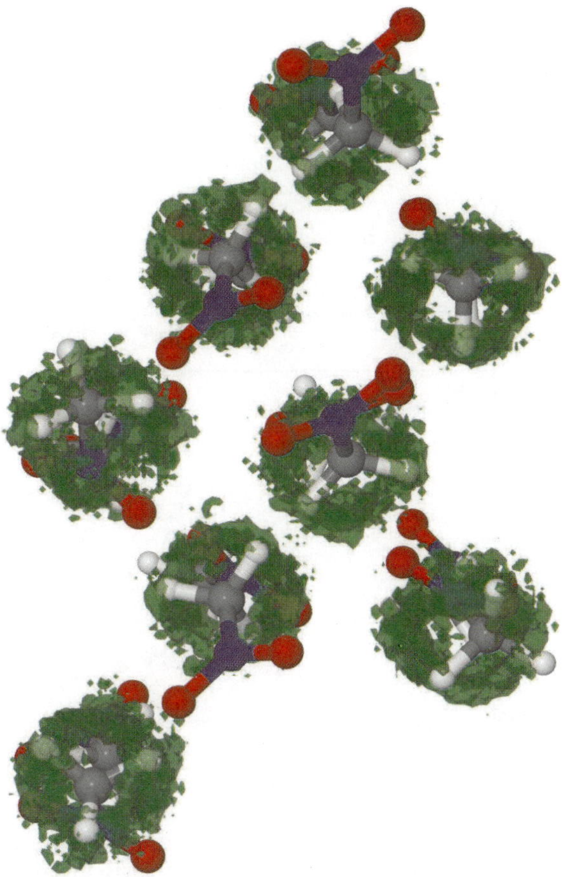

Fig. 3.22 Packing of the molecules in the crystal of nitromethane showing the free rotation of the methyl group around the C-N axis. The green surfaces represent the space region explored by the hydrogen atoms of nitromethane at the pressure of 1.1 GPa (phase I) during a Car–Parrinello molecular dynamics simulation.

isosymmetric phase transformation, in which the structure changes without variation of the space group [214, 231].

Using X-ray diffraction, the variation with pressure of the crystallographic parameters (lengths of the unit cell axes and angles between the axes) and of the unit cell volume can be obtained. In monoclinic crystals and, even more so, in triclinic crystals, the strain tensor that is obtained experimentally does not refer to the principal axes of strain [114, 200]. In monoclinic crystals, one of the principal axes coincides with the *b* crystal

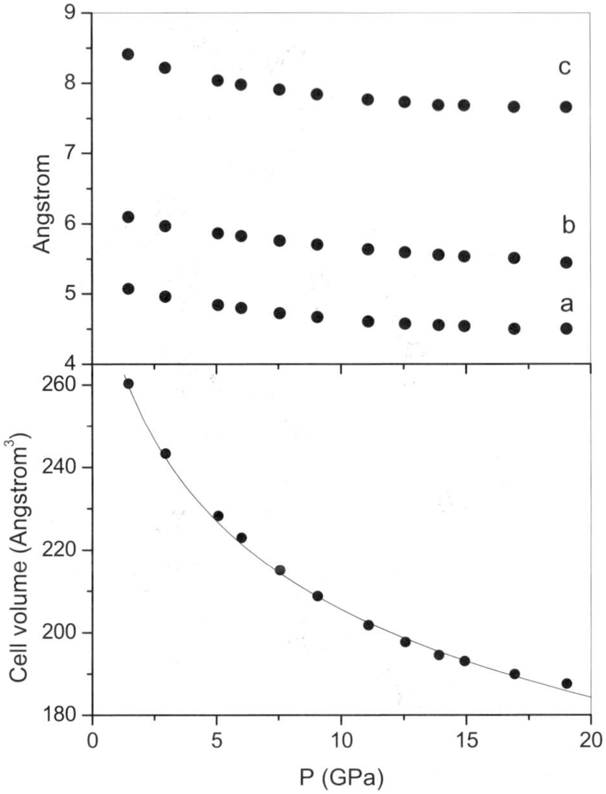

Fig. 3.23 Variation with pressure of the unit cell parameters (top) and of the unit cell volume (bottom) in the nitromethane crystal. Full line: fit to the Murnaghan equation of state.

axis, while the other two lie in the *ac* plane. Additional useful structural information can be obtained by transforming the experimental strain tensor to the principal axes [232–235] and correlating the directions of the principal axes with those of the relevant intermolecular interactions. This kind of analysis has actually been performed for several of the systems considered in this section [25, 211, 214, 215, 217, 224, 227, 228]. It can be of particular interest to follow the change in orientation of the principal axes with increasing pressure, as a probe of molecular rearrangements of the crystal packing.

The lattice parameters give a fairly good and complete picture of the effect of an external stress on the crystal structure and on the interplay

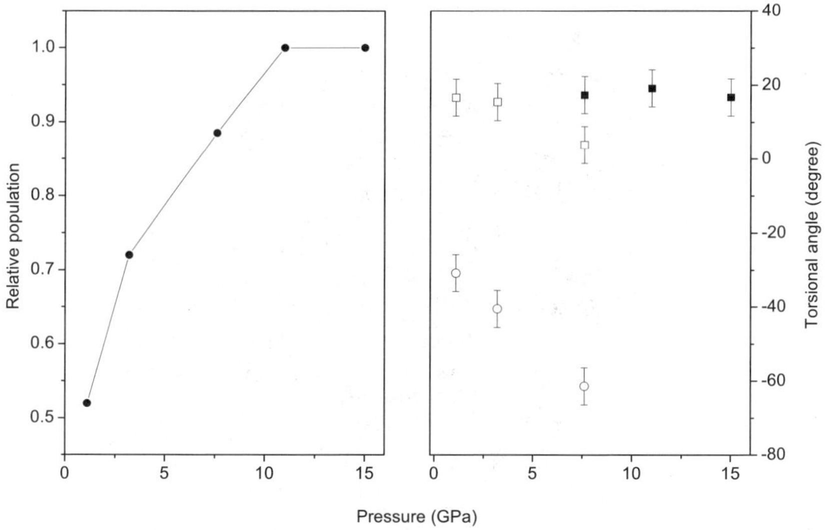

Fig. 3.24 Variation with pressure of the relative population of the eclipsed conformation (left) and of the torsional angle between the methyl and nitro groups (right) in nitromethane.

of intra- and intermolecular interactions. However, for crystals of lower symmetry (monoclinic and triclinic) and for special situations encountered in high-pressure studies, it is more desirable to resort directly to the strain tensor defined in (3.7). We extend the definition to also consider the finite Lagrangian (Green's) strain [113],

$$\eta_{ij} = \frac{1}{2}\sum_{m}\left(\frac{\partial u_i}{\partial rj} + \frac{\partial u_j}{\partial ri} + \frac{\partial u_m}{\partial ri}\frac{\partial u_m}{\partial rj}\right). \tag{3.41}$$

The lattice strain can be visualized through the strain ellipsoid [114]. As already noted, the lattice strain, not to be confused with the macroscopic strain, is a field strain and not a matter strain, and is not required to conform to the symmetry of the lattice. However, under hydrostatic compression (as in the case of thermal expansion and dilation), requirements are imposed on the strain ellipsoid. In crystals of cubic symmetry, the strain ellipsoid degenerates into a sphere while in crystals of hexagonal, tetragonal and trigonal symmetry, the strain ellipsoid has rotational symmetry around the unique axis. In orthorhombic crystals, the principal axes of the ellipsoid coincide with the crystal axes while, in monoclinic and triclinic crystals, the principal axes need not coincide with the crystal axes

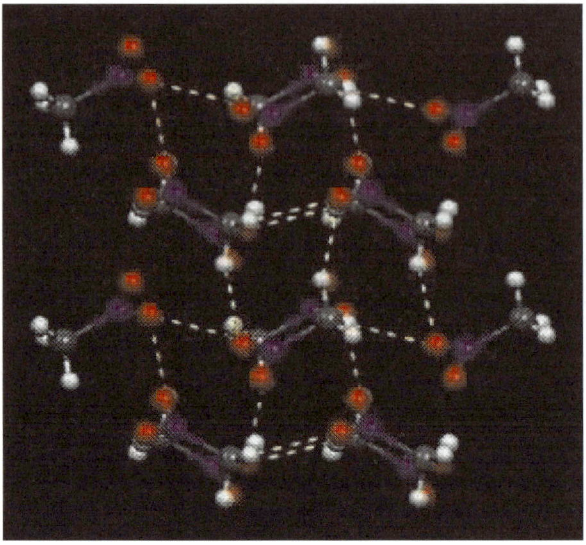

Fig. 3.25 Projection of the high-pressure crystal structure of nitromethane at 15 GPa showing the network of hydrogen bonds represented as yellow dashed lines.

and thus may contain some additional information on the structure and intermolecular interactions in the system. These requirements will not be obeyed in the case of uniaxial stresses, as may happen in a diamond anvil cell. The lattice strain can be defined in terms of the crystal parameters of the initial, zero strain, solid and those of the deformed solid as discussed in several places [113, 141, 197, 233, 236–241].

We consider the case of a homogeneous strain such that the initial and the deformed state are both perfect crystals, and the fractional coordinates in the crystal frames of all the atoms in the unit cell are unchanged after the deformation. We denote by (a,b,c) and by (a',b',c') the crystallographic reference frames of the initial and deformed crystal, respectively, and by R and R', the transformation matrices from the Cartesian to the crystallographic frames. Then if X and X' are the Cartesian coordinates of a point in the crystal prior and after the deformation, homogeneity of the strain implies,

$$R'X' = RX, \tag{3.42}$$

and it is seen that the Lagrangian strain is given by,

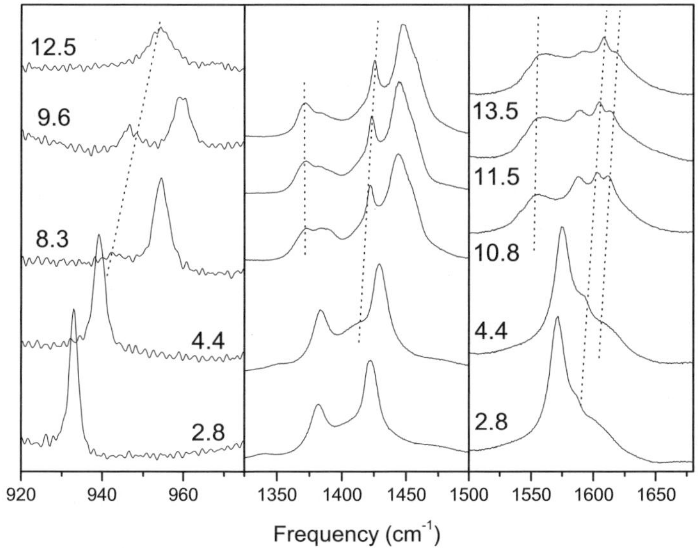

Fig. 3.26 Evolution with pressure (values in GPa) of some of the vibrational infrared modes of the nitromethane crystal. Dotted lines are a reference guide showing the new peaks intensifying with increasing pressure.

$$e = R'^{-1}R - I, \tag{3.43}$$

where I is the unit matrix. Upon symmetrization we obtain,

$$\Sigma = \frac{1}{2}(e + e^T) = \frac{1}{2}[R'^{-1}R + (R'^{-1}R)^T] - I. \tag{3.44}$$

For the finite Lagrangian strain one obtains,

$$\eta = \frac{1}{2}[e + e^T + e^T e] = \frac{1}{2}[(R'^{-1}R)^T R'^{-1}R - I], \tag{3.45}$$

which can be rewritten introducing the metric of the initial state,

$$G = (R^T)^{-1}R^T, \tag{3.46}$$

and of the deformed state,

$$G' = (R'^T)^{-1}R'^T, \tag{3.47}$$

as

$$\eta = \frac{1}{2} R^T [G' - G](R^T)^{-1}, \tag{3.48}$$

showing that the finite Lagrangian strain depends on the variation of the metric tensor between the final and initial state. Explicit expressions of the elements of the matrix of the infinitesimal Lagrangian strain e, have been given by Schlenker et al. [236]. In Figure 3.27, as an example, the linear strain along the principal axes is reported for the three polymorphs of glycine and the orientation in the *ac* crystallographic plane of the principal axes of the strain ellipsoid are shown. It can be seen that the direction of the principal axes changes with pressure or temperature as a response to the reorientation of some intermolecular interaction (hydrogen bonding in the present case).

The lattice strain defined above is only the external strain. In the homogeneously strained crystal a relaxation of the structure will occur to ensure that the system reaches the minimum structure. This will occur by a relaxation of the lattice, and the fractional coordinates of the lattice points will experience changes determined by what is called the internal strain. The problem has been discussed by Catti [242].

In many experiments using the diamond anvil cell, the external stress is uniaxial [237–241]. The applied stress,

$$\begin{pmatrix} \sigma_{11} & 0 & 0 \\ 0 & \sigma_{22} & 0 \\ 0 & 0 & \sigma_{33} \end{pmatrix},$$

can be decomposed in a component representing the average hydrostatic pressure,

$$\begin{pmatrix} \sigma_p & 0 & 0 \\ 0 & \sigma_p & 0 \\ 0 & 0 & \sigma_p \end{pmatrix},$$

and in the deviatoric stress,

$$\begin{pmatrix} -t/3 & 0 & 0 \\ 0 & -t/3 & 0 \\ 0 & 0 & 2t/3 \end{pmatrix},$$

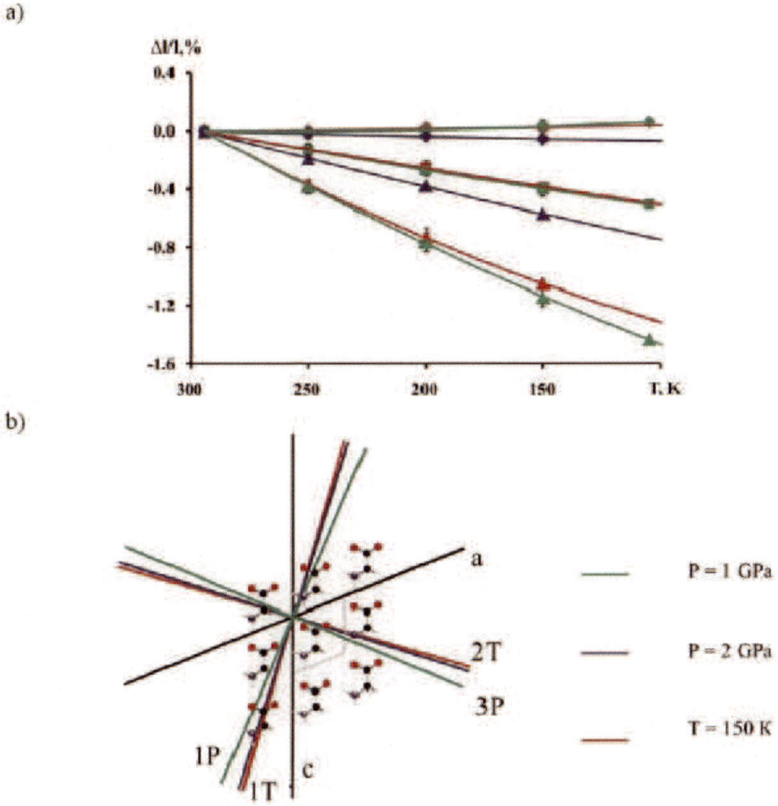

Fig. 3.27 (a) linear strain along the principal axes of the strain ellipsoid for the polymorphs α (red), β (green) and γ (blue) of glycine; (b) orientations of the principal axes of the strain ellipsoid on cooling and on increasing the pressure (reprinted with permission from E.V. Boldyreva et al. *Arkivoc* **12** (2004) 128).

where,

$$\sigma_p = \frac{1}{3}(\sigma_{11} + \sigma_{22} + \sigma_{33}), \tag{3.49}$$

and,

$$t = \sigma_{33} - \sigma_{11}. \tag{3.50}$$

In X-ray diffraction experiments of polycrystalline materials in the radial diffraction geometry [243–248], deviatoric stress is a source of important information on the texture of the composite material [249–251]. The strain produced by the average hydrostatic stress component, can be separated from the deviatoric stress, and the diffraction data can be analysed to sort out the single crystal elastic constants from studies of polycrystalline samples at high pressure. Deviatoric stresses reveal as deformation of the Debye rings and variation of intensities along the elliptical rings, depending on the orientation of the crystallites with respect to the direction of compression. In particular, intensity variations correlate with preferred orientations of the crystallites and with the texture of the polycrystal. It has been reported that at high pressures, non-hydrostatic stresses are produced, with distortion of the lattice. Unusual ellipsoidal diffraction patterns have been observed in quenched polymerized C_{60} [252] as a consequence of anisotropic deformation induced under non-hydrostatic compression. Anisotropic compression in solid argon [253] at high pressure has been reported, inducing strength and elastic anisotropy to the level that the melting point of argon exceeds that of iron above 30 GPa [55].

3.3 Parametric equations of state

In most cases, one is forced to resort to empirical equations of state which simply have the purpose of reproducing and fitting the experimental pressure-volume data or, in a more general view, may be rationalized with a connection to the intermolecular interactions at work in the system. These equations of state allow the determination of the bulk modulus K and its first and second pressure derivatives K' and K". If these parameters are known independently, they may serve to validate the fit of the equation of state to experiments.

A large variety of empirical equations of state have been proposed based on different approximations [254, 255]. It is worth mentioning, that the first empirical equation of state, the Tait equation of state (Eq. 2.2) for liquids and solutions, already reported in the previous chapter, was formulated over 100 years ago. It is important to mention that the Tait equation of state goes beyond the simple expansion of pressure in powers of the temperature and volume.

All the empirical equations of state are founded on the basic definition of the pressure in terms of the Helmholtz free energy F, expressed as a function of volume and temperature, at a constant number of molecules,

$$P = -\left(\frac{\partial F}{\partial V}\right)_T, \tag{3.51}$$

which defines the isothermal equation of state, on the assumption that the free energy simply coincides with the strain energy. One way to proceed further is to then find an expression of the free energy in terms of the volume or, since we are generally interested in non-infinitesimal strains, of the finite strain f and of the temperature. The equation of state in the form 3.51 is then obtained as,

$$P(V) = -\left(\frac{\partial F}{\partial f}\right)_T \left(\frac{\partial f}{\partial V}\right), \tag{3.52}$$

or, more simply,

$$P(f) = -\left(\frac{\partial F}{\partial f}\right)_T. \tag{3.53}$$

The coefficients of the expansion of the free energy (see Eq. 3.58) are expressed in terms of the bulk modulus,

$$k_0 = -V\left(\frac{\partial P}{\partial V}\right)_T, \tag{3.54}$$

and its variation with pressure through $k^{'} = (\frac{\partial k}{\partial P})_0$ and $k'' = (\frac{\partial^2 k}{\partial P^2})_0$. The simplest approximation assumes that the bulk modulus varies linearly with pressure,

$$-V\frac{\partial P}{\partial V} = k_0 + k^{'}P. \tag{3.55}$$

Direct integration then gives the Murnaghan equation of state [256],

$$P = \frac{k_0}{k^{'}}\left[\left(\frac{V_0}{V}\right)^{k^{'}} - 1\right], \tag{3.56}$$

or, in the inverted form,

$$\frac{V}{V_0} = \left(1 + \frac{k^{'}}{k_0}P\right)^{-\frac{1}{k^{'}}}. \tag{3.57}$$

The Murnaghan equation has been largely used but generally it is only able to reproduce experimental data for small compressions, i.e., $\eta \leq 0.9$ ($\eta = V/V_0$). In the case of nitromethane [24], the P-V data from X-ray diffraction have been satisfactorily fitted to the Murnaghan equation of state for compression up to $\eta = 0.67$, as can be seen from Figure 3.23. The limitation of the Murnaghan equation of state to small compressions also applies to more elaborate forms of the equation, obtained by expanding the bulk modulus to higher orders of pressure [257]. It is a common drawback of many empirical equations of state to be unable to give a finite strain energy at very large compressions and to correctly reproduce the limiting value of the bulk modulus at $P \to \infty$. This is actually not a serious problem for molecular crystals, which generally decompose at not very high pressures. In fact, apart from diatomics, triatomics and methane, the most stable molecular crystal reported so far is probably the benzene crystal that at 200 K, has been found to definitely maintain its molecular structure up to ~50 GPa before amorphization occurs [208]. Considering the monoclinic crystal phase of benzene ($P2_1/c$ space group), it has been reported that a pressure increase from 1 to 50 GPa produces a compression $V/V_0 = 0.7$. Aromatics are indeed quite stable molecular arrays. Recently, a high-pressure stability, similar to that found for benzene, has been demonstrated for the indole crystal [258].

At constant temperature we may expand the free energy in a Taylor series of the finite strain,

$$F = F_0 + c_2 f^2 + c_3 f^3 + c_4 f^4 + \cdots, \tag{3.58}$$

where f is the positive compression strain. The finite strain can be expressed in different ways according to a general form given by,

$$f_n = \frac{[(\frac{V}{V_0})^{-\frac{n}{3}} - 1]}{n}. \tag{3.59}$$

The free energy is defined relative to a reference state. The Eulerian strain f_E ($n = 2$) is relative to the strained (final) state, while the Lagrangian strain f_L ($n = -2$) is relative to the unstrained (initial) state. The different ways to represent the finite strain do not differ substantially at small compression. In Figure 3.28, the behaviour as a function of V/V_0 of the Eulerian, Lagrangian and of the so-called natural, or Henky, strain (the latter to be discussed in the following section), are reported. It is evident that the definition of the finite strain becomes important only at high compressions that, for the reasons already explained, are not of particular interest for molecular crystals. The Eulerian strain has been more largely used in high-pressure studies, since it ensure a better convergence of the free energy expansion.

In the limiting case of $n = 1$, the strain factor is simply related to the linear strain,

$$f_1 = \frac{l_0}{l} - 1. \tag{3.60}$$

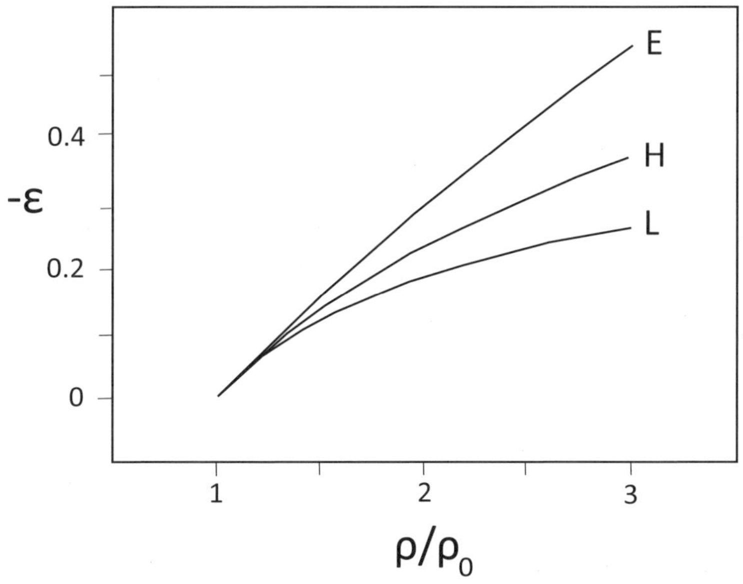

Fig. 3.28 Evolution as a function of the compression of the Lagrangian (L), Eulerian (E) and natural or Henky (H) strain.

Expanding the free energy to fourth order in the Eulerian strain, the fourth order Birch–Murnaghan equation of state is obtained [259],

$$P = 3k_0 f_E(1+2f_E)^{\frac{5}{2}} \left\{1 + \frac{3}{2}\left[(k^{'} - 4)f_E + \frac{3}{2}(k_0 k^{"} + (k^{'} - 4)(k^{'} - 3) + \frac{35}{9}\right] f_E^2\right\}. \tag{3.61}$$

Truncation of the energy expansion to third order in the strain gives a three-parameter (V_0, k_0 and $k^{'}$) equation of state (the coefficient of the f_E^2 term is equal to zero), implying for the second derivative of the bulk modulus a value,

$$k^{"} = -\frac{1}{k_0}\left[(k^{'} - 4)(k^{'} - 3) + \frac{35}{9}\right]. \tag{3.62}$$

The expansion can be truncated to second order, this leads to the condition,

$$k^{'} = 4, \tag{3.63}$$

a value of the first derivative of the bulk modulus close to the actual experimental values for many crystals and this leads, neglecting higher order terms, to the simpler equation,

$$P = 3k_0 f_E (1 + 2f_E)^{\frac{5}{2}}, \tag{3.64}$$

known as the Birch equation of state [259]. These, and other empirical equations, are judged on their ability to reproduce experimental P-V data [77, 121, 260] or to account for the values of the bulk modulus and its pressure derivative known from other sources.

A different equation of state has been proposed by Poirier and Tarantola [119, 261], derived from an expansion of the free energy in the natural or Henky strain defined as,

$$f_H = \ln \frac{\ell}{\ell_0}, \tag{3.65}$$

or, for hydrostatic compression as,

$$f_H = \frac{1}{3} \ln \frac{V}{V_0}. \tag{3.66}$$

The natural strain considers the instantaneous strain as the reference state. Considering the hydrostatic compression case, one obtains to fourth order,

$$P = 3k_0 \left(\frac{V}{V_0}\right) f_H \left[1 + \frac{3}{2}(k^{'} - 2) f_N + \frac{3}{2}(1 + k_0 k^{'} + (k^{'} - 2) + (k^{'} - 2)^2) f_H^2\right]. \tag{3.67}$$

In this case truncation to third order implies,

$$k'' = -\frac{1}{k_0}\left[1 + (k^{'} - 2) + (k^{'} - 2)^2\right], \tag{3.68}$$

and to second order,

$$k^{'} = 2. \tag{3.69}$$

The equation of state gives a representation of the response of the system to the application of an external pressure. The response is dictated by the interatomic interactions and, in particular, by the repulsive interatomic forces which become predominant at high densities. Therefore, an alternative approach to the equation of state, is to select an appropriate form of the interatomic potential and to obtain the internal energy of the solid therefrom, whose derivative with respect to the volume gives the pressure. An appropriate form of the intermolecular potential for molecular crystals is the Lennard-Jones 6-12 potential, already considered in Eq. 1.1 [5],

$$E(r) = -\frac{A}{r^6} + \frac{B}{r^{12}}, \tag{3.70}$$

or, more generally,

$$E(r) = -\frac{A}{r^m} + \frac{B}{r^n}, \tag{3.71}$$

where the constants A and B and the parameters m and n must be selected according to the type of solid and the interatomic potential. Alternatively, the repulsive part of the potential can be expressed in the Buckingham approach as an exponential in the interatomic separation [5].

Defining the constants in terms of the equilibrium zero-pressure interatomic distance and introducing the bulk modulus K_0, from the general form (3.51), one obtains the equation of state,

$$P = \frac{3K_0}{n-m}\left[\left(\frac{V_0}{V}\right)^{\frac{n+3}{3}} - \left(\frac{V_0}{V}\right)^{\frac{m+3}{3}}\right]. \tag{3.72}$$

It can be seen that taking n = 4 and m = 2, the Birch equation of state (3.64) is recovered showing that the finite strain empirical equations of state can also be given an interpretation in terms of interatomic interactions.

An equation of state derived from a model of interatomic interactions has been obtained by Vinet [262, 263]. Vinet assumes that the normalized interaction energy E^*, is expressed in terms of reduced interatomic separation r^*, with an exponential part accounting for the repulsive interaction,

$$E^* = -(1 + r^*)\exp(-r^*). \tag{3.73}$$

This leads to the Vinet equation of state,

$$P = 3k_0\frac{1-f_V}{f_V^2}\exp\left[\frac{3}{2}(k^{'} - 1)(1 - f_V)\right], \tag{3.74}$$

where the strain is defined as,

$$f_V = \left(\frac{V_0}{V}\right)^{\frac{1}{3}}. \tag{3.75}$$

The Vinet equation of state has the important advantage, beside its applicability to different type of solids, that it can be used at higher compressions, i.e., for $\eta \geq 0.6$, than the finite strain equations of state [264]. The Vinet equation of state simplifies further assuming,

$$k^{'} = 1, \tag{3.76}$$

and reducing Eq. 3.74 to a two-parameter (V_0, k_0) equation. By derivation, one obtains an expression for the second derivative of the bulk modulus,

$$k'' = -\frac{1}{k_0}\left[\left(\frac{k'}{2}\right)^2 \frac{k'}{2} - \frac{19}{36}\right] . \tag{3.77}$$

The Vinet equation of state has been denominated the universal equation of state in view of its applicability to several kinds of solids (metals, covalent, van der Waals solids) because of a supposed similarity of the repulsive interaction at short interatomic separation. The denomination of the Vinet equation of state as "universal" has been questioned by Holzapfel [124] with the argument that it fails to give the correct limiting value of the compression at very high pressures. Nevertheless, the equation has been applied with success in many systems with different bonding characteristics.

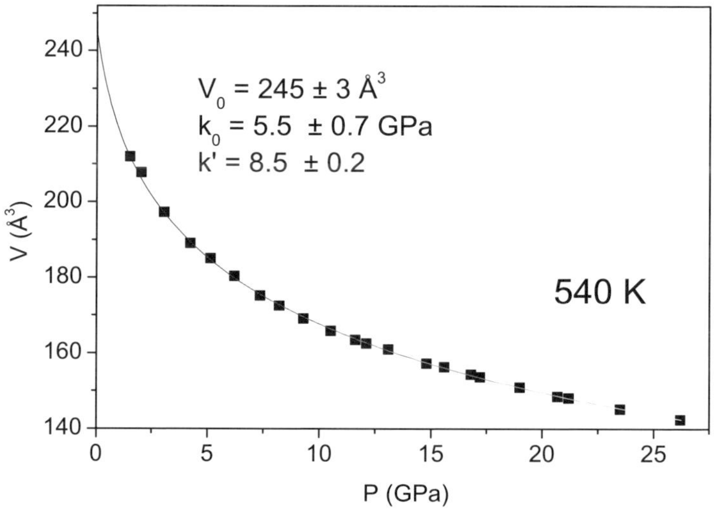

Fig. 3.29 Fit of the Vinet equation of state to the pressure-volume data of the monoclinic phase II of crystalline benzene.

The interaction potential leading to the Vinet equation of state has a central form implying that the units in the crystal have no internal structure. Therefore, the Vinet equation of state should not be applied to a system with internal degrees of freedom. However, in practice it has been applied with success to molecular crystals as well. As an example, the fit of the Vinet equation of state to the P-V data obtained from X-ray diffraction of the monoclinic $P2_1/c$ polymorph of benzene [208], is shown in Figure 3.29. It can be seen that the Vinet equation of state is able to reproduce, quite satisfactorily, the experimental pressure-volume data of crystalline benzene. This will obviously also depend on the accuracy of the data. In this respect, the fit can be equally satisfactory with different parametric equations of state. As a comparison, Figure 3.30 shows the Murnaghan, the Birch–Murnaghan, the Vinet and the Poirier and Tarantola equations of state, traced using the same V_0, k_0 and k' parameters obtained from the fit of the benzene P-V data. It can be seen that the equations qualitatively

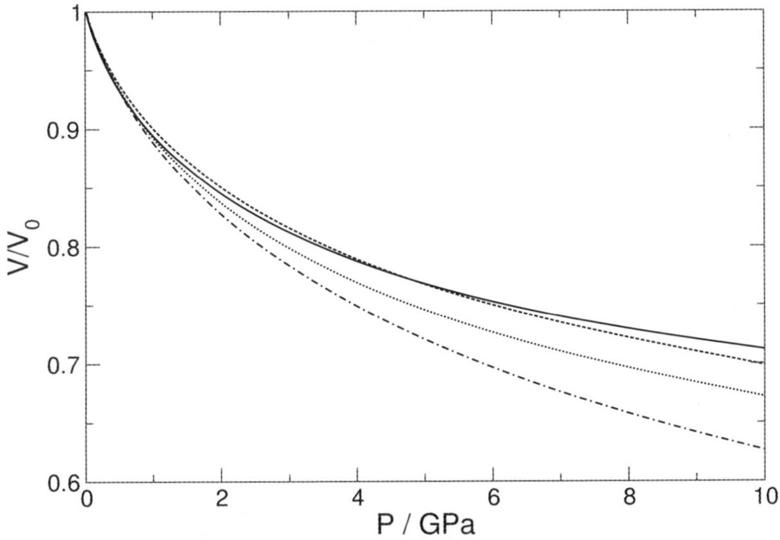

Fig. 3.30 Comparison of the Murnaghan (full), Birch–Murnaghan (dash), Vinet (dot) and Poirier–Tarantola (dot-dash) equations of state calculated with values of the parameters k_0 = 5.5 GPa and k' = 8.5.

agree at low compression but, as expected, due to the fact that for truncations to lower orders they require different values of the parameters, the fit of experimental data will give values of the bulk modulus, and its first derivative, depending on the equation selected.

Once the equation of state has been obtained at a given temperature T_0 it can be of interest to extend the equation at a different temperature T [265]. The simplest approach is to use the same equation, adjusting the parameters V_0, k_0, k'_0 ... for the new temperature. The definition of the thermal expansion coefficient α,

$$\alpha = \frac{1}{V}\frac{dV}{dT}, \tag{3.78}$$

gives the simplest isobaric equation of state and by integration we obtain,

$$V_0(T) = V_0(T_0)e^{\left[\int_{T_0}^{T} \alpha(T)dT\right]}. \tag{3.79}$$

As a first approximation α, which can be assessed independently, can be considered to be a constant or to depend linearly on temperature. The bulk modulus can likewise be considered to depend linearly on temperature,

$$k_T = k_0(T_0) + \left(\frac{\partial k_T}{\partial T}\right)_P (T - T_0). \tag{3.80}$$

As an average elastic constant, the bulk modulus has a temperature dependence similar to that discussed above for the elastic constants.

Chapter 4

High-pressure Technical Survey

A description of the experimental methods employed to induce and study chemical processes at high pressure, first requires the identification of the pressure range of interest. In fact, the pressure range and the modalities of high-pressure generation (static or dynamic) sharply define the nature of the chemical processes, and especially the related probing techniques. Most of the organic syntheses are performed below 1 Kbar, where a commercial technology offers reliable high-pressure reactors working on large volumes and at variable temperature. Going to the GPa range, the sample size decreases more and more with increasing pressure, whereas the experimental difficulties correspondingly grow. Here, the commercial instrumentation availability is noticeably reduced, and custom apparatus are generally designed and built according to the application. Large volume presses are employed for syntheses where the reactants are mainly inorganic solid materials, offering the advantage of synthesizing mm^3 size products using pressures as high as some tens of GPa and temperatures of thousands K. All these methods rarely allow *in situ* monitoring of the chemical reaction. The third method, in this schematic approach, is represented by the diamond anvil cell (DAC) technique. This method is extremely useful from a fundamental point of view, being the unique tool to understand chemical transformations occurring under higher pressure conditions (up and above the Mbar) such as in the Earth's interior. This technique successfully conjugates the possibility of monitoring the reaction *in situ* over a wide P-T range, the obvious drawback being represented by the sample dimension (a few hundreds of microns in diameter at 1 GPa, becoming a few microns in the Mbar range), which also defines the amount of recoverable product.

In this chapter our attention is essentially focused on the static compression techniques employed above 0.1 GPa, where most of the chemical

Table 4.1 Conversion of units of pressure

	Bar	Pa(10^5)	Atm	Kg/cm^2
Bar	1	1	0.986923	1.019716
Pa(10^5)	1	1	0.986923	1.019716
Atm	1.013249	1.013249	1	1.033226
Kg/cm^2	0.980665	0.980665	0.967842	1

processes described in the next chapters occur. In spite of the huge variety of apparatus presently employed to generate static high pressures for synthetic and characterization purposes in chemistry, physics, geology, life and material science, all are derivatives of four reference devices: the piston-cylinder, Bridgman anvil, belt and multiple anvils apparatus. In this framework, particular attention will be devoted to the DAC apparatus, which is central in recent developments to a plethora of *in situ* probing techniques, based on laser and synchrotron light.

4.1 The piston-cylinder device

According to the definition of pressure, as force acting per unit area, the simplest realization of a high-pressure device is based on the application of a mechanical force on a surface. However, in order to increase pressure on the material, its volume must reduce. This is usually accomplished by moving some of the confining walls surrounding the sample inward. A piston sliding into a cylinder is the simplest example of such a mechanism. The two basic requirements are related to the mechanical properties of the walls, which must be strong and relatively incompressible, and to the containment of the compressed material in order to avoid its extrusion. This is basically the approach employed in studying chemical processes, for pressures ranging from a few Kbar up to some GPa (conversion factors between commonly used pressure units are collected in Table 4.1).

Conventional chemical syntheses such as addition (Wittig, Michael), cycloadditions (Diels–Alder, 2+2, dipolar), ionogenic, substitution and polymerization reactions have been extensively studied with these high-pressure devices [266]. The interested reader is referred to the huge literature available in this field where a good starting point could be represented by refs. [56, 57, 59, 60, 267].

The simplest configuration of this high-pressure apparatus consists of a cylinder, where the sample is contained, in which slides a cylindrical piston which exerts the pressure on the sample. The sample is therefore compressed between the piston and the cylinder (closed-end cylinder geometry), or, in a different configuration, between two opposed pistons one sliding against the other, along the cylinder (open-end cylinder geometry). When solid substances are being studied nothing is required, in principle, to seal the sample if the clearance between the piston and the cylinder is small. Nevertheless, extrusion into the annular clearance between the piston and cylinder always occurs in liquids, soft solids and even harder materials, and can be prevented by placing, between the piston tip and the sample, one or more thin metallic disks or wedges with the same diameter as the piston. Under load, the disk expansion against the cylinder wall ensures sample sealing. Leak prevention in pipes and piston-cylinder devices was realized by Bridgman who developed a leakproof solid packing, based on the "unsupported area" principle [30]. According to this principle, the sealing pressure is higher than the internal pressure and this is realized through a reduction of the area supporting the internal pressure. Worth mentioning is the seal invented by P.W. Bridgman, which is a mushroom-shaped plug in which a soft ring of packing material is placed in the annular space between the stem and the chamber. A further reinforcement can be provided by conical shaping and additional gasketing support between the piston and the sample [268]. Despite the ultimate pressure attainable with this kind of device, which is in principle, determined by the tensile strength of the cylinder and piston material, sample size and cell geometries lower this value because of the deformation and rupture of the piston, plugs, seals and probing sections composing the cell. In addition, when very high pressures are generated, the expansion of the cylinder should also be considered. This expansion will not generally be compensated by that of the piston, thus determining extrusion of the material contained between the piston and the cylinder wall. Control of the clearance between the piston and cylinder can be realized by compressing the outside of the cylinder simultaneously with the progression of the piston in the cylinder. The piston and cylinder of the first high-pressure devices were constructed of steel, allowing maximum pressures in the order of 2 GPa. Cemented tungsten carbides later substituted steel, doubling the attainable final pressure. Tungsten carbide is indeed approximately three times stiffer than steel, with a Young's modulus of approximately 550 GPa, the workable material being shapeable in large pieces and possessing the highest compressive strength.

Because of the extensive employment for synthesis, there was an urgent need to generate high pressure simultaneously with high temperature.

3000°C at 1.5 GPa were reported at the beginning of the 20th century by Parsons, using internal electrical resistance heating (see Figure 4.1) together with a solid pressure-transmitting material, which also served as thermal and electrical insulation [269]. A piston-cylinder device with simultaneous pressure-temperature capabilities, considerably beyond those used by Parsons, was employed by Coes, Jr. to synthesize a new dense phase of silica, which was named coesite [270], but the apparatus employed for this synthesis was only described in 1962 [271]. It consists of an alumina cylinder, which is pressed using tungsten carbide pistons, by both sides of the alumina cylinder. The alumina cylinder insulates the two pistons from each other and heating is realized by resistive heating of a cylindrical graphite-heating tube, surrounding the sample chamber. Other variants of simple piston-cylinder devices for high-pressure-temperature studies have been later described [272–274], also including a device equipped with a teflon cell container, positioned in the piston-cylinder device to get a perfect sealing of the sample [275].

Laterally supported pistons are generally stiffer allowing the notable increase of the attainable final pressure. The first equipment consisted of three separate means to support the compression: binding rings radially supporting the inner chamber whereas bolted, or hydraulically loaded end plates, give axial support. In addition, the end of the piston is supported by a compressible, solid material such as KBr [276, 277]. Electrical leads, to power internal resistive heaters, temperature and pressure gauges could be introduced in the high-pressure vessel avoiding leakages or extrusion of the material, or rupture of the vessel components [278, 279]. Optical access is also often demanded for the *in situ* monitoring of the chemical processes occurring in the high-pressure chamber. The introduction of windows in high-pressure vessels is a major technical issue and different solutions have been adopted depending on the vessel characteristics and the purpose of window insertion. In any case, perfectly flat sets are required [280, 281] and their polishing is often pushed to the observation of a single interference ring when optical methods are used to measure the flatness. The stress gradient between the supported and the unsupported portion of the window is greatly reduced by decreasing the aperture of the window, with respect to the portion in contact with the seat, but the unsupported principle for sealing is hardly matched in the case of a very small inlet. Diamond windows would be the best choice in terms of mechanical resistance and transparency over a large portion of the electromagnetic spectrum, but the cost is very high. Other materials such as sapphire, Si or Ge can be used,

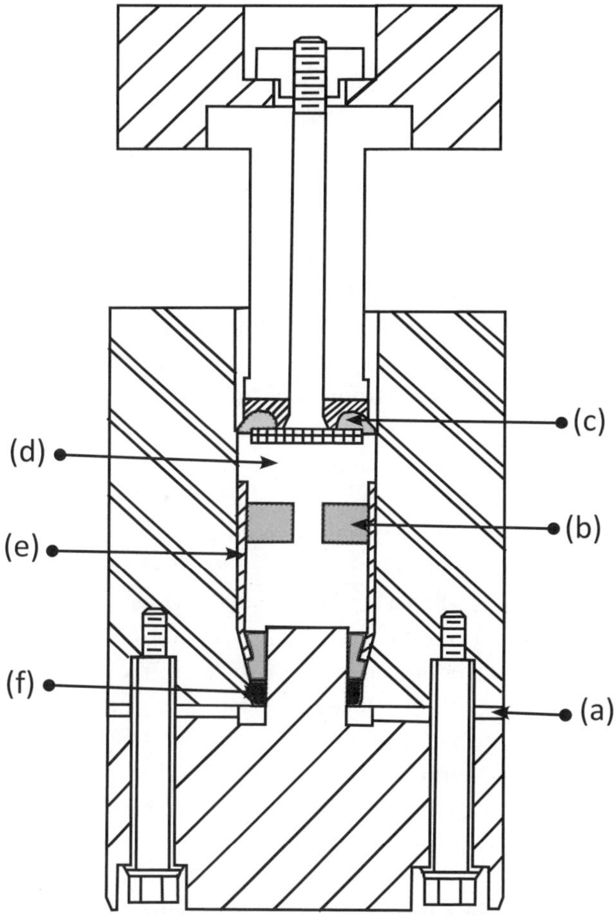

Fig. 4.1 The piston-cylinder device realized by C. A. Parson [269]: (a) presspahn insulator; (b) magnesia; (c) cup packing in brass and leather; (d) graphite; (e) asbestos and mica liner; (f) cup packing in rubber.

depending on the probing technique [282, 283]. Truncated conical windows were realized by Drickamer upon compressing NaCl into small cylinders of increasing diameter [284, 285]. Pressures up to 15 GPa could be sustained thanks to the frictional forces acting on them. The employment of optical fibres to access the high-pressure container is current practice, especially in industrial applications, to spectroscopically monitor the chemical reaction. The applications have been limited in these cases to a few Kbar as the fibre fragility represents one of the main drawbacks.

Friction represents an unavoidable issue which should be carefully taken into account both from a pressure generation aspect and for the implications to sample reactivity. In addition, friction is strictly bound to the hysteresis phenomena being related both to the sliding surfaces, with solid-transmitting pressure substances, and to internal friction effects which, in turn, depend on the pressure to which the solid under investigation is subjected. The friction at the interface between the piston and the cylinder prevents an exact determination of the pressure transmitted to the sample from the force applied by the piston, which is not completely transferred to the sample. When the force exerted on the piston is released, frictional forces will act with an opposite sign and the pressure within the sample will be higher than that calculated from force/area relationships. Frictional effects drastically increase with increasing the applied load, making accurate friction correction unreliable. Friction can be reduced by rotating the piston while pressure is applied [286]. The efficiency of this operation, which is quite good in fluid systems, is only partially effective when solid materials are used to transmit pressure because of an intrinsic hysteresis effect in many solids, especially when they are finely powdered. Besides technical aspects, friction also represents a mechanical initiator of a reactive process. A strong shear field exists on the piston tip. Here, the sample is in contact with the cylinder wall and, because of friction, a very rapid transformation of the sample could occur in this region, as compared with the transformation in the bulk of the sample. To avoid this kind of problem, hydrostatic high-pressure apparatus have been realized [287]. In these devices, a compressed fluid medium transmits the pressure to the sample through a piston, granting a hydrostatic compression. The pressure-transmitting fluid must be inert, stable and remain fluid up to the maximum operating pressure. The sample itself can be used as a compression medium, but normally this is not advisable, due to the risk of chemical contamination in the compression stages. The sample and pressure medium should therefore be kept separate. Besides different mechanical solutions to accomplish this separation, the sample under investigation can also be confined in an inner cell, to avoid the unintended effects on the reactivity of the metallic parts of the vessel in contact with the sample. Quartz cells equipped with a movable quartz piston, fitting perfectly with the cylinder [288], or cells made of polytetrafluorethylene (PTFE) have been reported [287]. The latter have been used to optically monitor the reaction; in fact, the sample is sealed inside the PTFE bag, remaining completely separated from the pressure medium, and the cell fits into the high-pressure autoclave in contact with the inner surface of the windows.

4.1.1 *Large volume piston-cylinder apparatus*

Synthesis of ultrahard materials, such as diamond and cubic boron nitride, have motivated the realization of large volume presses in order to increase the amount of product per compression cycle. Pressures of $\leq$5 GPa and temperatures up to 2000°C are required for these syntheses. In principle, these pressures can be realized with sample areas of cm radius size [289], the compression limits being defined by practical fabrication problems regarding the manufacturing quality of the stressed parts of the apparatus as their dimensions are scaled up. The selection of the materials is therefore a central aspect in these realizations in terms of quality, homogeneity, strength, workability and application [290].

Pressure is generated according to the previously described piston-cylinder approach. A massive metallic pressure vessel is filled with the fluid and is compressed through a piston, driven by a hydraulic ram. Sealing is provided through various types of Δ-ring and O-ring. High temperature is realized through the employment of resistive heaters. Reaction volumes of 40 cm^3 effectively absorbing 12000 ton of the total press force (50000 ton) have been reported in a toroidal large volume press [291], whereas 1000 cm^3 has been pressurized with 30000 ton in a belt apparatus [292].

Higher pressures can be generated through intensifiers which are always based on the piston-cylinder principle. The pressure is increased by gradually reducing the piston diameter through a sequence of compression stages. In this way, the size of the sample vessel can be increased, regardless of the geometry, because it can be separated from the pressure-generating unit [287]. Intensifying systems are generally more complex and expensive than piston-cylinder devices, but they can work on larger volumes. Normally, reactants are not allowed to come into contact with the pressure generation unit, in order to avoid corrosion of the vessel, catalytic effects and contamination of the reaction, but this is not possible in the case of continuous flow reactors. In some cases, reactants are introduced in the high-pressure vessel in some type of flexible package, compressed by a pressure-transmitting medium.

4.2 The opposed anvil devices

The opposed anvil concept, an evolution of the piston-cylinder principle, is generally associated with the work of Bridgman [293–295], but it should probably be dated back to the previous century because the first anvil apparatus was most likely constructed by Wartman in 1859 [296].

The opposed anvil principle is based on the magnification of pressure by compression of the sample between massively supported tapered pistons. These pistons, with the flat tapered tips pointing one against the other, are named anvils. In this set-up, the maximum pressure experienced by the sample largely exceeds the compressive strength of the anvil material [297]. The capability of the anvils in producing the highest static pressures should be found in the principle known as "massive support". Practically, when the small faces of two broad truncated cones (anvils) are forced together by an axial thrust, they are able to sustain a force per unit area much greater than that withstood by a right circular cylinder of the same cross-sectional area. This is possible because the axial thrust applied to the anvil faces, is redistributed into the greater circular area behind the faces.

Extensive description of the most important device based on this principle, the diamond anvil cell, will be given in the following sections, while the most representative opposed anvil devices are schematically shown in Figure 4.2.

In the Bridgman anvil device, the sample is compressed between anvils of tungsten carbide contained in steel-binding rings. The sample is embedded in salt disks in order to transmit as much pressure as possible to the sample in a hydrostatic manner. Lateral containment is ensured by a gasket of finely grained pipestone (catlinite), but other solids with similar frictional characteristics have also been used, such as talc and pyrophyllite, the latter material is still in use [298]. Later, these materials were substituted by metallic gaskets which ensure a good seal in high-temperature studies of liquids [299].

Several cells have been designed employing the Bridgman opposed anvils principle [300–302]. Pressures up to 20 GPa can be generated with this instrument and the limit is fixed by the extrusion of the gasket. In order to raise the attainable pressure, an upgrade of the Bridgman opposed anvil device was developed by Drickamer [303]. In its traditional design, the Drickamer cell is characterized by a thick containment ring, made of either hardened steel or tungsten carbide, therefore representing a kind of intermediate version between the piston-cylinder and the supported anvil

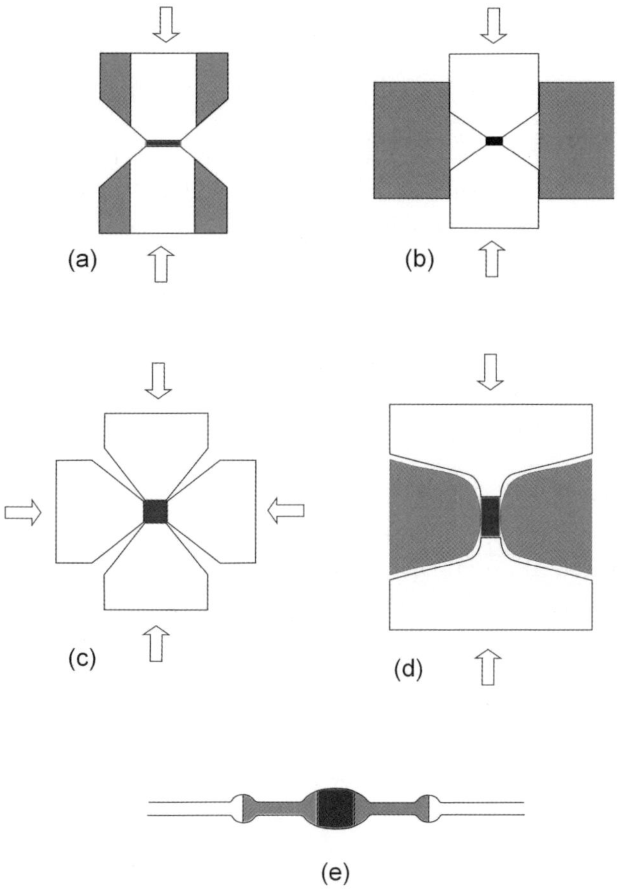

Fig. 4.2 Schematic representation of different types of opposed anvil devices: (a) Bridgman; (b) Drickamer; (c) girdle; (d) belt. Arrows indicate the applied load direction, grey areas in (a) and (b) are binding rings whereas anvils are always represented as white areas. In the belt apparatus (d) both white and grey areas are composed of concentric-binding rings surrounding the central conical piston (white area) or the cemented tungsten carbide conically shaped chamber (grey area). In (e) the sample area of the toroidal anvil is reported.

device of Bridgman. By inserting sintered diamond tips, pressures as high as 35 GPa could be reached. This instrument can also be equipped with windows of transparent material to allow optical measurements up to 10 GPa [284] at ambient or low temperatures [304]. Interest in the Bridgman anvils faded with the introduction of the diamond anvil cells but increased again with the availability of good quality sintered diamonds, which allowed reaching pressures up to 40 GPa on mm^3 size samples [305, 306]. Anvils made of nano-polycrystalline diamond have been recently employed both in supported and unsupported geometry [307].

Several other instruments were developed in order to reach higher static pressures and increase the sample dimensions [287]. In these devices, indicated as belt, girdle and profiled Bridgman anvil high-pressure apparatus, a radial support is continuously provided to the tapered pistons and pressures up to 20 GPa can be applied on mm^3 size samples, thus making these instruments appealing for the high-pressure synthesis of superhard materials.

The belt apparatus [308] consists of two conical-cemented tungsten carbide pistons pressed into each end of a conically shaped chamber. The pressure exerted by the piston is transmitted to the sample, contained in a metal tube, by pyrophyllite which serves as both a compression medium and, thermal and electrical insulator. Sealing is ensured by a sandwiched gasket realized through alternate elements of pyrophyllite and steel. The conical piston, as well as the chamber, are supported by hardened steel-binding rings. Pressures up to 10 GPa and temperatures in excess of 2000°C, realized through resistive heating, could be produced and maintained by this apparatus. Girdle devices use a conical piston profile [289]. The pistons are insulated from the belt or the girdle by a gasket assembly and thermocouples or electrical leads can be inserted through the gasket. They are particularly suitable for the synthesis of materials at high pressures and high temperature, but their assembly is quite complex with respect to the Bridgman anvil devices.

The two main profiled Bridgman anvil apparatus are the cupped and toroidal anvil cells. These instruments have a deep cavity in the centre of the anvils, which can host a larger amount of sample ($\sim$100 cm^3), with respect to the Bridgman and Drickamer cells, but in spite of this larger volume, the pressure limit is only slightly reduced. In the cupped type, the anvils have a cup-shaped profile with a semispherical volume depression, while in the toroidal type, the depression is formed by a central cavity and a circular concentric groove. In the toroidal cell, the sample is confined within the cavity, delimited by the profiled anvils and the gasket, which is squeezed in the volume of the toroidal groove. The role of the toroidal groove results in a reliable support for the gasket and smoothing of the pressure variation. This cell is comparable, in volume and pressure performance, to the multianvil apparatus but is very simple to assemble, compact and convenient, the main drawback being the absence of optical windows. Heater and pressure gauges can be placed in the sample cavity with the pressure-transmitting medium. Due to the spheroid shape of the sample cavity, an almost perfectly hydrostatic pressure distribution is achieved on compression and high-temperature studies are possible. The toroidal cell represents an improved version of the cupped anvil cell, due to the increased stability and higher load [309, 310]. This cell was the main device for the synthesis of superhard materials in the industry of the USSR, but did not diffuse into Western countries until the early 90s, because of doubts regarding the reliability of the functioning of the profiled anvils. Pressures up to 14 GPa can be reached with tungsten carbide anvils, and a small toroidal cell working on several hundreds mm^3 can sustain pressures of 11 GPa and temperatures of $\sim$2000°C, while larger volumes ($\sim$ 800 mm^3) can resist up to 9.5 GPa and 1800°C. Pressures of 8 GPa have been applied to a $\sim$ 200 cm^3 sample [311]. Several modified toroidal cells were developed and adapted to a large variety of techniques [281].

Among these the Paris-Edinburgh cell [312] (see Figure 4.3), an improved version of the toroidal cell, is an extremely powerful realization especially for the possible use in high-pressure neutron diffraction [313–315] and X-ray [316, 317] studies. The success of the Paris-Edinburgh cell is due to the reduced weight and its portability if compared with presses of similar performance and sample size. In the Paris-Edinburgh cell, the sample is compressed between anvils made of either tungsten carbide or sintered diamond allowing pressures to be reached well above 10 and 25 GPa, respectively. The first TiZr toroidal gaskets were fitted with grooves machined into the anvil faces to confine the sample, normally embedded in a solid pressure-transmitting medium. Complete encapsulation of the sample can be realized through the employment of two flanged hemispherical caps (see Figure 4.3). With this technical solution, the available pressure range for hydrostatic studies by using fluid pressure-transmitting media, was extended to 9 GPa, the freezing pressure of methanol:ethanol mixtures [318].

4.3 Multi-anvil devices

Larger volumes of solid samples can be compressed by using multi-anvil systems. A multi-anvil device is a high-pressure apparatus with more than one axis of loading and four or more anvils compressing the sample [319, 320].

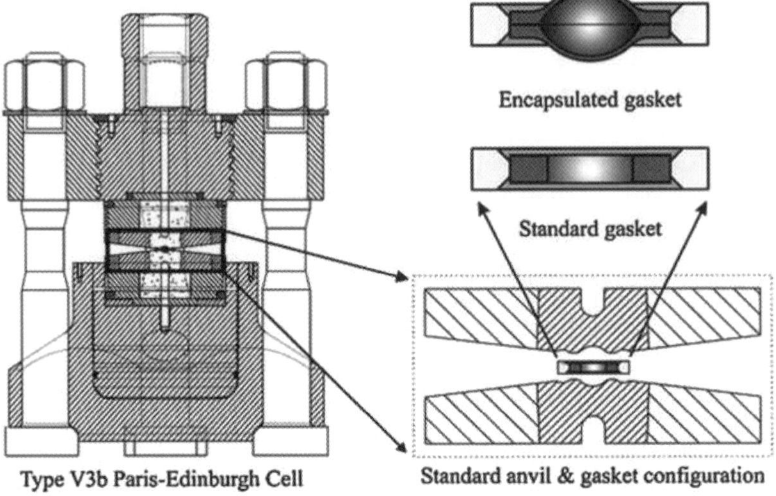

Fig. 4.3 The Paris-Edinburgh cell and two different types of gaskets (reprinted with permission from W. G. Marshall et al. *J. Appl. Cryst.* **35** (2002) 122, Copyright (2002), by the International Union of Crystallography).

The original arrangement dates back to the end of the 50s when H.T. Hall realized the first multi-anvil, high-pressure apparatus, a tetrahedral-anvil device capable of reaching pressures up to 10 GPa and temperatures as high as 3000°C [273]. Since then, this apparatus has rapidly evolved, and nowadays pressures of tens of GPa and temperatures exceeding 2000°C can be reached. They obviously represent an incredible tool for chemical applications, especially in view of reproducing Earth's interior conditions. Several reviews and books describe the characteristics, evolution and application of these devices (see Figure 4.4) [268, 287, 319–329].

The two opposing anvils with circular faces, characterizing the Bridgman design, were replaced by four anvils with triangular faces, independently driven by four hydraulic rams to compress a pyrophyllite tetrahedron. In the tetrahedral press (see Figure 4.4a), and in other multi-anvil devices as well, the principle of massive support given to the anvil faces is reduced with respect to the Bridgman anvils because the solid angle subtended by each anvil must decrease as the number of anvils used is increased. Nevertheless, this is compensated by the higher number of anvils that reciprocally support each other through the gasket. The number of anvils was later raised from four to six, giving rise to hexahedral devices compressing both trigonal-bipyramidal (Figure 4.4b) and cubic (Figure 4.4c) assemblies [320, 330].

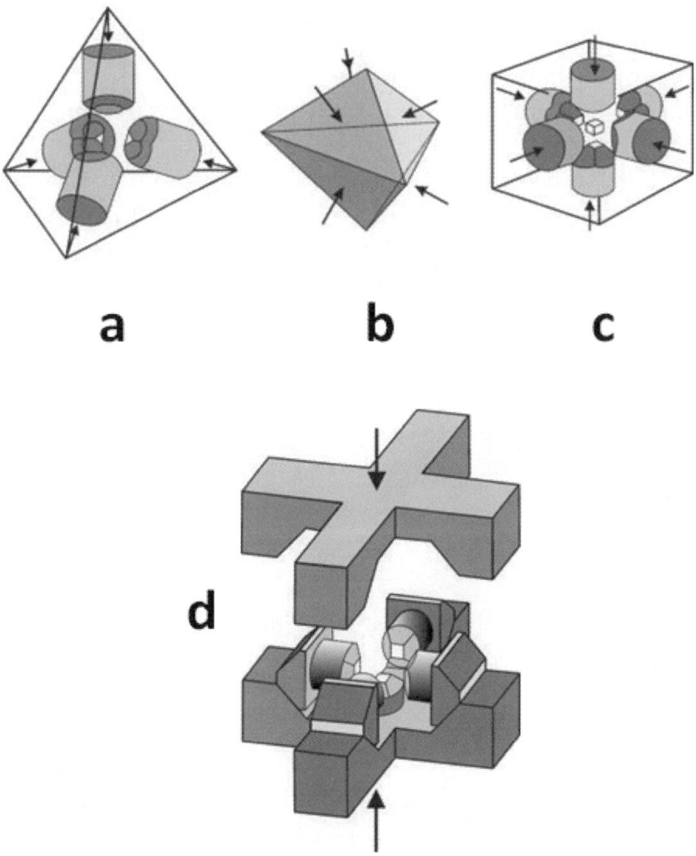

Fig. 4.4 Different multi-anvil geometries: (a) tetrahedral, (b) trigonal-bipyramidal, (c) cubic, (d) cubic DIA (reprinted with permission from H. Huppertz, *Z. Kristallogr.* **331** (2004) 122, Copyright (2004), Oldenburg Wissenschaftsverlag GmbH).

In order to reduce the number of rams as the number of polyhedral faces increases, different arrangements were realized to transfer a uniaxial applied force in the perpendicular plane. Particularly relevant is the cubic apparatus (DIA) developed by Osugi, and reported in Figure 4.4d, in which the cubic cell assembly is designed in a way that a uniaxial compression drives the movement of the wedges in the perpendicular plane, on which four anvils are mounted so that the cell assembly is hydrostatically pressed [331]. The top and bottom anvils are fixed to the upper and lower guide blocks, respectively. Lubrication and insulation between the guide blocks is ensured by teflon sheets and glass epoxy plates.

Pressure generation in excess of 10 GPa, the nominal limit of tetrahedral and cubic anvil devices, motivated the realization of devices with a greater number of anvils. An octahedral-anvil device was developed by Kawai [332], who used eight wedge-shaped tapering anvils, obtained by cutting a sphere along three orthogonal planes, passing through the centre and forming an octahedral space into which a cell assembly is placed (see Figure 4.5). Pressures up to 16 GPa could be reached by this device and the results

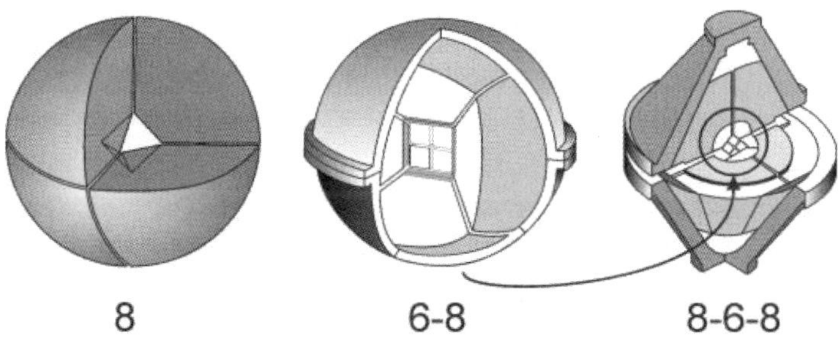

Fig. 4.5 Sections of the octahedral-anvil device (8) developed by Kawai [332], of the two-stage (6-8) and three-stage (8-6-8) devices (reprinted with permission from H. Huppertz, *Z. Kristallogr.* **331** (2004) 122, Copyright (2004), Oldenburg Wissenschaftsverlag GmbH).

seemed to support a type of experimental proportionality rule between the number of polyhedral faces and the attainable final pressure. This also led to unsuccessful experiments with 20 multipistons. Kawai et al. [333] later realized a two-stage evolution of this split-sphere device, in which six anvils compress a second stage, composed of eight cubic anvils of tungsten carbide. The truncated corners of the eight inner anvils define the octahedral cavity containing the sample assembly. The whole assembly, composed of six anvils of the external first stage and eight anvils of the internal second stage, was initially covered by two half shells of rubber and put into oil in a high-pressure vessel. Increasing the oil pressure, a homogeneous compression is applied to the first stage and then to the second stage [333]. This press is known as a 6–8 octahedral anvil device or KMA [334], whereas the eight cubes assembly is known as the Kawai cell [319, 335] (see Figure 4.5).

Further modifications have been introduced and compression can be mechanically achieved by means of three anvils fixed to the upper and lower blocks of the hydraulic press (Kawai-type apparatus) [336]. This configuration is probably the most successful among multi-anvil presses and new, improved versions are continuously realized [337]. Three-stage spherical devices (8-6-8) were also realized [322], whereas more complex arrangements, always based on the octahedral void, and consisting of up to 26 anvils were also tested [338, 339].

A modified multi-anvil apparatus was developed by completely redesigning the anvil housing as a cylindrical assembly of wedges contained by two massive rings. The driving motivation was the realization of a 6–8 octahedral anvil device, which could be adapted to hydraulic presses employed in piston-cylinder applications, thus representing an elegant and "cheap" solution [340–342]. This apparatus, named Walker-type by its inventor, has been widely used in petrology laboratories for reaching higher pressures with respect to piston-cylinder devices.

4.4 The diamond anvil cell

A marked change of gear in high-pressure research dates back to 1958 when the diamond anvil cell (DAC) was invented. New horizons in the study of the physical and chemical properties of materials, well beyond the P-T limitations encountered in the Bridgman era, were disclosed by this technical realization, thanks also to the wealth of *in situ* probing techniques which could already be applied to the sample in the DAC [33, 343]. The operational principle of this device is that of the opposed anvil press proposed by Bridgman, where the cemented tungsten carbide anvils are replaced by tiny diamonds. This apparently obvious substitution, diamond is indeed the hardest material known, has in the reduced sample size, the relevant drawback that likely delayed the exploitation of this material. The transparent diamond anvils opened the possibility of having visual monitoring of the high-pressure effects on the sample under examination [343], a really exciting novelty in the field, whose evolution was represented by the rapid growth of optical techniques specifically developed to monitor the sample behaviour under compression.

The first employment of diamond in a high-pressure device was pursued in the 50s at the University of Chicago (UC) to perform high-pressure X-ray experiments [344]. Pressures in excess of 2 GPa were produced in a device, called the "split-diamond bomb", obtained by producing a semi-cylindrical groove along the surface in contact with the two diamond halves in which, once the diamonds were clamped together, two pistons moved through driven by an external press to compress the sample. In this case, the diamond acted as a lateral sample container. This approach was followed by J.C. Jamieson who used a three carat single diamond, in which a cylindrical hole was drilled to host the sample allowing the collection of X-ray data on several systems at slightly higher pressure [345].

The next step, i.e., the employment of the diamond itself to press the sample, along with other great ideas, are only obvious in hindsight and seemed to happen by chance in 1958 at the same time in two different laboratories. In fact, whereas Jamieson and co-workers set up a DAC at UC using diamonds for X-ray studies at high pressure [346], Weir and co-workers did the same at the National Bureau of Standards (NBS) for IR absorption measurements [347]. The two devices differed in the way the diamonds were pressed together, a clamped DAC at UC whereas the NBS group realized the lever-arm system, and how the sample was probed: perpendicularly at UC and along the compression direction at NBS. This latter approach, which allowed the visual observation of the sample during compression represents, as nicely described by Bassett [343], the primary event preceding any analytical optical probe. The evolution of the DAC in the last 50 years has been continuous and thanks to a properly designed diamond anvil, the pressure has been raised in the Mbar range by simultaneously producing and controlling high temperature, even the more reluctant materials such as hydrogen have been properly confined and hydrostatically compressed, whereas different methods to determine the sample pressure have been developed, depending on the P-T-hν conditions and the probing technique.

The main technical and operational aspects characterizing the DAC technology are described in several reviews [348–355]. As previously stated, the DAC operation is based on the use of diamonds as the opposed anvil. The sample is placed between the small flat faces of two opposed brilliant cut gem quality diamond anvils, its lateral containment is ensured by a metallic gasket [356] and a force is perpendicularly applied to the large diamond faces by a thrust generating mechanism (see Figure 4.6). In general, one diamond is fixed on a backing plate, whereas the other is mounted on a piston or another plate thrust by the external force. The successful generation of pressures up to the Mbar range lies essentially in the surface reduction going from the plate, where the force is applied, to the small diamond culet in contact with the sample. The attainable pressure is therefore directly related to the diamond culet diameter and to the load applied which is, in turn, limited by breakage of the diamond culet. A qualitative inverse square relationship between the maximum pressure and the diameter of the diamond small culet is suggested [355]. Besides the diamonds, the backing plates represent the other critical area limiting the maximum useful pressure and for this reason strong materials are required for their realization.

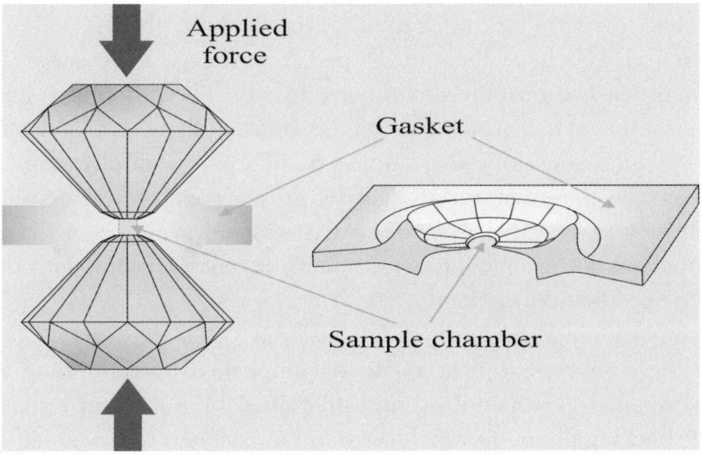

Fig. 4.6 Scheme of the DAC working principle: the force applied perpendicularly to the diamond faces is transmitted to the sample enclosed between the small tips of the diamonds and the metallic gasket.

Once the force is applied to the diamonds, if the stress to which the backing plates are subjected exceeds that needed for plastic deformation, the diamond alignment will be affected, likely resulting in diamond damage or even breakage. Materials with very high plastic deformation are therefore required. The force typically applied to the diamonds is in the order of 10^3–10^4 N originating a stress, given by the ratio between the force applied and the supported diamond area, exceeding 1 GPa. This requirement drastically reduces the materials possessing useful characteristics. The backing plates supporting the diamonds are often constructed of tungsten carbide [357, 358], a brittle material containing small concentrations of Co as a binder. Depending on the grain size and binder content, values of the compressive strength ranging between 3 and 6 GPa are typical for cemented carbide. These values are greater with respect to other materials also in use, such as Cu-Be alloy, with a small concentration of cobalt, and Inconel 718 being the most commonly employed. An extension of the stress range used can be realized by the employment of compressive stress components perpendicular to the applied stress. In the backing plate, a hole behind the diamonds should also be present to allow optical access. The hole dimension will determine the part of the diamond face supported by the plate, whereas the commonly employed conical-shaped optical access, terminating with this hole, will be cut at an angle giving the best compromise between large optical aperture and weakening of the backing plate.

Different DACs have been developed over the years to satisfy specific requirements of the probing techniques in use. The differences are essentially due to the way in which the force is generated and transmitted to the diamonds which generally also implies a different anvil-alignment design. In addition, the chosen materials also dictate the constructive details. Here, we will briefly review the types of DAC, reported in Figure 4.7, that have been employed so far, paying attention to recent developments generally related to specific applications.

In the National Bureau of Standards' cell [347], the force is generated through the compression of a spring accomplished by tightening a screw. The load applied is transmitted and magnified by a lever-arm driving the moveable backing plate, on which the diamond is mounted, against a similar fixed counterpart. Improved upgrades of this cell were later realized by Piermarini and Block, consisting first in the employment of a hemispherical mount for the diamond which is pressed by a movable piston driven by the lever-arm [359]. The hemispherical mount can be tilted in its socket to ensure the alignment of the diamond anvil. A further improved design was later realized for viscosity measurements, by adopting a yoke-type pressure plate in order that the load applied by the lever is parallel to the piston axis, thus minimizing wear and distortion in the piston-cylinder assembly [360].

In the DAC developed by Bassett [361], the force on the anvils is produced by a threaded gland, which moves a piston whose rotation is prevented by a pin in the cylinder engaging a groove in the piston. The piston acts on half-cylindrical rockers on which the diamond is mounted. The rockers ensure the tilting and translation of the diamonds for alignment purposes.

Mao and Bell produced the first DAC, providing access to the Mbar range [357]. This cell was based on fundamental devices characterizing

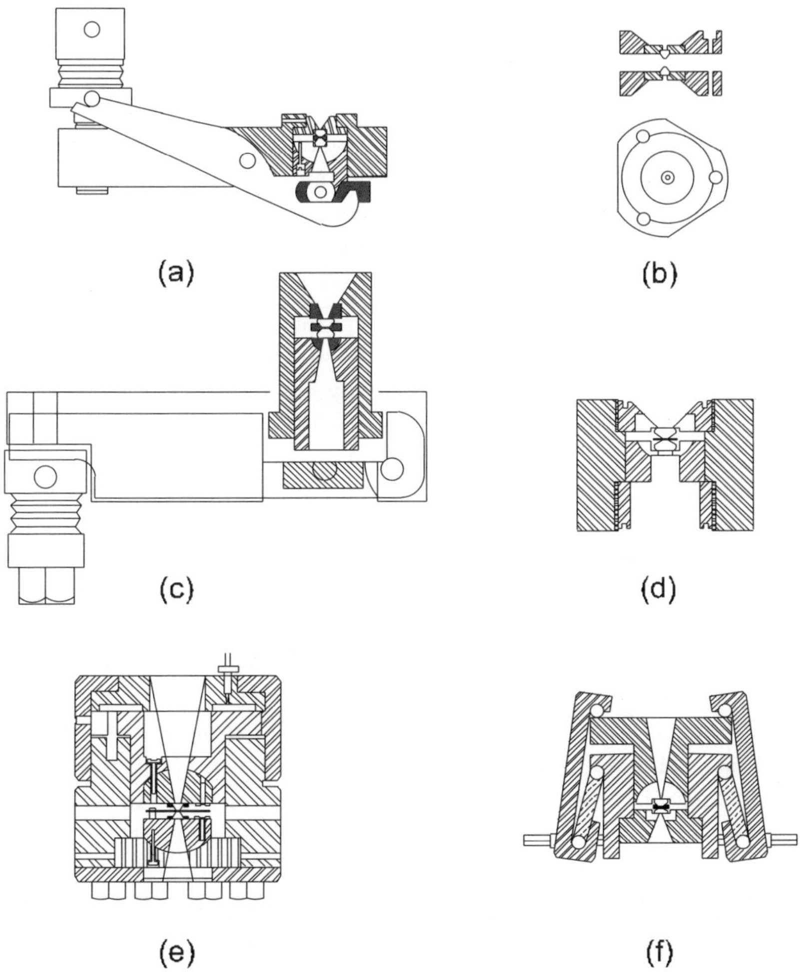

Fig. 4.7 Schematic design of the historically relevant and commonly employed diamond anvil cells: (a) National Bureau of Standards; (b) Merrill–Bassett; (c) Mao–Bell; (d) Bassett; (e) Membrane-type; (f) Syassen–Holzapfel.

the two previously described DACs. The force-generating mechanism is indeed the same employed in the Piermarini–Block cell, based on a Belleville spring-loaded lever-arm mechanism. The thrust here is applied to a long piston which, in turn, presses the half-cylinder tungsten carbide rockers, as in the Bassett cell. The well driven long piston-cylinder assembly and the employment of tungsten carbide for the diamond seats are the basis of the megabar limit being overcome [362]. The design of the Mao–Bell cell was recently adopted and modified to realize a rotational DAC (RDAC), where shear can be applied to the sample under pressures exceeding 50 GPa and temperatures exceeding 2000°C [363]. Shear deformation of the sample, realized through the rotation of one anvil relatively to the fixed one, was originally realized by Bridgman using tungsten carbide anvils [293]. Implementation of this approach in a DAC has been reported by several authors (see references in [363]), but none of them had a precise control of the rotation. With respect to the Mao–Bell design, a thrust bearing seated in a centring mount on the thrust block, facilitates the piston rotation which is driven by a gearing system whose movement can be computer controlled.

In the Syassen–Holzapfel DAC [364], the thrust for the long piston compressing the diamonds is generated by knee-type lever arms, whose folding drives the sliding piston. The thrust from the lever is obtained by closing the brackets through opposite threaded rods, which are synchronously turned by a simple gear-set. The back diamond is mounted on a hemispherical mount for alignment purposes. Beryllium seats, where the diamonds are hot pressed, are employed for X-ray diffraction measurements [365], whereas a single crystal sapphire mount was later adopted for Raman measurements [366].

Following the DAC evolution, the demand for probing the compressed sample using different techniques over wider P-T conditions has also dramatically grown. New DACs have been inserted in measurement apparatus or cryostats, where the available dimensions were extremely reduced. Attempts to miniaturize DACs based on the piston-cylinder design have brought the realization of cells of 2 cm in diameter and 2 cm in height, differing in the attainable final pressure and for the constructive (magnetic or non-magnetic) materials [367–371]. Pressures beyond 1 Mbar have been obtained with some of these cells [369–371], whereas non-magnetic materials have been employed first by Tozer [368], and recently by Gavriliuk et al. [371].

Although with different technical solutions, all the cells we have described so far share the piston-cylinder assembly to convey the force necessary to generate pressure on the sample. A different approach, not employing sliding elements, was adopted by Merrill and Bassett for the realization of a new cell [372] used for single crystal X-ray diffraction measurement at high pressure. The cell dimensions should be small enough to be mounted on a goniometer and include space requirements for a precession camera.

The force is here generated by tightening three trigonally arranged screws, pulling the stainless steel platforms on which beryllium disks supporting the diamonds are mounted. These disks have a diameter of about 12 mm and also serve as transparent windows for X-ray radiation. Replacement by stainless steel plates can be adopted, when only optical measurements are performed. The simplicity and compactness of this DAC is the reason of its great success. A quite recent evolution of this cell was realized by implementing piezoelectric actuators, parallel to the static load screws [373]. The pressure is statically set by the loading screws and then, by applying a suitable voltage, an expansive load is applied by the actuators, thus reducing the pressure on the sample, with a strain rate depending on the actuator response time and on the mechanical friction. Dynamic changes (Δt $\sim$0.1 sec) can be monitored by using appropriate techniques. Melting processes, phase transitions and chemical reactions could also be effectively monitored through this device.

In addition, the hydrothermal DAC, set up for compressing fluids and to study hydrothermal reactions over a wide temperature range (from -190 to 1200°C) [374], is based on the Merrill–Bassett DAC design [372]. The most significant differences consist of the Mo wires wrapped around the tungsten carbide seats to heat the diamonds, and ceramic barriers around the diamonds to maintain a stable high temperature. In addition, the sample can be cooled by putting liquid nitrogen into the brass chamber forming the DAC base or directly into the chamber containing the diamonds. Recently, an upgrade of this cell was realized with the purpose of improving the diamond alignment to minimize the risk of losing the sample, which is generally magnified by a temperature increase [375].

All the cells described so far are driven by mechanical devices based on the employment of screws, levers and threaded rods to generate the force. These types of cells have a very high mechanical stability and are suitable for constant pressure experiments, nevertheless they present some limits connected to the way in which the pressure, and then the applied force, is regulated. Changing the pressure could require the removal of the cell from the optical or X-ray measurement set-up, a time-demanding operation because of the realignment procedures. This is particularly important when the cell is contained in purged or evacuated vessels, such as cryostats, or instruments (FTIR spectrometers), situations which would benefit from a remote pressure control. To overcome this problem, technical solutions are represented by push-pull, linear and rotary motion high- and low-temperature vacuum feedthroughs, which can give access to the screw-based system regulating the sample pressure in the cell. Nevertheless, this would not be sufficient in cases where fixed thermodynamic conditions are required, a quite normal occurrence in kinetic studies of phase transitions, as well as of chemical reactions, and fast pressure adjustments necessary to compensate volume variations would be mandatory. For example, in the study of polymerization reactions, a pressure drop due to the contraction of the sample volume follows the polymer formation. If not compensated, this pressure variation can affect the kinetic evolution of the process or even stop the process itself [376]. Finally, pressure adjustments in mechanically based systems would not be as fine as required by the process. For instance, many chemical reactions of simple unsaturated molecules take place from tenths to a few GPa [377–379], thus requiring rapid pressure adjustment in the order of one or few kilobars. This is particularly relevant in the early stage of the reaction when the process is faster. All these kinds of requirements are satisfactorily accomplished by pneumatic devices such as the membrane DAC [380–382]. Here, the force on the diamond is applied by inflating a metallic membrane, diaphragm or bellow, by means of a fluid medium. The membrane dilatation pushes one diamond against the other, this mechanism being typically driven by a piston-cylinder movement or by rods. Pressure can be released using the same principle by deflating the membrane. Helium is usually used as a pressurizing gas due to its low liquefaction temperature, which prevents any condensation of the gas inside the membrane down to 4 K while performing low-temperature experiments or simply in some loading procedures. The helium pressure inside the membrane is regulated through a thin high-pressure stainless steel capillary (~1 mm in diameter) connected to a gas cylinder. When cooling or heating the cell, the helium pressure inside the membrane must be carefully monitored and adjusted in order to maintain the desired pressure inside the cell. Pneumatic DACs allow a fine and remote control of pressure, so that pressure can be changed by leaving the cell on the measurement bench and no mechanical contacts with the cell are required, thus not altering the sample position to the probing beam. The helium supply can, in addition, be computer controlled, thus making the pressure change fully remotely automated. This is a significant advantage when performing any optical, synchrotron light-based and low-temperature experiment, or when monitoring kinetic evolutions. This system is highly suitable whenever fine-tuning of the pressure is required, as is the case with phase transitions, equilibrium processes, crystal growth and processes, such as reaction kinetics, where compensation of pressure due to volume contraction is mandatory.

A hybrid cell attempting to combine the advantages of a multi-anvil press with those of the DAC was recently produced [383]. The issue which motivated the realization of this device was basically the increase of the sample volume to the multimegabar range. Pressure increase in the DAC requires a reduction of the diamond culet dimensions and then a reduction of the sample volume. This cell is essentially a miniaturized cubic press (c-DAC), where the load is remotely applied through a synchronized screw mechanism to six pistons on which trimmed pyramidal ~1/2 carat diamond anvils are mounted. The size of the square culet (200 x 200 μm^2) of these six diamond anvils, sets the size of the cubic sample. The sample is contained among the diamonds taking advantage of a single empty metal cube whose faces have been drilled to gasket the six diamond tips.

Since the first versions of the DAC, many developments and improvements were introduced [384–389] such as the metallic gasket technique for the generation of hydrostatic pressure [356], the ruby fluorescence method for pressure calibration [390–392], new media for hydrostatic pressure transmission at low [393] and ultrahigh (Mbar) pressure [394, 395], making the DAC the most widely used high-pressure device up to hundreds of GPa. The success of the DAC is essentially due to the extraordinary versatility and compatibility with many investigation techniques. The mechanical and optical properties of diamond made the DAC a suitable device for spectroscopic studies using electromagnetic radiations ranging from X-rays to radio frequencies [396, 397] (with the only exception being UV and soft X-ray, due to the absorption of the diamond between 5 and 5000 eV [287, 348, 349, 355]) and for electrical conductivity [398, 399], magnetic susceptibility, rheology, elasticity and neutron diffraction and scattering as well [396]. The DAC has allowed the investigation of high-pressure phenomena in an astonishing and unpredictable range of combined extreme pressure and temperature conditions, ranging from 0.1 up to 500 GPa [400], and from liquid helium temperature [401, 402] (or down to a few mK in some cases [403–406]) up to 5000 K [407, 408] using laser heating techniques. From a chemical point of view, the great advantage in using the DAC is represented by the possibility of maintaining the required P-T conditions for the desired time allowing the kinetic, and then thermodynamic, study of a chemical reaction. The small sample dimensions did not prevent experimental measurements because many modern investigation methods can be performed even with a very reduced amount of sample. As a matter of fact, the DAC increased by more than one order of magnitude, the experimentally accessible pressure range, making this variable, much more effective in changing the density of matter, as easy to control as temperature.

4.4.1 *Diamonds*

The mechanical and optical properties of diamond make it the best choice to build anvils able to produce the highest pressures and allowing *in situ* monitoring of the pressure-induced phenomena, using an incredibly rich variety of probing techniques. Diamond is an allotropic form of carbon, consisting of tetrahedrally covalently bonded carbon atoms. This arrangement makes it the hardest known substance. Hardness measured by the sclerometry method (scratch at a constant indenter load) has been reported to be 137 and 167 GPa for the (100) and (111) faces [409] of face-centred cubic diamond (space group Fd3m). Another very rare crystalline form called lonsdaleite, or hexagonal diamond in reference to its hexagonal lattice symmetry (space group $P6_3/mmc$), exists. First principles calculations have found lonsdaleite to be harder than diamond [410].

The mechanical properties of a material are always quantified through strength in terms of compressive, tensile and also shear strength. The precise tensile strength of diamond is unknown, however, strength up to 60 GPa has been reported [411], but calculations have shown that it could be as high as 90–225 GPa, depending on the lattice quality and orientation, being highest in the normal direction to the cubic face (100) and smallest along the cube diagonal (111) [412]. The compressive strength is one of the most important parameters of a material, in view of its employment as an anvil in a high-pressure device. The opposite anvil design roughly corresponds to uniaxial compressive deformation conditions because axial loads are generally applied to anvil tips. Recent first principles calculations [413] provide compressive strength values of -223.1, -469.0 and -470.4 GPa along the (100), (110) and (111) directions, respectively, in fair agreement with previous results [412, 414, 415].

Strength and hardness are not the only properties which make diamond suitable for high-pressure studies. As already mentioned, diamond is transparent to parts of the electromagnetic spectrum other than visible, such as infrared, near-ultraviolet and hard X-rays, with energy greater than 10 KeV. Diamonds are classified according to their purity (presence of nitrogen or boron impurities), which is strongly reflected in their optical properties [416]. Diamonds can be schematically classified as type I (98%) and II (2%) according to the presence of nitrogen impurities. Type I diamonds contain nitrogen impurities (~0.05–0.2%) which can be present as aggregates (type

Ia), the so-called platelets, giving rise to anomalous X-ray diffraction peaks, or as single substitutional site impurities dispersed throughout the crystal (type Ib). These diamonds present a characteristic IR absorption band centred at about 1200 cm^{-1} which is missing in type II diamonds (see Figure 4.8). This broad band is indeed due to the activation of symmetry

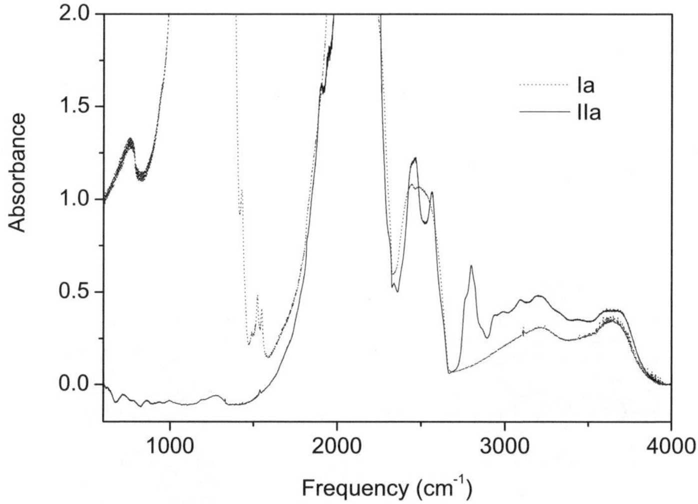

Fig. 4.8 Comparison of the mid-infrared absorption spectra of a couple of approximately 2.4 mm thick Ia and IIa diamonds.

forbidden one-phonon absorption bands, due to the breakage of the translational symmetry by nitrogen substitution. Type IIa diamonds, which are almost impurity free, in fact only possess the infrared absorption between 1800 and 2200 cm^{-1}, which is due to the symmetry allowed two-phonon excitation. When diamonds are nitrogen-free but contain boron impurities they are classified as IIb and are p-type semiconductors. Types Ib and IIb are overall only about 0.2% of all natural diamonds. More complex is an attribution of the optical response of diamonds in the visible and UV region, due to the different effects that the distribution of impurities has on the electronic properties. In general, type IIa diamonds are colourless and transmit below 230–240 nm whereas type Ia diamonds transmit only below 320 nm. Diamond is also an excellent material for X-ray studies because of its transparency to X-rays, and the low atomic number and therefore very low absorption. In addition, the employment of single crystals and the proper cut reduces the X-ray attenuation especially for low-energy X-rays.

Finally, the electrical and thermal properties of diamond are also relevant in view of high-pressure applications. Diamond is a good electrical insulator having a resistivity of 10^{16} ohm·cm [417], unless substitutional boron impurities are present. Diamond is also a good heat conductor; thermal conductivity has been reported to change from about 22 W/(cm·K), for natural diamond, to 33.2 W/(cm·K) for synthetic crystalline diamond with 99.9% ^{12}C [418].

Diamond anvils are obtained from single crystal, gem quality and defect-free stones with the flat culet anvil set parallel to the (100) plane. Anvils are generally 16-sided standard cut stones whose dimension depends on the DAC employed and on the desired final pressure. The diameter of the larger anvil face usually ranges from 4.5 (1/2 carat) to about 2 mm (1/8 carat), whereas that of the culet from 50 to 700 μm (see Figure 4.9). The greater

Fig. 4.9 Typical diamond employed for the DAC, the ruler gives the dimension.

the ratio between the two surfaces, then the greater the magnification of the force applied by the small diamond tip and subsequently the greater the pressure applied to the sample. Pressures in the Mbar range are normally produced [389] by using bevel or double-bevel culet geometry (see Figure 4.10), in order to provide increased sample pressure and stability for a given force applied to the diamond tables.

The type and mechanical requirements of the backing plates, procedures for the mounting and alignment of the diamonds are extensively

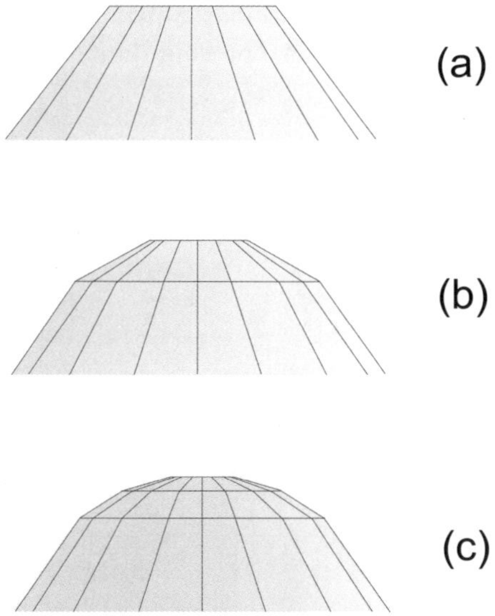

Fig. 4.10 Examples of different diamond tips employed to reach increasing final pressure (from a to c) in experiments with the DAC: (a) standard, (b) bevel, (c) double-bevel.

described in several reviews and books (see for example ref. [355]). All these aspects are, in fact, fundamental in producing and maintaining static high-pressure conditions. Diamond failure often occurs due to improper mounting and alignment. Epoxy glue or soft metallic rings, like copper, are usually employed to fix the diamond onto the backing plates. These are often hemispherical or hemicylindrical rockers which allow a fine adjustment of the anvils with regard to parallelism and centring. All of these operations are usually made by direct observation of the diamonds through the optical axis. While centring is usually obtained by translation of the anvil supports, parallelism of the culets can be realized by tilting these supports through screws and observing the disappearance of the resulting interference fringes with white illumination from the rear, while looking along the optical axis of the cell [419]. This method allows tuning of the parallelism in terms of the distance between the faces given by $\lambda/2$. A perfect alignment is fundamental for reaching very high pressures without damaging the anvils. A different mounting assembly, and consequently a new diamond design has been recently proposed [420]. Hard steel seats are designed to host conical anvils which are therefore laterally supported (see Figure 4.11). The advantages consist of better and reliable alignment, larger aperture and, in particular, remarkably smaller anvils (typically 0.1 carats). Conical support should also prevent damage of the seats in case of diamond failure.

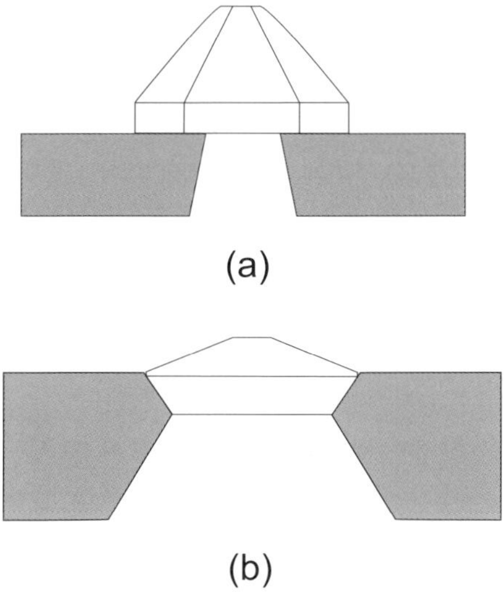

Fig. 4.11 Different backing plates and different supporting geometries: (a) classical flat and (b) conical support.

A synthetic Ib diamond backing plate (1.5 x 5.5 x 5.5 mm^3), having the plate surface (100) orientation, has been employed to support the diamond [421]. The plate is mounted on a tungsten carbide hemisphere sitting on a hole of 4–5 mm in diameter. This incredibly large optical access allowed X-ray measurements up to a maximum scattering angle of 34° and pressures beyond 2 Mbar.

Besides misalignment and improper mounting, structural defects and cracks can also affect the diamond performance and ultimately result in breakage on varying the pressure. Birefringence is normally employed for making a qualitative assessment regarding the uniformity of the dislocation density inside the crystal and then to evaluate the mechanical characteristics of the stone. Birefringence enhances with increasing size and amount of strain, cracks and faults in the crystal, thus compromising the mechanical characteristics of the anvil. Therefore a common criterion in the identification of good stones is a low birefringence magnitude (2 x 10^{-5}) [422].

Another qualitative criterion commonly employed for the selection of stones intended for anvil realization is the fluorescence. The number of lattice defects that can give rise to luminescence is incredibly high (more than 100) and only a portion of them has been analysed in detail [416, 423, 424]. Luminescence can also be due to internal stress or impurities other than nitrogen and boron. According to the extremely variable type and concentration of defects and impurities in natural diamonds, selection of the stones becomes mandatory before planning an experiment in which the probing technique is based on the employment of a UV and visible laser source, such as any linear and non-linear optical spectroscopy. Luminescence can indeed prevent the identification of the desired signal from the sample. Conversely, this is in principle, a negligible problem when using probing techniques such as infrared spectroscopy or X-ray based techniques. A useful criterion for the selection of the stones is the observation, and eventually the comparison, of the intensity of the second-order Raman spectrum of diamond around 2500 cm^{-1} with respect to the fluorescence background [425].

Quite often, a reduced diamond thickness would be desirable to decrease the attenuation of the probing beam (for example, in infrared or soft X-ray experiments). Perforated anvils employed as backing plates of miniature anvils (0.01 carats) have been proposed [426].

Larger natural anvils are extremely expensive and usually more fragile than smaller ones, due to the increased probability of containing defects. This can somewhat limit the search for new greater pressure limits parallel to an increase of the sample dimensions. The continuous improvement in the techniques to synthesize large diamonds now allows the attainment of high-quality single-crystal diamond. Incredibly high growth rates are now possible by microwave plasma chemical vapour deposition (MPCVD) techniques leading to single crystal transparent diamond [427] with properties comparable to those attainable by the more expensive high-pressure, high-temperature syntheses that, besides the cost, are also more limited in size and in the control of the impurities . As a matter of fact synthetic diamonds fabricated from defect free and isotopically pure ^{12}C atoms show very low luminescence, even under compression [428, 429], providing extraordinary clarity and extending the range of optical measurements [430, 431].

This presentation clearly shows that the employment of diamond presents some limitations, although unavoidable when pressures beyond some tens of GPa are required. Microscopic sample dimensions and availability and cost of the stones are likely to be the principal factors which limit the employment of diamonds and the related high-pressure technology to only a few specialist laboratories. In addition, the intrinsic spectral signals of diamond interfere with probing techniques in not negligible portions of IR, visible, UV and soft X-ray spectral regions. Materials other than diamond have been therefore investigated in view of their possible employment as anvil in a high-pressure device. Cubic zirconia, sapphire and moissanite are by far the most studied, and a comparison of the properties of interest for high-pressure applications is reported in ref. [432].

Gem cut, single crystal cubic zirconia allowed pressures as high as 16.7 GPa to be reached [433], with a routinely achievable pressure of 10 GPa when solid media were investigated [434]. The advantages of using cubic zirconia are especially evident in IR absorption spectra, giving access to the 5 μm spectral region which is hidden in diamond by the two-phonon allowed transition, and in Raman scattering experiments where no fluorescence is observed in the typical internal vibrations region by using the 514.5 nm excitation line. Another feature is also of interest: cubic zirconia is stable at high temperatures, allowing studies above 800°C without any particular precaution such as those adopted to prevent oxidation of diamond.

The employment of sapphire as a possible material in opposed anvil cells dates back to the 60s. Cost, hardness (2000 kg/mm^2 in Knoop hardness), good UV and IR transparency are the main characteristics of interest for high-pressure studies. In particular, the UV transparency is much better than that of diamond, because a good transmission is achieved down to 144 nm, thus allowing excitation in the energy scale of chemical bonding, which is extremely interesting in the study of photo-induced chemical transformations. A cell combining a sapphire anvil with a tungsten carbide one, was employed to realize a pressure just above 10 GPa [435]. Extended studies of the mechanical strength of sapphire were performed in view of the possible high-pressure applications [436]. Making use of sapphire balls ($\sim$1 cm in diameter) as anvils, pressures above 10 GPa were realized [437]. Pressures as high as 25.8 GPa were produced by using beveled sapphire anvils [438]. The low luminescence is another interesting feature of sapphire, which makes it particularly suitable to study wide band gap materials [439–441]. Moreover, it can be used in large volume cells for neutron scattering and electrical measurements.

Moissanite is probably more interesting than cubic zirconia and sapphire for anvil manufacturing in high-pressure studies. Moissanite is hexagonal silicon carbide with hardness (3000 in Knoop scale) almost double than sapphire and cubic zirconia. Anvils have been produced using synthetic gem quality, single crystals with the *c* axis parallel to the compression direction allowing maximum pressures above 50 GPa [442, 443]. Besides hardness, the appealing characteristics of moissanite concern the optical properties, which are somehow complementary to diamond. None of the four sharp Raman lines overlap with the diamond first- and second-order peaks [442, 443]. Also of interest is the greater transmittance of moissanite, about one order of magnitude, with respect to diamond in the region between 1900 and 2300 cm^{-1} [444]. Moissanite anvils can also be employed both in laser as well as resistive heating high-pressure experiments. Moissanite limits the possibility of performing X-ray experiments due to the strong absorption below 20 keV. Thermal conductivity is also high and being a wide-gap insulator it can be used in high-pressure electric conductivity measurements. Owing to the large dimensions of moissanite, single crystal anvils can be scaled up to three orders of magnitude larger than diamond anvils, allowing the study of mm^3 size samples. This obviously can open the pressure range up to 50 GPa allowing a plethora of new probing techniques.

4.4.2 *Gasket*

The introduction of the metallic gasket [356] represents a milestone in high-pressure research. It is indeed an unavoidable component of the DAC and its use marks the beginning of quantitative measurements of strongly compressed samples. The metallic gasket has several critical functions. It ensures sideways encapsulation of the sample giving rise, together with the diamonds, to the high-pressure chamber; in addition, the gasket provides a lateral support for the tips of the anvils. The gasket is a metallic foil having a typical initial thickness of 100–250 μm. The shape of the foil can change depending on the cell geometry.

When the gasket is compressed by the two opposite diamonds, extrusion of the metal occurs because of its plastic deformation. Friction between the metal and the anvil surface, shear strength of the metal and thickness of the metal layer are the relevant parameters in determining sealing efficiency, stability of the sample and pressure gradient [445]. Two limiting cases are also illustrated to give a practical method to follow the sample stability on increasing pressure. If the metal extrudes outwards and the gasket hole dimensions increase with pressure, the so-called thick gasket regime is identified. Here, the gasket support may become insufficient and asymmetric with respect to the loading direction, resulting in assembly failure. Conversely, if the gasket is thinner and the metal extrudes inwards, determining a reduction of the gasket hole as the pressure is increased, the gasket is essentially stable. As a general rule, the thickness of the gasket should be thin enough to fit the latter situation.

In general, the gasket is therefore first indented to a typical thickness of 50–30 μm by applying pressure to the diamonds, which were previously carefully aligned, with the metal foil in between. In this way, the gasket deforms plastically and symmetrically on both sides, and an imprint-shaped crater, depending on the diamonds shape, is left (see Figure 4.12). The extruded material is fundamental for preventing failure of the anvils, due to the concentration of the stress at the edge of the anvil culets [348], in addition, it is also effective in supporting the material between the diamonds, thus increasing the pressure for a given gasket thickness by several GPa [445]. A further reduction of the stress can be obtained by using bevel diamonds [362].

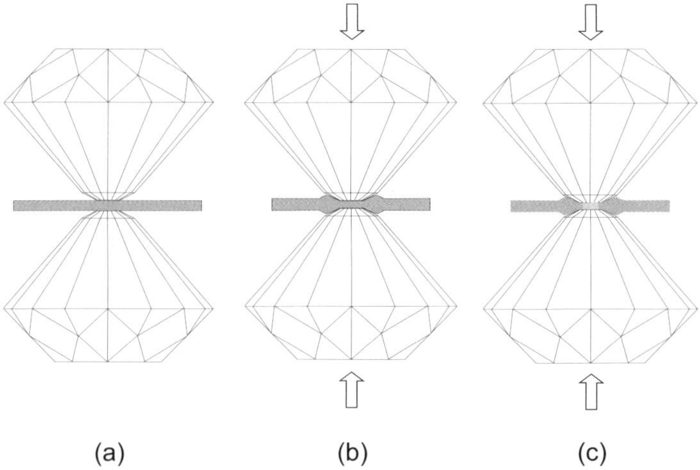

Fig. 4.12 Gasket preparation procedure: the metal foil employed for the gasket is placed between the diamonds (a); is deformed by the anvils (indenting) through the application of pressure to the diamonds (b); the sample volume (central transparent area in (c)) is produced by drilling the indented gasket in the centre and again positioning the gasket between the diamonds for the sample-loading operations.

The sample cavity can finally be realized by drilling a cylindrical hole centred, within micrometric precision, with respect to the indented area of the gasket throughout the indented thickness. The centring is particularly important to guarantee stability to the sample area when the load is applied.

The hole is produced through the employment of drilling machines, electro-erosion or laser drilling systems. The latter are increasingly employed in view of the possibility to drill materials other than metals, such as diamond or ceramics. The 1.06 μm emission of nsec pulsed solid state lasers is usually employed for this purpose. Drilling accuracy, centring and reproducibility allow the production of high-quality holes with diameters as small as about 20 μm, mandatory for experiments in the Mbar range. High-quality holes, in terms of roundness and edge finish, of smaller diameter (10 μm) have been produced through the employment of a 30 keV Ga ion beam [446].

X-ray transmission experiments allowed the monitoring of the diamond deformation as a function of pressures in the multimegabar regime [447]. The topography of the single-beveled diamond surface, reveals a strongly inhomogeneous pressure distribution with a pronounced peak at the centre, which decreases going to the edges of the gasket as is shown in Figure 4.13. For this reason, it is advisable to use a gasket hole to culet area ratio as small as possible, in order to ensure the most homogeneous compression on the whole sample. Also remarkable is the progressive flattening of the diamond surface, which assumes a kind of cup profile at the highest load (300 GPa on the centre of the diamond surface) with the bevel angle reversed at the edge.

Suitable materials for gaskets should simultaneously possess a large yield strength, a large friction coefficient on the diamond and a considerable ductility. High strength stainless steel and Cu-Be are diffusely employed for many purposes, but in the very high-pressure regime and at high temperature, rhenium is more suitable because of its higher yield strength. Boron and beryllium gaskets have also been used in order to provide optical access for X-ray measurements [448]. Laminated gaskets formed by the superposition of very thin metal foils glued together and then pressed, have also been realized for electrical measurements in the DAC, owing to the possibility to safely bring wires into the sample region [449]. The gasket material must also be inert with respect to the sample, in order to control the pressure-induced reactive processes without interference or contribution from catalytic effects due to the metallic gasket. This issue is not uncommon and is particularly relevant with fluid samples or in high-temperature studies, where contamination and corrosion of the gasket will likely bring sample loss and possible diamond failure. To prevent this kind of problems, isolation of the sample from the gasket through a gold ring has been reported [450, 451]. Gold powder is employed to fill the sample chamber after the indented gasket has been drilled, then it is pressed between the diamonds

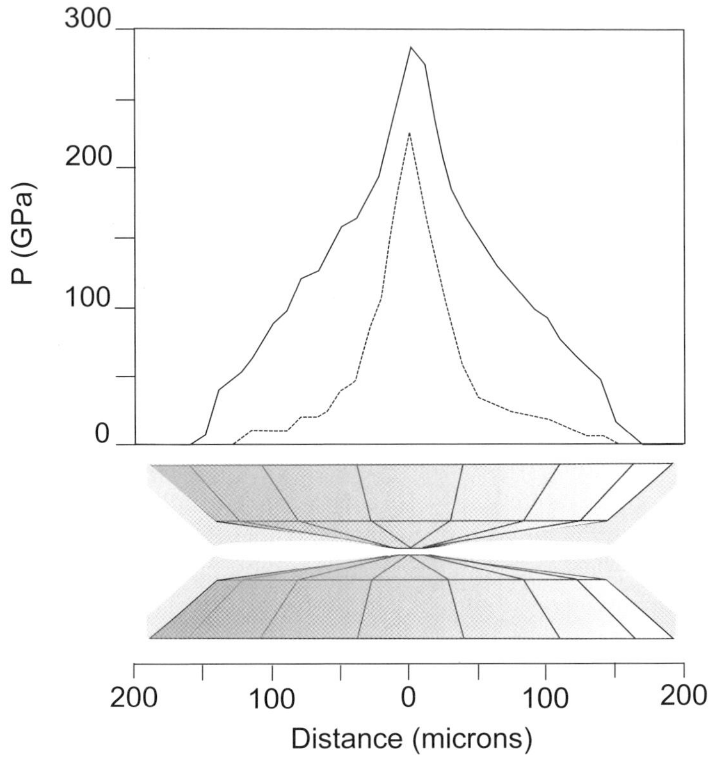

Fig. 4.13 In the upper part, the pressure distribution between bevel diamond culets under two different loads is shown. The sample is confined in the central region, about 10 μm in diameter, the remaining culet diamond surface being covered by the gasket. Below, the tip deformation at the maximum load (grey area) with respect to the ambient pressure profile is shown. (Modified from R.J. Hemley, et al. *Science* **276** (1997) 1242–1245. Reprinted with permission from AAAS).

and finally drilled again using a lower diameter tool.

Bulk metallic glasses, such as the amorphous metallic $Pd_{40}Ni_{40}P_{20}$ alloy, have also been suggested as interesting materials for gasket preparation in X-ray diffraction experiments in the Mbar range because of their excellent mechanical properties, together with the absence of sharp X-ray diffraction peaks [452].

Gaskets with several parts optimized for different specific properties can also be realized. Inert materials, like MgO or Al_2O_3, can be inserted in the metallic gasket in order to provide the electrical insulation necessary for allowing the introduction of electrical leads into the sample chamber [396].

Diamond-coated gaskets have been realized by filling the crater, produced in the preindentation procedure, with diamond powder (0.5–1 μm size) and a small amount of epoxy. This mixture is then compacted by again applying pressure to the diamonds and then drilled to produce the sample chamber [453, 454]. By exploiting the greater shear strength of the central flat area of the gasket, thicker samples are stabilized at high pressure in comparison to non-diamond coated gaskets of the same metal [454]. This solution is also extremely interesting for laser-heating experiments to minimize the thermal conduction from the heated sample to the diamond anvils.

4.4.3 *Loading techniques*

The sample preparation sometimes represents the most difficult part of a high-pressure experiment and in general, difficulties coincide with the loading procedure. Liquid and solid samples at ambient conditions are generally easily loaded, the main difficulties likely being related to the inert atmosphere created in order to avoid contamination of the sample. Glove boxes are generally employed in these cases, the only difference with ordinary sample manipulation in a controlled atmosphere is the need for a microscope to monitor the sample during the loading procedure. A glove box is also the basis of a cryogenic-loading technique for gaseous materials, that crystallize at temperatures higher than liquid nitrogen, which has been recently set-up [455]. Successful loading without removing the pressure gauges or compromising the gasket stability should be checked. Gaseous samples are obviously more complicated to load and several techniques are available depending on the thermodynamic and chemical properties of the sample. This aspect is particularly relevant in view of the common necessity to load these materials, employed as pressure-transmitting media, to ensure as far as possible, hydrostatic compression conditions for the sample under examination. Types and characteristics of several pressure-transmitting media will be discussed in more detail in the next section.

The general principle for loading a gaseous sample in the DAC is to condense it at low temperature or high pressure, in order to have enough matter when the pressure is applied to seal the sample and prevent the gasket closure. This is commonly achieved by three different methods: condensation in the liquid phase at low temperature (cryogenic loading), low-temperature crystallization by direct spraying onto the diamonds or loading of compressed gas (0.1-0.2 GPa) (gas loading).

In the cryogenic loading, the low temperature is usually realized by means of cryostats or liquid nitrogen thermal baths. The DAC is accommodated in the stainless steel tank, usually designed to operate up to pressures of several tens of bar [456]. The tank can be powered by electrical, optical, mechanical and capillary feedtroughs or even windows for having direct optical access. Any cell in principle, can be adjusted inside the tank, being a crucial difference in the way of applying pressure in order to seal the sample between the diamonds. In the loading procedure, the cell is inserted into the condensation chamber with the diamond slightly spaced ($\sim$ 150 μm), then the container is purged or possibly evacuated. In general, the sample gas atmosphere is kept cooling in the tank, maintaining an overpressure to avoid contamination due to the possible presence of leaks in the tank. Gas is continuously supplied during condensation. At this stage, the liquid starts to flood into the container and as the sample chamber is filled, the DAC can be closed by sealing the sample between the diamond anvils. Depending on the DAC employed, different ways to seal the sample have been set-up. In general, mechanical feedthroughs are employed to drive screws or threaded pistons. Easier by far is the employment of cryo-loading devices with pneumatic DACs where a capillary, bringing the helium gas to the membrane, is easily brought inside the vessel allowing a fast and reliable sealing of the sample.

The relatively simple cryogenic loading is probably the easiest solution when the loading of fluids with melting points above the liquid nitrogen temperature is required. Nevertheless, the study of hydrogen, a major issue in physics, as well as the need for loading helium or neon as a pressure transmitting media, pushed the realization of gas-loading techniques capable of producing a sufficiently dense gas inside the sample chamber. For instance, the same density of liquid helium at ambient pressure is obtained at 0.12 GPa and room temperature. The first high-pressure loading of helium was performed in 1979 [395]. Another great advantage of this loading method is also represented by the capability to load gas mixtures [457]. In fact, owing to the different boiling temperatures, a gas mixture decomposes during cooling and hence is impossible to load with the cryogenic technique.

The working principle of a gas loading apparatus is based on the production of a sufficiently dense gas inside a vessel, where the unsealed cell is contained. In general, single or double stage compressors allow the production of gas pressures up to 0.2 GPa. The high density is needed to prevent the gasket closure once pressure is applied to seal the sample. The sealing is achieved by applying an excess pressure on the diamonds. Depending on the DAC type, different solutions have been set-up to close the cell. For mechanical cells the gas-loading systems are constructed in such a way as to clamp the DAC by tightening the screws which drive the diamond movement, using a gear assembly located inside the high-pressure vessel [353, 458, 459]. Versatile gas-loading apparatus capable of accommodating various types of cells have been developed [460, 461]. In addition, these systems also have more sophisticated systems driving the piston to apply pressure to the diamond. A geared motor, placed externally to the vessel, drives a shaft which is coupled to the piston acting on the diamonds through a nut, thus the piston is advanced by rotating the shaft [460]. Recently, a system in which on the pushing end of the piston, which drives the sealing of the sample through the diamond clamping, a force sensor is placed so that the force applied to close the cell can be precisely adjusted [461]. When membrane DAC are employed, the sample sealing can be ensured by applying a small overpressure to the membrane. The small pressure difference between the membrane and the vessel must be maintained during the pressure release, otherwise there is the risk of losing the sample or even breaking the diamonds, making this operation rather delicate. Recently a clamping method based on a shaft moving a rotary drive to seal the cell has been realized. The most innovative aspect of this realization is the possibility to visually monitor the loading procedure through a CCD camera [462]. An apparatus for loading gases in a Paris-Edinburgh press has been recently realized [463]. Here, the loading is performed in an external vessel where the anvils, contained in a clamp, are placed. Once the loading is completed, the clamp is closed and transferred into the Paris-Edinburgh cell. The clamp is provided with the necessary apertures for the scattered neutron beams.

In the case of loading of special gases (flammable, toxic or explosive) or mixtures of species having very different melting and boiling points, both standard cryogenic and gas-loading techniques may not be suitable. In these cases, condensation of the desired material (small hydrocarbons, for example) directly onto the diamond surfaces have been performed [376]. To increase the probability of filling the gasket hole as the condensation occurs, advantage can be taken of the indium dam technique [464]. A thin indium ring fixed, around the sample chamber, should ensure the capture of as much as possible of the sample, sprayed between the diamonds. Once the dam is fixed, the cell is cooled by a liquid nitrogen bath in an inert atmosphere or by a close-cycle cryostat. When the temperature of the cell is close to the condensation temperature of the sample, the sample gas is blown through a small capillary, inserted between the slightly spaced ($\sim$ 150 μm) anvils. In order to prevent the obstruction of the capillary due to sample condensation, a very weak flow of helium at room temperature is maintained through the capillary while the temperature decreases. Once the condensation has occurred, the cell can be closed by sealing the sample through the application of load to the diamonds. The employment of a cryostat is mandatory when the condensation temperature is below 80 K. The cell is mounted on the cold tip of a close-cycle cryostat and the end of a capillary is placed at $\sim$2 mm from the surface of the diamonds. The capillary is connected through a flange to the deposition line. The deposition of the gas, or of the components of a mixture [465], can be followed by observation through a microscope and once completed, the cell can be screwed by a mechanical feedthrough. In all these cases, the employment of a membrane DAC facilitates the remote sample sealing.

4.4.4 *Compression media*

The study of chemical and physical properties of materials, as a function of pressure, should be ideally performed in conditions in which the stress applied is homogeneous and the sample does not undergo any shear stress due to the environment. In order to achieve these conditions, the sample should be surrounded by a hydrostatic medium which, by definition, does not have shear strength. Non-hydrostatic conditions give rise to an inhomogeneous strain in crystalline samples, which is revealed by broadening of the diffraction peaks or of the relevant spectral, both vibrational and electronic, signatures. Non-hydrostatic stresses may therefore alter the evolution with pressure of lattice parameters, as well as of vibrational frequencies, providing misleading data regarding phase transitions, or preventing accurate determination of the relevant parameters of the EOS. In addition, shear stress can severely affect the chemical stability of a substance by inducing or, more generally, lowering the onset pressure of a chemical reaction. For this reason fluids or, as diffusely observed in the case of large volume presses, low-shear strength solids such as AgCl or NaCl have been employed as pressure-transmitting media. Relevant information to perform a high-pressure study is therefore the choice of the most suitable pressure-transmitting medium, depending on the desired pressure range, chemical requirements and loading procedure. Knowledge of the hydrostatic limits of the possible materials is therefore of interest. The hydrostatic pressure range has been determined for many fluids (see Table 4.2) by studying the

Table 4.2 Hydrostatic and quasi-hydrostatic pressure limits for the materials commonly used as compression media. me: 4:1 mixture of methanol-ethanol, mew: methanol-ethanol-water mixture 16:3:1 from ref. [466]

medium	P_{hyd}	P_{q-hyd}
He	11.6	50
Ne	4.7	16
Ar	1.2	9
H_2	5.7	50
N_2	2.4	13
me	10.4	10.4
mew	14	14

pressure distribution across the sample, using the broadening of the R1 ruby fluorescence line [393]. These authors showed that a 4:1 methanol-ethanol mixture remains hydrostatic at room temperature up to 10.4 GPa, a higher pressure with respect to the 1:1 pentane-isopentane mixture (7 GPa). By adding a small amount of water (16:3:1 methanol-ethanol-water), the hydrostatic range extends to 14.5 GPa [467], a pressure which coincides with the glass transition and therefore above this limit, large pressure gradients are present in the sample. Fast pressurization allows supercompressing the mixture and homogeneous non-hydrostatic conditions up to 35 GPa have been claimed [287, 468]. Silicone fluid is claimed to behave as comparably well, as a pressure-transmitting medium, as the 4:1 methanol:ethanol mixture below 30 GPa, being much better at higher pressures [469]. These conclusions are however controversial because comparative studies with argon have shown a substantial non-hydrostatic contribution above the glass transition [470]. The strong IR absorption, the quite important reactivity at high pressure under laser irradiation and the perturbation of the chemical environment when a reaction occurs, make all these materials unsuitable for studies of high-pressure reactions.

Rare gases are by far the best pressure-transmitting media because of their stability, chemical inertia, high solidification pressure and employment at low temperature [348]. In addition, hydrostaticity also persists in the solid phase, in fact, even though helium solidifies at 11 GPa, a homogeneous distribution of pressure across the helium sample has been measured up to 60 GPa [471] and nearly hydrostatic behaviour has been reported up to 120 GPa [472]. Neon freezes at ambient temperature at 4.7 GPa but is reported to be hydrostatic up to 16 GPa [473]. Both these gases also have the additional advantage of presenting a low X-ray absorption also giving a low background, therefore representing the best choice for X-ray diffraction measurements. Nevertheless, neon is to be preferred in high-temperature experiments because of the diffusion of helium into the diamonds at high temperature, which strongly affects their mechanical stability. Two other commonly employed fluid-transmitting media are argon and nitrogen. The main advantage of using these two materials is the ease of performing the loading as low-temperature liquids (see previous section). Argon is reported to be quasi-hydrostatic well above the melting point [474], not observing deviatoric stress in X-ray diffraction studies of layered ionic crystals up to ~30 GPa [470]. Comparative X-ray diffraction studies of the EOS of ^{12}C and ^{13}C diamonds suggest that N_2 is nearly hydrostatic up to 22 GPa [475]. Based only on the observable splitting of the ruby lines, which is a simplified and questionable criterion, xenon is reported to be a hydrostatic pressure medium up to 55 GPa [476].

The reliability of the hydrostaticity limit as obtained by ruby fluorescence line-broadening measurements has recently been discussed because of the ruby stiffness (bulk modulus ~254 GPa [477]), which makes ruby less sensitive to non-hydrostatic stresses than the molecular compounds employed as pressure-transmitting media [478]. These authors used another method of detecting non-hydrostatic stresses in the pressure media, consisting of the study of the sharp diffraction peaks of quartz single crystals. The smaller bulk modulus (37.12 GPa) [479] with respect to ruby, makes the peak parameters extremely sensitive to small non-hydrostatic stresses [480]. Using this method remarkable lower values have been determined especially for Ar (1.9 GPa), N_2 (3.0 GPa) and silicon oil (0.9 GPa).

4.5 High- and low-temperature techniques

One of the factors contributing to the success and diffusion of the DAC technique is strictly related to the possibility to independently change pressure and temperature. This peculiarity allows a detailed and precise analysis of a broad part of the P-T diagram, where a wealth of phenomena of interest for different disciplines occur. This is obviously extremely appealing from a chemical point of view, due to the possibility to study reactions at constant pressure as a function of temperature or, conversely, at constant temperature as a function of pressure, so that the relevant thermodynamic parameters can be obtained. For this reason, the coupling of the DAC technique to cryogenic or heating methods has been developed independently by many different groups. Temperatures ranging from a few mK up to thousands of K have been produced simultaneously, at pressures in excess of tens or hundreds of GPa. From a chemical point of view, the possibility of conjugating high pressure with variable high-temperature conditions is far more attractive, especially in view of studying geochemical processes, than low-temperature studies, which mainly concern crystallographic-oriented studies. Low- and high-temperature methods will be described in the following sections separating the two common approaches employed in high-pressure research to generate high temperatures, the resistive and laser heating methods.

4.5.1 *Low-temperature devices*

Many different solutions have been proposed in the literature to couple almost all the different types of DAC to commercial cryostats, especially for optical and X-ray diffraction measurements. Extensive descriptions are provided in different books [281, 287]. In addition, specifically designed DAC are now commercially available in view of their employment in low-temperature applications. The main difficulties of coupling a DAC with a cryostat is related to the larger distance between the focusing and collecting optics, and the sample, which complicates optical measurements. In fact, the achievements of low temperatures requires thermal insulation so that mechanical elements sorrounding the cell are necessary, such as thermal shields ($T \leq 20$ K) and vacuum chamber with relative windows. Long working distance optics are therefore necessary. For this reason specific cryostats, where the focusing and collecting optics are also contained in the vacuum chambers, have been realized both for Raman [481] and IR [482] studies. The system realized for IR absorption measurements is rather peculiar because IR measurements in the DAC are normally limited, unless when synchrotron light is employed, in the utilization of IR microscopes, where the available space for the cell is generally extremely limited, not allowing cryo-apparatus. Specifically designed thin flux-cooled DACs can be mounted on common microscope stages and translators and coupled to short working distance high-magnification optics. The system described in ref. [482] uses the sample chamber of an FTIR spectrometer as a cryo-chamber. A membrane DAC is mounted on the cold tip of a cryostat (close-cycle or flux) and placed between the focusing optics (Cassegrain objectives or ellipsoidal mirrors) of the IR beam. Windows (CaF_2 or CsI) are employed to isolate the cryo-chamber from the rest of the instrument (see Figure 4.14). Far infrared spectra at low temperature have also been measured with this apparatus [483, 484]. In general, membrane DACs are better adapted to cryostats because the remote control is here easily realized via a simple vacuum feedthrough, allowing the helium gas to be brought to the membrane. Nevertheless, mechanical cells now do not require heating to change pressure, and are commonly employed, with the pressure control allowed via a diaphragm, gear box or lever-arm assembly driven from outside.

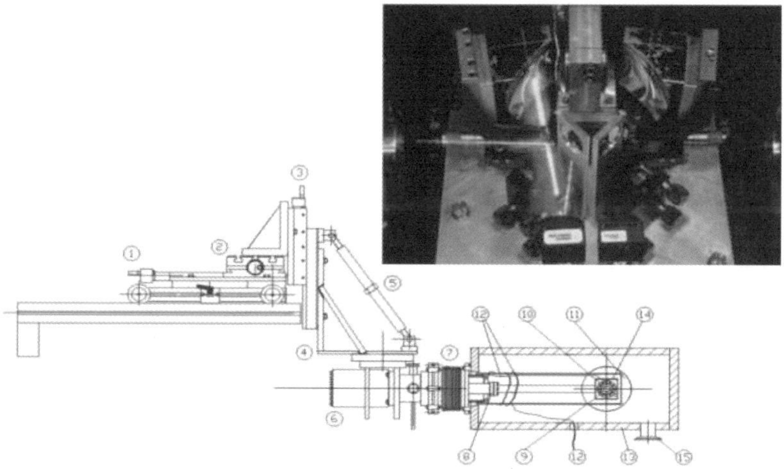

Fig. 4.14 Schematics of a close-cycle cryostat coupling to an FTIR spectrometer as described in ref. [482]. The beam-condensing apparatus is realized with ellipsoidal mirrors and employed to focus the IR beam on the sample contained in the DAC is shown in the inset.

A number of different kinds of cryostats are in use, and fall into two categories, He or N_2 flux or close-cycle cryostats, the former being far more commonly employed. Compact He flux cryostats are more easily adapted to different experimental set-up, but generally they are limited in the possibility having radiation shrouds and shields so that they can be routinely operated at typical temperatures, on the sample, between 10 and 20 K. Large He flow cryostats can reach down to 2K. Close-cycle cryostats allow samples to reach temperatures $\geq$10 K, but present the important drawback of sample oscillation ($\sim$10 μm). Magnetic measurements have been performed up to 10 GPa [485] using a DAC in which the piston is pneumatically driven by a membrane bellow, pressurized by liquid helium. The main peculiarity here is that by using a ^{3}He-^{4}He dilution refrigerator [486], temperatures as low as 0.03 K could be obtained.

4.5.2 *Resistive heating*

Temperatures as high as several thousands of degrees can nowadays be obtained on samples contained in a DAC. These temperatures can be obtained by using both laser or resistive heating techniques. Laser heating, which is currently the preferred heating technique used in combination with the DAC, presents however a few drawbacks, consisting of large temperature gradients across the sample, and in the difficulty in handling this technique below 1300 K [487], where large temperature errors are also made by employing the black body radiation fit [488], the usual temperature determination technique in laser heating experiments. Resistive heating of the sample can be achieved by placing the wires where the electrical current passes through and around the DAC, the diamonds' support or just around the diamond and gasket assembly. This is the best heating method to achieve homogeneous temperature distribution in the sample, even though important drawbacks, such as the graphitization of diamonds or the change with temperature of the mechanical properties of the gasket and even of the cell, should be considered. This obviously poses precise limits to both the maximum pressure and temperature that can be obtained. These limitations can be avoided by heating the sample internally.

Internal resistive heating can produce thousands of degrees onto the sample, with the remarkable advantage that diamonds and the rest of the cell remain at a much lower temperature. This is realized through the employment of fine wires of an inert high-melting metal, well isolated by the diamonds. Iron [489–491] and tungsten [492] wires were first used either as the heater or as the sample, and temperatures of several thousands of degrees have been measured by fitting the emission to a black body radiation. A quite complex assembly of the internal heating apparatus has been reported by Zha et al. [493] for the study of non-electrically conducting materials. Here, an electrical insulating layer is sandwiched between two gaskets in which a hole, fitting the diamond surface, has been drilled. The non-metallic gasket is produced by a mixture of diamond powder and MgO compacted by pressurization between the supporting gasket. A rhenium strip is employed as the heater and a laser-drilled hole is produced in the strip to contain the sample. Thermal insulation from the diamonds is realized by packing low-thermal conductivity powders around the strip, whose choice is dictated by the kind of optical, physical and chemical properties required.

External resistive heating of the DAC can be realized through different approaches which can also be combined to produce higher temperatures. In all cases, the uniform heating of the sample is achieved but, as previously mentioned, diamond anvils are exposed to the risk of oxidation, graphitization and chemical reaction with the gasket material, whereas the mechanics of the cell can be seriously affected by the high temperature. The limit temperature, fixed by diamond failure even in vacuum conditions and when using an inert gas flow, is about 1500°C. In order to avoid diamond oxidation and graphitization, the volume containing the heaters and the anvils is enclosed in an inert gas atmosphere (Ar with 1% hydrogen gas).

The heating of the DAC can be realized by placing it inside an evacuated oven [494]. Since all the components of the cell are exposed to thermal stress, an important mismatch between the different materials composing the cell can take place. Careful choice of materials with low-thermal expansion coefficients should be made. Tungsten carbide, boron carbide, rhenium, inconel and udimet 700 are generally used [494, 495]. Despite the difficulties in combining optical measurements with the use of a furnace, Raman spectra have been recorded [494, 496]. Similarly to the external furnace, several solutions have been reported in the literature where resistances were wrapped around the body of the cell.

Heating can also be achieved by passing a high electrical current through molybdenum wires wrapped around the tungsten carbide seats, which support the diamond anvils [374, 497]. The sample temperature is claimed to be very uniform. Alternatively, the cavity around the diamond anvils is heated by suitably designed furnaces [495, 498–500]. This approach is by far more effective and temperatures up to 1000 K can be obtained at several tens of GPa. By combining this method with an additional external heater, the temperature stability and the heating performances can be improved. Temperatures largely in excess of 1000 K can also be obtained at quite high pressures (72 GPa) [501].

Temperature can be measured by means of different thermocouples, some of them also operating above 2200 K [281]. Temperature sensors should be placed as close as possible to the sample, in order to minimize errors in the temperature measurement. This drawback can, to a certain extent, be overcome by calibrating the thermocouple with some material of a known transition temperature [502, 503]. Comparative measurements between two thermocouples placed inside the sample chamber and glued onto an anvil, as close as possible to the sample, provided an almost constant 5% difference between the two sensors up to 500 K [504]. For very high temperatures, the spectroradiometry of the black body emission or the intensity ratio between the corresponding Stokes and anti-Stokes Raman bands should be used (see the laser heating section). Diamond and cubic boron carbide are very promising materials for this purpose because they have an intense Raman signal and they are chemically inert in most conditions. From the experimental point of view, measuring the Raman intensity in terms of absolute values in order to compare the Stokes and anti-Stokes regions is not trivial, due to different instrumental sensitivities and different luminescence of the diamonds in the two regions. In principle, the ruby luminescence line shift could also be used as a temperature gauge if pressure is known from any other technique, but practically for temperatures exceeding 400°C, the decrease in the signal-to-noise ratio limits the use of the ruby for this purpose.

4.5.3 *Laser heating*

The synthesis of diamond by irradiating graphite with a ruby laser in a DAC by Taro Takahashi and William A. Bassett [505] opened, in the late 60s, access to an almost unexplored portion of the P-T space. Laser heating in a few years became a reliable tool for high-pressure and high-temperature studies [408, 506]. Geophysics [507] and material science [31] are the disciplines that have greatly benefited by the development of this technique. The structural transformations and chemical reactions occurring in the Earth's deep mantle and core demanded controlled and reliable techniques to increase temperature under high-pressure conditions. As an example, in order to determine the temperature distribution in the core, a precise characterization of the iron and iron-rich compounds is mandatory. This search motivated several different experiments regarding the structure and melting of iron under pressures exceeding 2 Mbar, and temperatures above 4000 K [407, 508–511]. In addition, the synthesis of attracting extended materials from simple molecular crystals, such as the energetic polymeric form of nitrogen [512], the silica-like HT-HP structures of CO_2 [513], or superhard nitrides of C, B, Si and Ge expected to be close to diamond in terms of hardness [514], is a direct result of the high-temperature conditions realized through the laser heating technique. Temperatures of 6000 K and pressures well above 2 Mbar are produced at equilibrium conditions while maintaining independent control of the two variables [515]. Achievement of such high temperatures without damaging the diamonds is due to the stability of the diamond with respect to the graphitization above 10 GPa [281].

Sample heating is achieved by focusing high power (~100 W) cw infrared lasers. These powers are necessary because the diamond anvils absorb a significant part of the beam (~10% for a IIa type and 30–40% for a Ia type diamond anvil [516]). Pulsed lasers have also been employed, presenting the advantage of very high peak power, and then less average power being required with respect to cw lasers, with reduced heating of the cell due to the reduced time in which the sample remains at high temperature. Nevertheless, measurement of the peak sample temperature requires more complex electronics even though time-averaged measurements have been proposed [517]. Both solid state (Nd-YAG and Nd-YLF) and gas (CO_2) lasers are commonly employed, the choice being driven by the sample absorptivity. Both YAG and YLF lasers emit around 1 μm (1.064 μm Nd-YAG, 1.053 μm Nd-YLF) whereas CO_2 emits in the middle IR region at 10.64 μm. The former are employed with metallic samples, or metallic absorbers to heat the sample. The CO_2 laser is instead employed with materials, such as minerals, transparent oxides or even molecular systems, exploiting the absorption of lattice phonons or, in the molecule case, internal vibrations. Due to the different emission wavelength, different optics are also required with the two laser sources. Normally, BK7 glass can be used with YAG and YLF lasers, whereas ZnSe or CaF_2 optics are necessary when a CO_2 laser is adopted. Transparent samples can also be efficiently heated up by placing a small piece of a metal foil (Re, Pt, W), with a thickness of 5–20 μm, in which one or more small holes are laser drilled [518]. The laser is focused onto the foil which acts as an efficient laser absorber and conveys heat to the transparent sample, which is contained in the small holes (see Figure 4.15). The employment of the foil guarantees more stability of the sample with respect to powders.

Due to its extremely large thermal conductivity, diamonds act as nearly infinite heat sinks. For this reason, the sample should be thermally insulated from the diamonds and embedded in a chemically inert pressure medium with low strength and low-thermal conductivity (sapphire, MgO, alkali alides). Also, rare gases, especially argon, are often employed in spite of their high compressibility which reduces their thermal insulation power at high pressure because of the greatly reduced thickness.

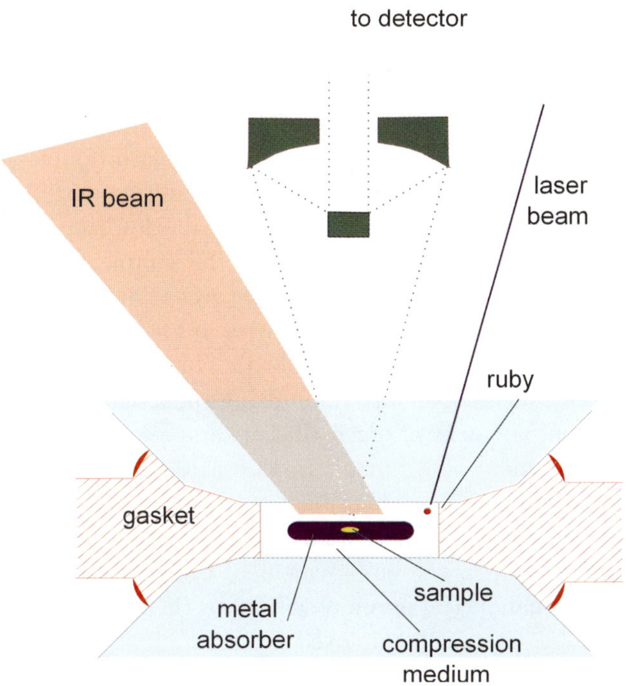

Fig. 4.15 Schematic representation of the sample assembly in a laser-heating experiment. The IR laser beam is directed onto the absorbing metal immersed in a compression medium also acting as thermal insulator. The absorbing metal can be the sample itself or can be drilled to contain the material under study. Thermal emission from the sample is registered and analysed for temperature determination. In this scheme, pressure is obtained by the ruby fluorescence technique (see next section).

The reactivity of the medium surrounding the sample, and of the sample itself, is a major issue in laser-heating experiments. Their reactivity with the gasket materials and with the diamond can be greatly enhanced by the extreme pressure and temperature conditions. For example, this aspect prevented the study of attracting materials such as hydrogen and alkali metals, even though protection of the gasket by gold [451] or compressed ceramic powder [501] allowed the extension of the melting line of hydrogen up to 800 K.

Two important drawbacks of the laser heating technique concern the large temperature gradients existing across the sample and the indirect temperature determination through the sample emission. The latter is indeed affected by the interpretative emission model and by temperature fluctuations, due to intensity changes of the laser source. In spite of the thermal insulation, large temperature gradients are present both in radial and axial directions. Besides the high-thermal conductivity of diamonds, the Gaussian intensity distribution of the commonly employed TEM_{00} laser mode also contributes. In order to overcome this problem, a double-sided laser-heating technique has been adopted [519]. A multimode Nd:YAG laser having a profile with a relatively flat power distribution (boxcar like) is split and focused simultaneously on both sides of the sample. The intensity profile makes it possible to heat broader sample areas (20–50 μm), thus decreasing radial temperature gradients; as a matter of fact, uniform radial heating can also be realized by defocusing the beam. In addition, the laser focusing from both sides succeeds in also minimizing the axial temperature gradient allowing the employment of much thicker samples.

Temperature measurement is a difficult task in a laser-heating experiment because of the small sample dimensions and in particular the inhomogeneous temperature distribution. In addition, due to the high temperatures produced in the sample, the employment of ordinary temperature gauges such as conventional optical pyrometry [506] or thermocouples [503] is only sporadic. By far the most employed technique to determine the sample temperature is based on the measurement of the sample thermal emission over a wide range (600–900 nm) of the visible region, the temperature being obtained by a fit to the Planck profile of the black body emission [520] (see Figure 4.16):

$$I(\lambda) = \frac{(2\pi c^2 h)\epsilon\lambda^{-5}}{e^{(hc/k)/\lambda T} - 1}, \qquad (4.1)$$

where $I(\lambda)$ is the measured intensity, ϵ the emissivity, c the velocity of light and h and k are Planck's and Boltzmann's constants. Two important approximations characterize this approach, the emission measurement in a quite limited spectral range and the assumption that the emitting system is an ideal black body, which implies an emissivity value independent on wavelength and equal to 1. The greater the spectral range analysed the more reliably the results fit the black body emission, nevertheless, chromatic aberration should be considered when setting up the optics for the spectroradiometry analysis [521]. Corrected temperature values with respect to reference metal-melting temperatures are obtained when the data are fitted to a grey body emission, i.e., a material having $\epsilon < 1$, but still independent of wavelength, corrected for the experimental values of the spectral emissivity [488]. As already mentioned, temperature fluctuations are essentially due to the power instability of the laser source and given that the emission power is directly proportional to the fourth power of the emitting sample temperature (Stefan–Boltzmann law), it results that a fluctuation of 2–3% in the laser output may result in temperature fluctuations of ~200 K at about 2000 K [522, 523]. Temperature stabilization has been realized by measuring the sample emission with an auxiliary system which drives an electro-optical attenuator. When the thermal radiation from the sample changes, the servo-controller varies the voltage applied to the liquid-crystal cell of the electro-optical attenuator, thus changing the polarization of the laser beam which is further selected by a calcite polarizer. The incident power of the laser is therefore continuously readjusted in order to keep the sample emission constant. In this way, standard deviations of 8 K were obtained at 3050 K [488].

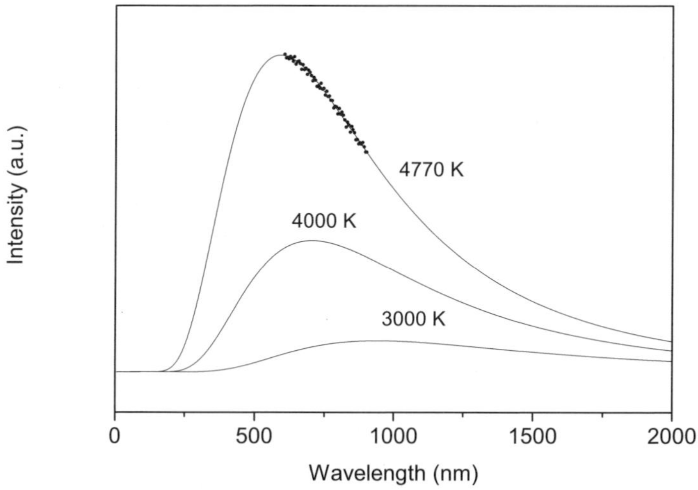

Fig. 4.16 Sample temperature determination by fitting the emission data from the sample (symbols) to Eq. 4.1.

Two-dimensional imaging of the hot sample permits determination of the real time-temperature distribution across the hot spot, instead of only relying on a small point or an average measurement. This method has recently been developed by simultaneously analysing different spatially correlated images of the sample, in narrow spectral ranges, so that chromatic aberrations are practically eliminated. The temperature and the emissivity of the sample, within the grey body approximation, are mapped in two dimensions thus yielding precise measurements of temperature gradients, their temporal evolution as a function of time, and also the possible dependence of emissivity on wavelength and on temperature [524, 525].

Sample temperature determination from the intensity ratio of Stokes and anti-Stokes Raman bands has also been reported [518]. The intensity ratio of these bands only depends on temperature, regardless of the material and is determined by the Boltzmann distribution. Adoption of this method in laser-heating experiments on CO_2, realized through a metal foil absorber, revealed an overestimate of the sample temperature measured by the black body emission fit, the emitted light being measured for the temperature determination most likely dominated by the metal foil emission [526]. This difference, extremely relevant ($\sim$400 K) below 40 GPa, decreases at higher pressure because of the larger thermal conductivity of both the sample and the metal foil.

Analogously to the Stokes/anti-Stokes intensity ratio, the principle of detailed balance has been employed in nuclear resonant inelastic X-ray scattering experiments to determine the sample temperature [527, 528]. Measuring the ratio of the phonon creation and annihilation peak intensities, represents a reliable temperature determination because the nuclear resonance cross section depends only negligibly on temperature. The temperatures measured in this way agree nicely with values determined from the thermal radiation spectra fitted to the Planck radiation.

4.6 Pressure measurement

In the previous section we reviewed how difficult the accurate temperature determination in a high-pressure environment is. Analogously, difficulties are met when a reliable pressure measurement has to be performed. The relevant findings of the intense technical and conceptual efforts made during the 60s for the realization of a high-pressure scale are exhaustively detailed in a review by Decker et al. [529]. At low pressures, the calibration may be based on direct measurements of pressure by measuring the load and the surface on which it is applied. The primary pressure gauges employed are the mercury manometer, whose employment is limited to pressures of a few hundred bars, and the free-piston gauge which has been successfully used up to 2.6 GPa. Although the use of a piston-cylinder system, with either piston packing or a solid-medium pressure environment, represents the best approximation to a primary scale, at pressures above 2.5 GPa the friction between the piston and cylinder or packing, and the internal friction associated with the solid-medium environment, can affect the load-pressure curves. A great accuracy (10 ppm) is required in the determination of the effective area, and on corrections for friction and for buoyancy of the piston [287]. The reported accuracies vary from 0.02% at 0.8 GPa to 4% at 10 GPa.

In analogy with the construction of the International Practical Temperature Scale, employment of fixed points secondary gauges allows the realization of a practical pressure scale. Fixed pressure points are liquid-solid or solid-solid transformations in pure substances. Besides the existence of a polymorphic transition, several additional conditions should be satisfied. First of all, the parameter of interest should exhibit a clear detectable change at the transitions. Observations have regarded changes in electrical resistance, volume, refractive index, absorption and reflectance, but also direct probing of solid-solid phase transitions by X-ray and neutron diffraction, differential thermal analysis, magnetic properties and ultrasonic velocities have been performed. Among these parameters electrical resistance and volume measurements have been by far the most employed to identify the transition points [281, 287, 530]. The material under examination should be attainable with very high purity and should be chemically stable. The phase transitions, in order to be of interest, should occur sharply both in time and in pressure, and the pressure transition should depend smoothly on temperature. Accuracies are high below 4 GPa where the melting of mercury (0.76 GPa), bismuth I-II (2.55 GPa) and Tl II-III (3.67 GPa) phase transitions have been employed. At pressures up to 10 GPa, other phase transitions in Cs, Ba and Bi have been used, whereas above this pressure the phase transitions of interest are those between Sn I - Sn II (10 GPa), Fe-α - Fe-ϵ (12.6 GPa), Ba II - Ba III (14 GPa) and Pb I - Pb II (12-16 GPa). Once a set of fixed points has been chosen, an interpolation method to determine intermediate pressures is necessary. In order to obtain P-V-T curves (pressure, molar volume and temperature) accurate measurements of V are necessary. Dilatometry can provide this information, but X-ray diffraction is one of the most accurate methods presently used for determining the molar volume of a crystalline solid. Sonic velocity methods yield data from which the first derivative of volume, with respect to pressure (compressibility) or the reciprocal of compressibility (bulk modulus), can be obtained.

Secondary pressure gauges can be identified by relying on any physical parameter which varies monotonically with pressure. Even though optical absorptions, from the UV to IR, and refractive index measurements have been used as secondary pressure gauges, the change of electrical resistance of a metallic wire has been by far the most employed method. The selection of the metal is made by considering high sensitivity of resistance to pressure and a low one to temperature, purity, ease of manufacturing, chemical and physical-chemical stability, high degree of reproducibility from gauge to gauge (i.e., low sensitivity to chemical composition and manufacturing techniques), linear or nearly linear response of resistance to pressure and absence of phase transitions in the pressure range of interest. Manganin wires are generally employed for this purpose. The term manganin refers to a family of copper-manganese-nickel alloys. In high-pressure studies, commercial manganin with 84% Cu, 12% Mn and 4% Ni are generally employed. Nevertheless, as the pressure rises above 3 GPa, the calibration curve of manganin becomes progressively less reliable. The most employed interpolation technique is the NaCl scale, based on the theoretical equation of state of NaCl which also has the advantage being used as a reliable calibrant at elevated temperatures.

Dynamic shock measurements represent another method of measuring pressure. When a material undergoes a shock wave compression, pressure-volume-internal energy (P-V-E) data along the Hugoniot, a curve in the P-V plane reached by shocking a material at defined P-T conditions, can be obtained. These results can be employed to calculate P-V relations along an adiabatic or an isothermal line, the latter being commonly denoted as a reduced shock wave isotherm [531–535], more useful for comparison with static measurements, by using an appropriate equation of state. Experimentally the P-V relation can be obtained from the particle velocity (U_p) and the shock velocity (U_s) through the Hugoniot equations. These direct measurements provide a primary standard pressure calibration and can be used for the calibration of secondary standards of more practical use. As an example, the comparison of molar volume ratios of gold from shock wave data, X-ray diffraction measurements and the computed equation of state, also using compression data of NaCl as a reference, allowed the enhanced reliability of Au as an internal pressure calibrant over a broad range of pressures (0–200 GPa) and temperatures (300–3000 K) [503, 536, 537]. The EOS of gold is of particular interest because it is commonly employed as a pressure standard in high-pressure X-ray diffraction experiments. An attempt to reconcile the discrepancies among different experimental results has recently been performed by taking into account uniaxial stress contributions [466]. In addition, primary pressure scales have been derived by combining measurements of elasticity and density [538], and from high-precision X-ray diffraction data and Brillouin spectroscopy on MgO in hydrostatic conditions up to 55 GPa [539]. Here, the elastic wave velocities are determined under compression using Brillouin scattering, whereas the measurement of the specific volume is obtained by X-ray diffraction. Sound velocities are related to the volume values through the measurement, in both experiments, of the ruby R-line wavelength. The volume-dependent isothermal bulk modulus is obtained from the sound velocities, and from its integration the pressure is finally obtained.

Also, theoretical EOS are employed to interpolate and extrapolate fixed pressure data, but are also useful in determining pressure changes due to temperature differences. Density functional theory (DFT) calculations can provide the derivative of the bulk modulus which is then combined with the experimental values of reference volume (V_0) and bulk modulus (B_0), and corrected for the zero-point motion to give P(V).

One of the reasons which makes the DAC technique so successful and explains its widespread application, in combination with many different probing techniques, is due to the possibility of determining, *in situ*, the sample pressure by the ruby luminescence method [386, 390–393, 473, 540, 541]. In this method, the pressure is determined by the wavelength of a specific emission line of ruby which can be excited by appropriate visible laser lines, whose selection depends on the pressure range. Ruby sensors are in the form of small single crystals prepared by synthetic ruby, with a Cr^{3+} concentration in a specific range (3000–4000 ppm), through an annealing process requiring several days [542]. The ruby chips employed for high-pressure experiments with the DAC have typical dimensions ranging between 1 and 10 μm. An extensive and exhaustive description of the physical properties of ruby and their implications in the framework of its use as high-pressure sensors has recently been provided [543]. Ruby is a variety of corundum, the mineralogical term indicating α-Al_2O_3 (alumina), containing Cr^{3+} ions as substitutional impurities of Al^{3+}. The crystal structure of corundum can be regarded as a hexagonal closed packed oxygen lattice with Al ions occupying two-thirds of the octahedral interstitial sites. Neglecting the lattice distortion, the symmetry of these octahedral sites is given by the O_h cubic group, but the oxygen layers are not perfectly hexagonal given that each Al ion has three near oxygen atoms, whereas the other three are more distant by about 5%. The Al ion displacement along the hexagonal c-axis direction reduces the point symmetry to C_{3v}. In addition, the relative arrangement of oxygen atoms in the layers adjacent to the metal is such as to determine the loss of mirror symmetry, thus reducing the point symmetry for Al to

C_3. When Cr^{3+} is introduced it results in a small expansion of the host lattice because of its slightly larger ionic radius, so that the Cr^{3+} ions are not exactly coincident with the positions occupied by the aluminum ions [544, 545]. Powder [477, 546] and single-crystal [547–549] X-ray diffraction experiments indicate that the α-Al_2O_3 lattice deforms almost isotropically under hydrostatic pressure. Both computational and experimental data suggest that α-Al_2O_3 is not the thermodynamically stable room temperature form at pressures exceeding 100 GPa, but X-ray diffraction studies regarding the observation of α-Al_2O_3 up to 175 GPa are reported [546]. Recently, X-ray diffraction studies revealed a phase transition of ruby to a *Pbcn* structure, analogous to Rh_2O_3(II), at 96 GPa and high temperatures (~1200 K) [550]. The ruby fluorescence spectra of the quenched samples at ambient conditions show a significant red shift and line broadening.

The optical absorption and emission properties of ruby have been extensively studied because its energy levels are well described by the ligand field theory [551–553], thus representing a good model of d^3 systems, and because of its use as an active medium in the first solid state laser. The fivefold orbital degeneracy of the d orbitals is removed by the octahedral crystal field so that the 3d electrons of Cr^{3+} are hosted by the three low energy t_2 orbitals, which point in-between neighbouring ions, whereas the two e orbitals, pointing toward the neighbours, are empty. The energy level scheme deriving from this electron arrangement [553] is shown in Figure 4.17.

The lowest free-ion multi-electron state, in the weak crystal field limit, is the ^4F high-spin state. The ligand field and interactions between configuration terms determines a splitting of the states. The ground state is thus a high-spin state $^4A_2(t_2^3)$, followed in terms of increasing energy by $^2E(t_2^3)$, $^2T_1(t_2^3)$, $^4T_2(t_2^2e)$, $^2T_2(t_2^3)$, $^4T_1(t_2^2e)$ excited states. Two broad bands dominate the absorption spectrum, corresponding to the transitions classified as U ($^4A_2(t_2^3) \rightarrow {}^4T_2(t_2^2e)$) and Y ($^4A_2(t_2^3) \rightarrow {}^4T_1(t_2^2e)$). Thermal relaxation, following the excitation process, populates the lowest excited low-spin ^2E state, whose corresponding transition from/to the ground state is labelled as R. The trigonal distortion related to the repulsive interactions between neighbouring Al ions, in combination with spin-orbit coupling, originate an additional splitting of the energy levels. The two levels in which the $^2E(t_2^3)$ emitting state is split, differ by 29 cm^{-1}, whereas the splitting of the ground state (0.38 cm^{-1}) is generally neglected. As a consequence, two spin forbidden R emission lines, generally indicated as R1 and R2, at 694.25 and 692.86 nm (14404 and 14433 cm^{-1}) at 300 K and ambient pressure [542],

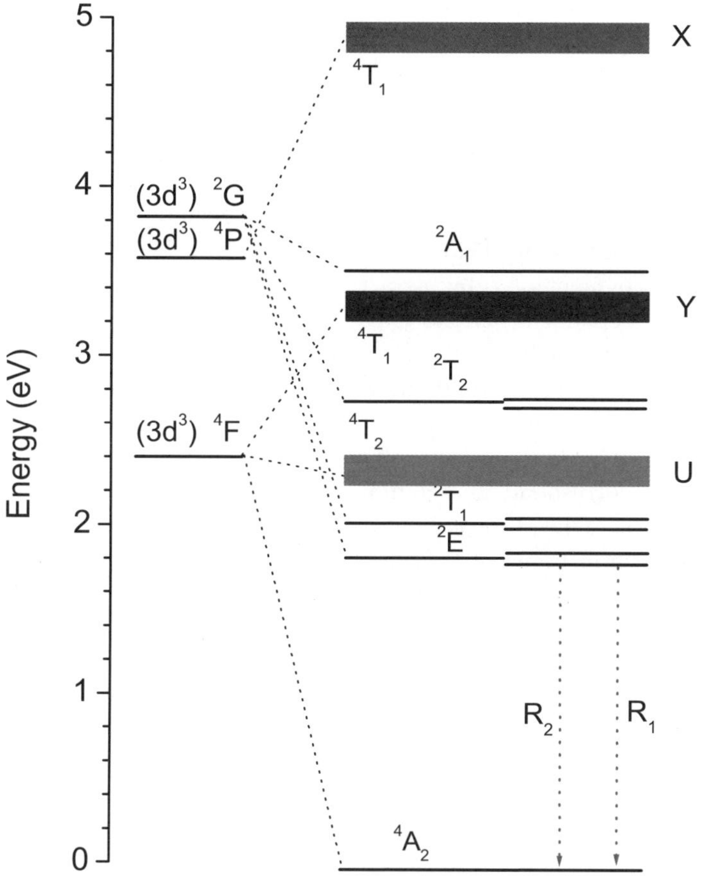

Fig. 4.17 Energy levels of Cr^{3+} in ruby. From left, the free-ion states in the weak crystal field limit split (dotted lines) because of the ligand field effect due to the cubic crystal field. Finally, on the right side, further splitting due to the trigonal crystal field and spin-orbit interaction.

dominate the emission spectrum persisting up to very high pressure [554]. The emission activity is due to the relaxation of the spin selection rules for the mixing of the doublet states with the quartet states via spin-orbit interaction. The fluorescence lifetime is strongly affected by the compression, exhibiting a non-linear increase with rising pressure [555].

The energy of the R lines depend on the Cr concentration [542, 556], temperature and pressure and, at least at very moderate pressures [557, 558], these effects are independent of each other. The temperature effect is by far more important than that of doping. Cooling determines a blue shift and a narrowing of the R lines [559], in addition, the intensity of the R2 line progressively decreases with decreasing temperature, following the Boltzmann distribution which also applies to excited states because equilibrium conditions are established within the two states by one-phonon processes. The ambient pressure-temperature evolution (4–350 K) of the R1 and R2 peak frequencies and linewidths has been reproduced within the Debye model of phonons [543, 559].

The employment of ruby, as a pressure sensor, is however related to the pressure-dependent shift to longer wavelengths of the R_1 and R_2 ruby fluorescence emission lines. The characteristics that make these sensors particularly valuable are the small linewidths, high fluorescence quantum yield and remarkable pressure shift. Furthermore, they maintain these characteristics with increasing pressure. The red shift, of the R1 and R2 emission lines, do not correspond to a common evolution of the excited states with pressure. In fact, the absorption spectrum exhibits a considerable blue shift [560], as shown in Figure 4.18, corresponding to the energy increase of the U and Y transitions, explained in terms of an increase in the octahedral crystal field strength. This blue shift of the strong absorption bands of ruby, dictates the wavelengths that can be used to excite the ruby fluorescence [555]. He-Ne (632.8 nm) or Kr (647.1 nm) red emission lines can be used at low pressure. The harmonic of Nd-YAG lasers (532 nm) can be used up to 30 GPa, whereas its absorption vanishes at higher pressures and shorter wavelenghts are needed. Above 100 GPa, the R1 emission also decreases drastically because of the reduced efficiency of the pumping to U and Y states [555], however excitation to the B states ($^2T_2(t_2^3)$) is suggested to increase the fluorescence, making it measurable up to 2.5 Mbar [561]. Also, direct pumping to the R state has been proposed for multimegabar applications [562]. On the contrary, the energies of the ^{2}E states depend on two different free-ion levels (^{4}F and ^{2}G), which respond in opposite ways to the octahedral crystal field change, thus roughly compensating the effects on the energy levels.

The effects of both static and dynamic uniaxial stress on the R ruby lines have been revised by Syassen [543]. Linewidth and splitting of the R lines are often taken as a reference of the quasi-hydrostatic character of the environment, but sharp ruby fluorescence lines and constant splitting with increasing pressure do not necessarily indicate the absence of deviatoric stress in the sample. The effects of controlled deviatoric stress on the R lines have been discussed in ref. [563].

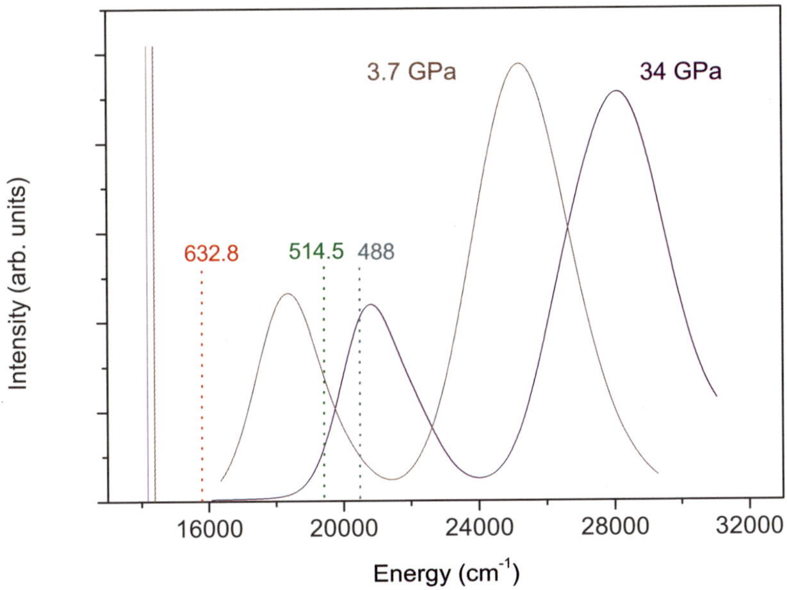

Fig. 4.18 Absorption and emission spectra of ruby at two selected pressures (the same colour for absorption and emission is used to identify the pressure value). The two absorption bands corresponding, in order of increasing energy, to the $^4A_2 \rightarrow {}^4T_2$) (U) and $^4A_2 \rightarrow {}^4T_1$ (Y) transitions remarkably blue shift with pressure, whereas the R1 emission line (sharp lines at lower energy) moves to lower energy. The position of some laser lines employed for pressure calibration are also reported.

Calibration of the ruby fluorescence emission by the equation of state of sodium chloride, shows a linear shift of the R lines up to about 20 GPa [392, 564]. An averaged value over the available experimental data of the modulus A_0, usually employed to characterize the linear shift of the R1 line with pressure, and defined as,

$$A_0 = -\frac{dP}{d(ln\nu_0)} = \frac{dP}{d(ln\lambda_0)}, \tag{4.2}$$

is given in ref. [543] as 1870(30) GPa. Above 20 GPa, the R line shift becomes substantially non-linear in pressure, as first reported in ref. [540]. Later, the quasi-hydrostatic calibration was extended up to 80 GPa [473]. This pressure scale became known as the quasi-hydrostatic scale with an accuracy of 1–2% at 50 GPa [539]. Application of this pressure scale in the megabar regime has also been performed [51, 565], but several reasons pointed to a re-examination of the pressure scale. Namely, the larger pressure achievable with the DAC, the employment of reliable hydrostatic compression media (He), improved experimental probing, both static and dynamic, and computational techniques. The first refinement of the ruby pressure scale was performed by Holzapfel, on the basis of the analysis of previously published X-ray diffraction data on diamond and Ta and low-pressure ultrasonic studies [566]. Several other studies based on shock data analysis [532], parallel X-ray diffraction for volume determination and Raman experiments on diamond [567], and X-ray diffraction of metals under hydrostatic conditions [568], brought new light to the calibration issue by including new data, and the following calibration relations were revisited or proposed:

$$P = \frac{A}{B}\left[\left(\frac{\Delta\lambda}{\lambda_0}+1\right)^B - 1\right], \tag{4.3}$$

$$P = \frac{A}{B+C}\left[\exp\left[\frac{(B+C)}{C}\left[1-\left(\frac{\nu}{\nu_0}\right)^C\right]\right] - 1\right], \tag{4.4}$$

$$P = A\left(1-\frac{\nu}{\nu_0}\right)\left[1+B\left(1-\frac{\nu}{\nu_0}\right)\right]. \tag{4.5}$$

In all cases, P is expressed in GPa and λ_0 (ν_0) is the wavelength (wavenumber) of the R_1 ruby fluorescence line at ambient pressure and at the reference temperature, typically room temperature. Equation 4.4 corresponds to Eq. 4.3 in the limit for $C \to 0$. The relevant parameters according to different authors are reported in Table 4.3. The comparison of these calibration relations is shown in Figure 4.19, from which a quite good agreement up to 150 GPa results.

A correction for the temperature dependence of the R emission lines is taken into account by including a temperature-dependent $\Delta\lambda_T$ term in Eq. 4.3. This term is given by a third-order polynomial:

$$\Delta\lambda_T = C_1(T-T_0) + C_2(T-T_0)^2 + C_3(T-T_0)^3, \tag{4.6}$$

with T_0 being the reference temperature and C_1, C_2 and C_3 a set of coefficients. The temperature correction becomes important above 100 K [574], where a 6 K variation is equivalent to 1 Kbar in terms of ruby fluorescence shift [575, 576]. In spite of the reliability of the ruby pressure scale over a wide P-T range, the significant temperature coefficient C_1 ~0.007 ÅK^{-1}, the thermal line broadening responsible for the overlapping of the R_1 and R_2 bands above 400–500 K, and the decrease in the fluorescence intensity

Table 4.3 Value of the parameters employed in the different non-linear calibrations of the ruby R1 spectral shift under pressure. The coefficients A, B, and C refer to Eqs. 4.3–4.5

	A (GPa)	B	C	
Eq. 4.3	1904	7.665		[473]
Eq. 4.3	1871	10.06		[532]
Eq. 4.3	1904	9.5		[568]
Eq. 4.3	1873.4	10.82		[569]
Eq. 4.4	1820	14.0	7.3	[566]
Eq. 4.4	1845	14.7	7.5	[570]
Eq. 4.5	1860	7.75		[571–573]

around 1 Mbar, define the limits of the pressure determination by the ruby fluorescence peak. Provided pressure is measured with another method, the isobaric frequency shift of the ruby fluorescence with temperature can be estimated [576].

For several other pressure calibrants working on the same principle as the ruby, the frequency shift with pressure of selected fluorescence lines, have been reported. To overcome the broadening and the quenching of the ruby luminescence for pressures beyond 100 GPa and temperatures exceeding 700 K, the pressure shift of the $^7D_0$-$^5F_0$ fluorescence line of Sm^{2+} in a SrB_4O_7 matrix was calibrated [577, 578]. This line remains intense, sharp and well isolated from other lines; it is negligibly affected by non-hydrostaticity; and both the peak energy and the linewidth have a weak dependence on temperature. These properties make this sensor superior to ruby for high-pressure high-temperature experiments. Similar features also characterize Sm:YAG, which shows an intensity comparable to that of ruby with the advantage that the pressure-induced frequency shift is temperature independent [579]. This makes Sm:YAG eligible for high-temperature and high-pressure experiments without using any temperature correction. In addition, the Sm:YAG fluorescence is narrower than that of ruby and can be detected just above 3 Mbar, a limit dictated by diamond fluorescence which overlaps that of Sm:YAG [580]. Other similar low-pressure ($P \leq 20$ GPa) high-temperature sensors which have been characterized are YAG:Eu^{3+} [581] and SrFCl:Sm^{2+} [582].

The employment of vibrational pressure gauges is also quite common, both in infrared absorption and Raman spectroscopy. Here, the pressure is determined by the frequency shift of a selected vibrational band under compression, generally calibrated through the ruby pressure scale. As far as the

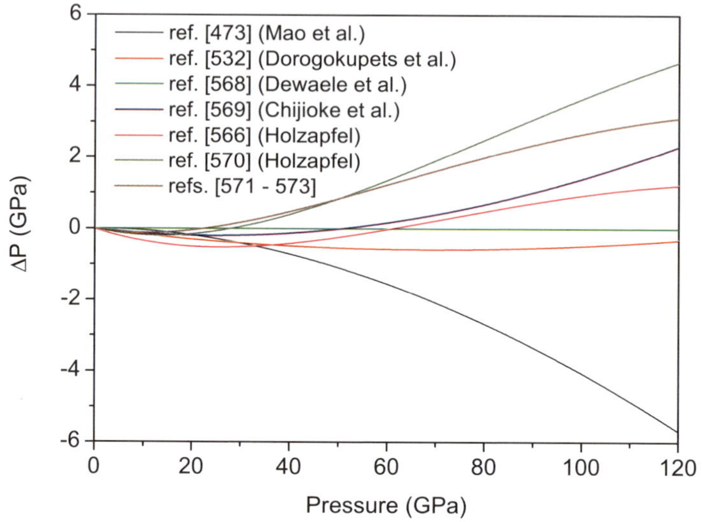

Fig. 4.19 Comparison of the non-linear calibration lines reported in Table 4.3. Pressure differences are computed with respect to Eq. 4.3 with the parameters from ref. [568].

Raman experiments are concerned, it is worth mentioning the employment of N_2 [583], which is particularly useful whenever nitrogen can be used as a compression medium. Recently, the pressure shift of the first-order Raman band of diamond was calibrated by different methods up to 4 Mbar, and the pressure to diamond phonon frequency relations were provided to be used as an optical pressure determination method. The stress-induced removal of the triply degenerated F_{2g} optical phonon mode, determines a Raman spectrum characterized by steep edges both on the high- and low-frequency sides, corresponding to the spectra of the culet and the table faces of the anvil, respectively. This method, first proposed by Hanfland et al. [584], produced calibration curves up to 140 GPa under hydrostatic conditions, by combining volume measurements using X-ray diffraction and frequency pressure shifts [567], up to 220 GPa using ruby as a pressure sensor [585], up to 300 GPa [586] and up to 410 GPa [587] using the Pt equation of state. Also, the pressure evolution of the Raman phonon of ^{13}C isotopic-enriched diamond chips has been calibrated to be employed as a pressure sensor [588]. More recently, the calibration has been extended up to 156 GPa using thin ^{13}C isotopic-enriched diamond layers grown on top of the diamond, using ruby fluorescence and the equation of state of copper as secondary pressure standards [589].

Also, cubic boron nitride (c-BN) was proposed as a high-pressure high-temperature pressure sensor [493]. Between the two intense Raman modes at 1054.7 cm^{-1} (TO) and 1305 cm^{-1} (LO), the TO mode has been calibrated using the EOS of gold up to 35 GPa and 600 K [590].

The ruby fluorescence method allows a pressure determination in a time scale ranging from one to a few seconds, thus allowing the following of a solid state reaction, which is normally slower, and compensate possible pressure changes due to volume changes. Nevertheless, in some cases exposing the sample to few mW of the visible laser beam, employed to excite the ruby fluorescence, can be sufficient to induce undesired photochemical reactions [591]. In these cases, the sample evolution with pressure is frequently monitored by IR absorption spectroscopy, usually employed for kinetic purposes, and pressure can be determined by the pressure shift of a selected vibrational band of materials employed as pressure gauges. The marker peak, besides fulfilling the requirement of stability, chemical inertness and significant intensity and frequency shift with pressure, must be isolated and should not overlap with the sample bands. Polyatomic ionic salts are the best candidates in matching these requirements, thanks to the strong transition dipole moments of the antisymmetric vibrational modes. Small amounts of the selected salt are typically used as a dilute solid solution in a transparent matrix. The asymmetric stretching vibrations of NO_2^- and NO_3^- have been successfully used as pressure calibrants in the infrared region [592]. In particular, the antisymmetric stretching mode of the NO_2^- mode has been characterized up to 50 GPa and between 100 and 400 K (see Figure 4.20) using 1–3% solid solutions of $NaNO_2$ in NaBr [376, 593]. $MgCO_3$ is also reported to be a useful vibrational gauge for pressure in infrared spectroscopy due to the pressure behaviour of the asymmetric C-O stretching vibration of the CO_3^{2-} ion [594].

As previously discussed, the ruby fluorescence scale has been calibrated mainly taking advantage of X-ray diffraction data against the isothermal equation of state as derived from the shock data of several metals. Claims of pressures largely exceeding 4 Mbar [595, 596], determined by extrapolating the R ruby emission up to to 550 GPa, were later reconsidered on the basis of a comparison between ruby fluorescence and the X-ray data of iron showing that a 100% error can affect this extrapolation [387]. Commonly employed materials for pressure calibration using volume data extracted by X-ray diffraction measurements are Au [597], NaCl [598], Re [599], Pt [600] and MgO [539]. The diffusion of this method is essentially due to the improved reliability of the last generation synchrotron sources for X-ray

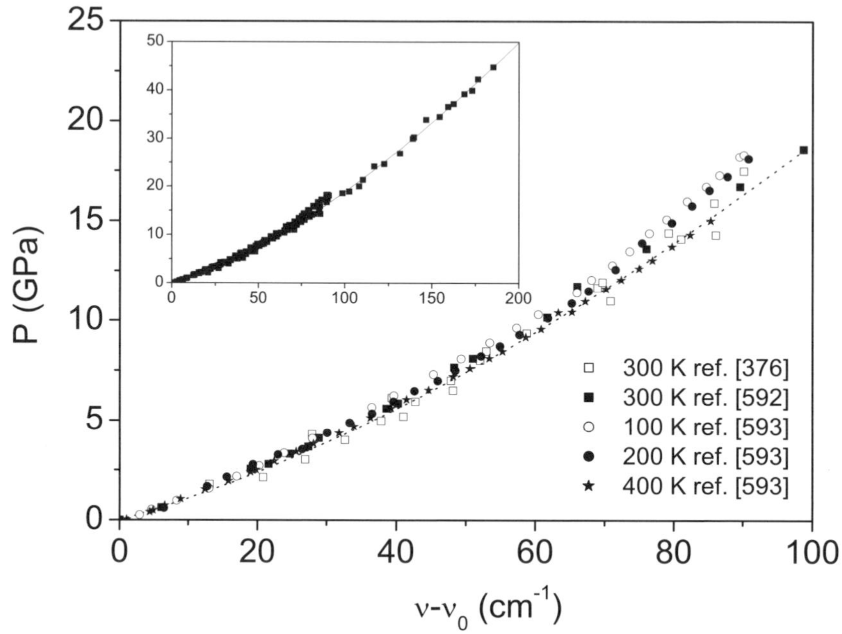

Fig. 4.20 Pressure shift of the NO_2^- asymmetric stretching mode at different temperatures. The shift is nearly independent of temperature up to 10 GPa, whereas above this pressure, the fitting coefficients change with temperature [593]. The inset contains the same data and additional data point up to 50 GPa and 300 K [376].

diffraction measurements and also to the possibility of performing these measurements when no optical access is available. Recently, metal (Fe, W and Pt) nanoprobes have been employed in X-ray absorption and diffraction experiments, together with a focalized submicron beam, to improve the spatial resolution in order to get precise pressure gradient measurements [601].

4.7 Probing techniques based on electromagnetic radiation

A general description of the probing techniques more commonly employed to monitor the sample response to compression in diamond anvil cells are reported in a number of reviews and books [281, 287, 348, 602]. Nevertheless, the rapid technological developments in realizing larger synthetic diamonds, by improving synchrotron as well as laser sources, as well as faster and more sensitive detection systems make an up-to-date photograph of the high-pressure probing techniques very difficult. In addition, high-pressure studies are rapidly diffusing into new communities and disciplines, so that research topics and probing techniques, traditionally extraneous to high-pressure research, add to the consolidated applications of high-pressure devices and in particular, of the diamond anvil cell. We will try here to summarize the most recent achievements and experimental novelties pursued through the employment of electromagnetic radiation and neutron sources in combination with the DAC.

4.7.1 *Optical spectroscopy*

Optical spectroscopy is historically the most adopted approach to study samples, under pressure, contained in the DAC. Only the development of new dedicated high-pressure set-up, within third generation synchrotron sources made the use of X-ray comparable to that of optical spectroscopic techniques in high-pressure research. As a matter of fact, one of the stronger motivations to adopt diamonds as an anvil-constructing material, was the possibility of providing optical access to the sample. Experimental details of the optical studies in the DAC have been reported in so many review and books (see, for example, ref. [281]) that any additional presentation is superfluous. Here, only the basic principles will be reviewed, and the most recent applications presented.

4.7.1.1 *Infrared spectroscopy*

Infrared absorption spectroscopy is an incredibly powerful method with which to gain insight into the structural properties of condensed phases by studying the phonon spectra. In addition, thanks to the fast and broad band acquisition allowed by the Fourier transform (FT) method, the sample transformations with time can be followed. The frequency evolution, with pressure, of phonons allows the Grüneisen parameters to be computed and elastic properties derived through quasi-harmonic models [603, 604]. In addition, the IR information together with that attainable by Raman spectra can be interpreted, when molecular systems are under investigation, according to symmetry arguments as provided by the Group Theory. Site occupancy and crystal phase symmetries can sometimes be inferred by the full vibrational information, but phase changes and phase transitions are more easily detected. In addition, the possibility to acquire spectra over a wide frequency range (typical mid-IR spectra are between 500 and 4000 cm^{-1}) and every few seconds, gives access to kinetic studies of phase transitions [605] and chemical reactions [378]. This is also possible because, as will be explained in the following sections, generally, all the sample in the gasket is probed at once so that quantitative information is provided by the spectra. In fact, the absorbance of each band is proportional to the corresponding amount of absorbing material, therefore its change with time is a direct measurement of the transformation process. In addition, IR light is ideal to study reactive processes because these frequencies are far away from any one- or two-photon electronic excitation, thus offering the unique possibility to disentangle pressure and photo-induced effects.

Infrared spectroscopy is also a powerful investigation technique for studying charge delocalization processes, thanks to the possibility of probing the sample through reflection and transmission spectra. High-pressure reflectivity measurements can also be performed on strongly absorbing samples and, if performed over a wide frequency range, allow the determination, through Kramers–Kroenig analysis, of the optical constants of the system. The analysis of infrared reflectance spectra provides, according to the Drude model, the plasma frequency which gives a measure of carrier concentration. This makes the measurement of infrared reflectivity a very powerful method to detect and characterize pressure-induced metallization in semiconductors [606], as well as in molecular systems [607].

In spite of the enormous importance of infrared spectroscopy to gain insight into structural properties, transformations and chemical stability of the samples under investigation, its employment in high-pressure studies is far more limited than Raman spectroscopy or X-ray diffraction techniques. The reason should be searched for in the non-trivial technical solutions necessary to couple the DAC to commercial FTIR spectrometers. By maintaining the imaging geometry of standard instruments, the sample acts as a diaphragm and only $\sim 10^{-4}$ of the incident light is transmitted, whereas reflectance losses, for normal incidence on the four diamond faces, determine an additional reduction of two orders of magnitude. The large refractive index of diamonds, interference fringes and diffraction from the gasket (typical dimensions larger than 100 μm) are additional complications that also limit the low frequency accessible range ($\bar{\nu} \geq 100$ cm^{-1}). Most of these problems are avoided or strongly reduced when a synchrotron infrared source is employed. This is indeed 2–3 orders of magnitude brighter than a conventional globar source, thus allowing smaller sample regions to be probed [608]. As an example, measurement of the H_2 vibron was recently performed at different temperatures up to 360 GPa [609]. Measurements in the Mbar range are really a major challenge due to the sample dimension ($\leq$10 μm), which precludes the employment of a different source. Nevertheless, synchrotron light-based techniques, including IR spectroscopy, suffer from limited access making in-house systems of primary importance. Many solutions, essentially differing in the focusing optics, have been adopted over the years. The first attempt to interface a DAC with an FTIR spectrometer was made using ellipsoidal mirrors to condense the IR beam also making the measurement of spectra in the far IR region ($\bar{\nu} \geq$25 cm^{-1}) possible [610]. An on-axis Cassegrain objective was used to obtain kinetic data, as a function of pressure and temperature, of the thermal decomposition of 1,3,5-trinitrohexahydro-1,3,5-triazine [611]. A similar set-up was used a few years later, succeeding in extending the frequency (20–5000 cm^{-1}) and temperature (80–300 K) ranges [612]. Light cones with different geometries have been employed to interface the DAC to a bolometer [613, 614], whereas a combination of a parabolic mirror before and a light cone after the DAC, allowed attainment of the first far IR spectrum above 6 GPa [615]. However, on-axis Cassegrain objectives are by far, the most employed in the more recent realizations, bringing to the realization of large aperture membrane DAC [616], and to more sophisticated set-up where the DAC is inserted in the evacuated sample chamber of an FTIR spectrometer, allowing remote control of temperature (20–700 K) and pressure (P$\leq$ 100 GPa) between 80 and 20000 cm^{-1} [482]. A slightly modified design was recently adopted to build an external microscopic optical bench, operating only in the mid-infrared, but also allowing reflectivity measurements [617]. A home-made dispersive set-up, based on Cassegrain optics, was employed to measure the low-temperature mid-infrared spectrum of solid para-H_2 up to 191 GPa [618].

4.7.1.2 *Raman and Brillouin spectroscopy*

The pioneering studies concerning the application of Raman and Brillouin scattering techniques to probe samples inside the DAC are nicely reviewed by Jayaraman [348], together with the first decisive improvement which made the Raman technique, by far, the most employed in high-pressure studies, especially as far as molecular systems are concerned [45]. Both Raman and Brillouin scattering are related to an inelastic scattering process of photons by phonons, magnons or simply by the variations of the dielectric constant due to translational and orientational motions of the molecules. In this two-photon process, where energy and momentum are conserved, the scattered photon has a different energy, which is greater, or smaller, than that of the incident photon depending on an annihilation, or a creation, of a phonon during the interaction between radiation and matter. Creation or annihilation of a phonon correspond to Stokes and anti-Stokes processes, respectively. The energies involved in Brillouin and Raman scattering processes are rather different, thus also determining relevant technical differences. The Raman experiments are directed to the analysis of the processes whose energy signatures are typically greater than 5–10 cm^{-1}, including the rotational and vibrational information in gases, liquids and solid phases, lattice phonons in crystals, but also the orientational dynamics in liquids. Brillouin experiments are on the contrary, dedicated to the study of processes having energies below 5 cm^{-1}. Because the response is close to the elastic Rayleigh scattering, a much higher resolution is necessary in Brillouin scattering experiments. A Fabry–Perot interferometer is coupled to the laser source to select the wavelength of the transmitted light according to the relation $2nd = m\lambda$, where n is the refractive index of the medium contained between the two plane mirrors at distance d, and m is an integer. Therefore, the constructive interference is obtained by varying the optical path nd, by either changing the distance between the mirrors or the refractive index of the medium. The quality of a Fabry–Perot interferometer is described by two parameters; the ratio between the maximum and the minimum intensities transmitted (contrast), and by the ratio between the distance of two successive transmission peaks and the full width at half-maximum of a transmission peak (finesse). The two parameters provide the capability of measuring small and close signals, respectively, with respect to the Rayleigh peak. Multipass Fabry–Perot systems allow for contrast values up 10^{10}.

Brillouin scattering is a suitable method to determine the density of fluids from measurements of sound velocity [619–622], or the complete set of elastic constants of a crystal by suitably choosing the experimental geometry [623–625], or for monitoring changes in the elastic properties of an amorphous material as pressure is applied [626]. The attainment of this information, in experiments with the DAC, is difficult and not always possible because of the limited interaction geometries between the laser beam and the material under study. Quite often, the optical access is along the diamond axis making only the backscattering geometry possible, thus limiting, due to the selection rules, the accessible phonons. Platelet geometry [623], consisting of a modified DAC in order to have symmetric conical openings offering wide aperture angles (~100°), have been employed in several studies [133]. Knowledge of the crystal orientation, the rotation of the cell around a specific axis, and the selection of incident and scattering geometries, also make the observation of the transverse modes possible, as reported, for example, in α-quartz [624], hydrogen sulphide [623] and argon [625].

As far as the Raman technique is concerned, a backscattering or nearly backscattering geometry is adopted. In general, both the focusing of the laser beam and the collection of the scattered light from the sample are realized through the employment of long working distance microscope objectives, which also allow the employment of cryostats or resistive heating apparatus. Analysis of the scattered light is performed by single or multiple stage monochromators, with configurations chosen depending if high spectral resolution or high stray light rejection is desired. Within this common and very general framework, sophisticated improvement of the Raman systems are realized, depending on the experiment for which they are designed. A recent challenging achievement, especially from the sample preparation and treatment point of view, finalized by Raman spectroscopy, is the measurement of the melting line of hydrogen for pressures exceeding 1 Mbar and temperatures close to 1000 K [627].

Raman spectroscopy with pulsed mode operation was recently demonstrated to be extremely efficient in discriminating the Raman signal from background radiation [628]. Fluorescence from the sample is a quite common occurrence in molecular systems with strong conjugation, or in reactions involving hydrocarbons. In addition, diamond fluorescence can also be important and depends on the stone quality, and the stress conditions, such as that generated in the vicinity of the diamond tips at very high pressures. The duration of these processes is much longer with respect to the instantaneous Raman response, therefore the use of a pulsed laser source and a synchronized gated acquisition makes it possible to slice the response from the sample suppressing, depending on the minimum gate width, most of the long-living background signal. Suppression factors of four orders of magnitude are claimed.

4.7.1.3 *Non-linear optical techniques*

As mentioned at the beginning of this section, the diffusion of the DAC across different disciplines demands new techniques to probe the compressed sample. In the case of molecular materials, systems which are central in this book, an urgent need is represented by the improvement of *in situ* diagnostic techniques which could allow the characterization of fast subpicosecond processes, such as the coordination and hydrogen bond dynamics in liquids, nucleation of crystalline phases, or the static and dynamic behaviour of electronic excited states as a function of pressure. The latter topic is particularly relevant in view of an understanding of the primary steps of purely pressure-induced or photo-induced high-pressure reactions [378, 591]. In fact, the change of the electronic properties tuned by pressure, or the selective electronic excitation at high pressure, have been identified as responsible of reactive events [465, 629–631], making the characterization of the electronic dynamics under pressure mandatory. The direct characterization of the dynamics by pump-probe or transient absorption experiments, using ultrafast laser spectroscopic techniques, is still a major challenge. Difficulties arise from the sample dimensions because of the active volumes required in non-linear optical processes, and from the need to overlap on this volume different laser beams, making very high spatial beam quality mandatory. In transient absorption experiments, the reduced sample thickness also represents an important drawback which would require intense laser fields, which in turn, induce non-linear processes in the diamond or sapphire anvils which dominate the sample response. Pulse energy of a few μJ are generally safely used, not inducing diamond damage, with laser pulse duration ranging from tens of femtoseconds to tens of picoseconds. To our knowledge, transient aborption experiments in the DAC are only sporadic. Femtosecond transient absorption measurements have been reported for pressures below 1 GPa, to investigate energy relaxation processes in strongly absorbing materials such as a laser dye (LDS698) [632] and β-carotene [633]. Two different relaxation processes have been characterized in the LDS698 case, a fast intramolecular process which is not affected by pressure and a slower ($\sim$1 psec) intermolecular process, which instead decreases with rising pressure. Particularly interesting in the β-carotene case is the observation of the lifetime lengthening of the S_1 state with rising pressure, explained in terms of a viscosity increase which dominates the decrease, with compression of the $S_0 - S_1$ energy gap.

Coherent Raman scattering techniques have been successfully applied since the end of the 80s to the study of the pressure effects on the vibrational relaxation dynamics. Time-resolved stimulated Raman scattering (SRS) was applied to study the relaxation of the breathing mode of benzene, in the liquid and crystal phases, up to 5 GPa [634]. After this study, both frequency and time-resolved coherent anti-Stokes Raman spectroscopy (CARS) experiments were performed on different molecular systems. CARS is a coherent Raman technique in which the energy difference of two different beams (ω_1 and ω_2) is tuned around a vibrational resonance ($\omega_{vib} = \omega_1 - \omega_2$). The two beams are focalized on the sample at an angle fixed by the momentum conservation (phase-matching condition). Interaction of these fields through the third-order non-linear susceptibility ($\chi^{(3)}$) of the material, gives rise to an output field at frequency $\omega_{as} = 2\omega_1 - \omega_2$ and spatially defined by the wavevector $k_{as} = 2k_1 - k_2$ (phase-matching condition).

In frequency resolved experiments ω_2 is tuned in the vicinity of the vibrational resonance, whereas in time-resolved CARS experiments, both ω_1 and ω_2 are fixed and the ω_1 beam is split, and one of the two beams is directed to a delay line, allowing the coherence induced in the vibrational state at $\omega_{vib} = \omega_1 - \omega_2$ to be probed at different delay times. Vibrational relaxation was studied directly in the time domain, in the case of the Ω^+ bound state in the ν_1;$2\nu_2$ Fermi resonance region of crystalline CO_2 [635], and for different modes in carbon disulphide and naphthalene [636, 637]. In all the cases, an increase of the vibrational dephasing, likely due to the opening of new relaxation channels because of the pressure modification of the phonon density of states, was suggested. Frequency resolved CARS spectroscopy was first employed to study nitromethane [638], and more recently to investigate frequencies and linewidths of the Raman active internal modes of solid nitrogen up to 22 GPa, providing information about the structural properties across different solid-solid phase transitions [639]. The geometrical constraint posed by the phase-matching condition, facilitates the discrimination of the CARS signal from the sample, and makes this technique particularly interesting for studying samples under extreme conditions of temperature and pressure, where black body and fluorescence background radiation dominate [640]. Another interesting characteristic of CARS spectroscopy is the possibility to extract information about the electronic properties by the analysis of the non-resonant component of $\chi^{(3)}$. This term contains all the $\chi^{(3)}$ terms with non-resonant denominators and can be considered as a constant contribution in the tuned frequency region, but as one, or a combination, of the laser frequencies employed in the experiment approaches an electronic resonance, the CARS lineshape is altered (see Figure 4.21). Using this approach, choosing $2\omega_1$ close to the lowest electronic transition, the band gap of deuterium has been obtained by measuring the CARS vibron lineshape up to 187 GPa [641].

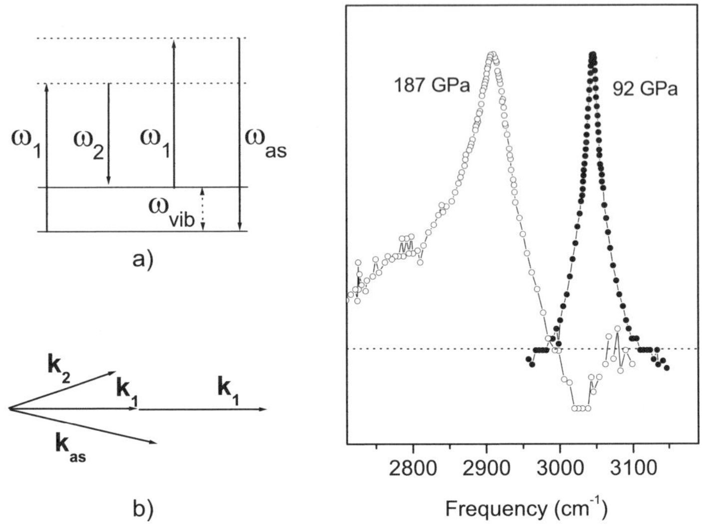

Fig. 4.21 Diagrammatic illustration of the energy conservation (a) and phase matching (b) requirements in a CARS experiment. On the right side, the deuterium vibron CARS spectrum at two different pressures are shown in a logaritmic scale (from ref. [641]). The lineshape modification on increasing pressure accounts for the activation of resonant term of $\chi^{(3)}$ by two-photon absorption.

Access to the electronic states through multiphoton absorption processes is extremely relevant in view of the diamond band gap, which precludes the direct probing of the electronic transitions of most of the simplest molecular systems, and the red shift of the absorption edge with increasing pressure, so that optical studies under high-pressure conditions must compete against the narrowing transparency of the diamond window. Two-photon (TP) absorption experiments have recently been employed to characterize the electronic changes of some model aromatic systems, gaining insight into the mechanisms of the pressure-induced reactivity. In these experiments the absorption is indirectly revealed through emission measurements at constant frequency, generally the peak fluorescence maximum, while a continuous scan of the excitation wavelength is performed. This can be accomplished through an optical parametric generator, which provides tunable light from the UV to the IR, thus allowing collection of excitation profiles up to 6.2 eV [631]. Relations between the structural arrangement and electronic properties have been derived by these studies, resulting fundamentally important understanding of the pressure-induced reactivity, or stability, of these systems. The structural excimer formation evidenced in benzene [631] allows an unified picture of the purely pressure [642] and photo-assisted [643] high-pressure reaction to be drawn. A pressure driven strengthening of the hydrogen-bonding network is instead derived by the TP fluorescence spectra of indole [644] and pyridine [645]. Particularly relevant is the pyridine case, where the fluorescence intensity increases by about six orders of magnitude upon crystallization at 3 GPa, and is explained by the inversion of the $n\pi^*$ and $\pi\pi^*$ excited states, likely due to the growing involvement, with rising density, of the lone pair of the N atom in intermolecular H-bonds.

4.7.2 *X-ray diffraction*

Presently X-ray based techniques are probably mostly applied in high-pressure research, also taking advantage of synchrotron radiation which allows the selection of high flux and brilliant (intense and collimated photon beams) specific wavelengths, from microwaves to hard X-rays. High-pressure dedicated beamlines are already present in most of the 3rd generation synchrotrons which are widespread around the world, while 4th generation sources are on the horizon. The use of X-ray techniques in high-pressure research dates back [646] well before the birth of the DAC. The main problems encountered when a DAC is placed in the X-ray beam are related to the diamond transmission, which dramatically reduces below 20 KeV to vanish just below 12 KeV, and to the Compton scattering from the diamonds, which gives an unavoidable background, one order of magnitude larger than the diffracted signal. Correction is usually performed using the background intensity, collected from the empty DAC, with no applied pressure, since no noticeable changes in the diffuse scattering of the diamonds is observed with increasing pressure. Recently, an alternative method has been proposed, in which diffraction is carried out at energies such that the Compton background is discriminated by coherent scattering, exploiting the absorption coefficient of barium, the primary composing element of image plates [647]. Other limitations are represented by the cell aperture, which dictates the accessible diffraction angle (2θ), but X-ray transparent seats can overcome this problem allowing a large angular opening for X-ray scattering. For instance, beryllium has been widely employed in single crystal studies [372, 448], boron seats were used both for single crystals [472] and for liquid [648], and more recently c-BN seats, the second hardest known material and possessing an X-ray absorption coefficient close to that of diamond, have become commercially available [649].

Energy dispersive X-ray diffraction (EXD) experiments are performed by analysing the diffracted radiation at a fixed angle which defines, according to Bragg's law, the product of the energy of a diffraction peak and its Bragg spacing. The efficiency and the spatial resolution of the detector is crucial in the analysis of the diffracted photons; in addition, the reduced scattering volume resulting from the small sample thickness inside the DAC, especially at very high pressure ($\leq$10 μm), coupled with the fine collimation required to probe the sample, made the synchrotron source indispensable in the experiment. To overcome the limitations, due to the single direction in which a sample in the DAC can be probed if X-ray transparent gaskets are not employed, rotation methods of the DAC have to be set-up [650]. Using this technique, a wealth of information in the megabar and multi-megabar range regarding the structural data (EOS, phase transitions) of different model systems, such as simple molecules [651], rare gases [652], metals [653, 654] and silica polymorphs [655, 656], have been obtained.

In angle dispersive X-ray diffraction (AXD) experiments, the sample is probed by a monochromatic beam and the entire diffraction cone is registered by an image plate. From this bi-dimensional plot (Debye–Scherrer rings), the intensity of the diffraction peaks as a function of the 2θ scattering angle, the required plot for Rietveld analysis, is obtained by integrating the intensity distribution over each entire ring. Integration of the entire diffraction cone limits, with respect to the energy dispersive technique, the texturing effects and allows immediate evaluation of the crystalline or polycrystalline quality of the sample. The resolution here is determined by the X-ray monochromator and by the focusing geometry. This is by far, the most applied technique to the study of the structural properties of both polycrystalline and single crystal samples under high-pressure conditions. Improvement of the X-ray optics, in principle the only limitation to the beam focus dimensions, allowed routine experiments in the multimegabar range, with focal spots of a few μm. The control of the focal spot enables the probing only of the sample volume heated by a focused laser, so that unambiguous evidence of melting can be gained by the diffraction pattern. Using this approach, a long standing debate about the melting line (P$\geq$1 Mbar and T$\geq$5000 K) of tantalum was solved [657], and precious information about a possible partial melt at the mantle base, through the study of the peridotite melting line (P$\geq$1 Mbar and T$\geq$4000 K), was obtained [658]. However, focus dimensions of a few microns are still unable to resolve sample dishomogeneities in stress, crystal quality, orientations, and chemical compositions, which can occur at the submicron level. Recent advances succeeded in obtaining focal spots between 250 and 600 nm [601], thus making submicron X-ray scanning of compressed samples possible. Providing lateral access through the employment of X-ray transparent Be gaskets, different crystal orientations can be probed in the case of single crystals or textured polycrystalline samples [659]. In this framework, of particular interest is the possibility to study single crystals by performing controlled angular scans around, in general, two azimuthal angles (0 and 90°) around the X-ray beam, which allowed the solution of complex and controversial crystal structures [660–662].

The information that can be gained by X-ray diffraction concerns the long range arrangement of the sample; on the contrary X-ray absorption spectroscopy accounts for the local coordination and electronic structure of an X-ray absorbing atom, thus being capable of revealing the fine structural effects of pressure application [663]. The structure of the absorption spectrum is rather complex, consisting of an absorption edge occurring when the energy of the incident photons corresponds to the excitation energy of a core electron to a continuum state, i.e., the energy required to produce a photoelectron. Thus, the energies of the absorbed radiation at these edges correspond to the binding energies of electrons in the K, L, M, ... shells of the absorbing elements. The emitted photoelectron is backscattered by the neighbouring atoms, giving rise to the interference pattern located at energies immediately above the absorption edge. Depending on the energy, the X-ray absorption spectrum is divided in two main regions: near edge absorption structure (XANES), which includes the energy range between ±10 eV with respect to the edge energy, and extended X-ray absorption fine structure (EXAFS), starting at approximately 50 eV and continuing up to 1000 eV above the edge. XANES gives information about the coordination and oxidation state of the absorbing atom, whereas EXAFS probes the local structure, giving information about the number and chemical identities of near neighbours, up to distances given by the mean free path of the photoelectron in the condensed matter, typically between 5 and 10 Å. Information about atomic coordination (XANES) and interatomic distance (EXAFS) evolution, as a function of pressure, have been derived on amorphous and crystalline GeO_2 [664]. Because of the EXAFS selectivity to a particular element and its short-range order, the technique is extremely suitable for the study of disordered systems, such as amorphous solids or liquids and how their properties can be altered by the application of pressure [665–667]. There are important limitations to conducting EXAFS experiments in a DAC, consisting firstly in the opacity of the diamonds which prevent access to the low-energy absorption edge, and secondly in the strong Bragg reflections from the single crystal diamond anvils which affect the quality of the EXAFS spectra [668]. Employment of nanocrystalline diamond anvils has been demonstrated to be extremely effective in avoiding the Bragg reflections from single crystal diamond anvils, permitting the measurement of the pressure evolution of the mean square relative displacement for the first neighbour shells of crystalline Ge [669].

Inelastic X-ray scattering (IXS) is a powerful spectroscopic tool which allows both the energy and the momentum transfer to be controlled over rather wide regions, making the study of quite distinct physical and chemical problems in condensed matter possible, including phonon-like excitations as well as electron spectroscopies at valence and core level. Exhaustive description of the fundamentals and of some high-pressure applications of IXS have been recently reported [670, 671]. In an inelastic X-ray scattering process, the energy and momentum lost by the incident photon are transferred to intrinsic excitations of the material under study thus providing, through the analysis of the scattered photons, information about those excitations. Historically, the study of phonon dispersion in condensed matter has been pursued through neutron spectroscopy. The dynamic structure factor S(Q,E), which is the space and time Fourier transform of the particle density correlation function, where Q is the momentum transfer, can be experimentally determined providing important information on the collective dynamics. The employment of neutrons is due to the sufficiently weak neutron-nucleus scattering cross section, which allows a large penetration depth, an energy comparable to that typical of phonons, and a momentum which gives access to the whole phonon dispersion (several tens of nm^{-1}), in contrast to Brillouin or Raman scattering, which only allow the acoustic and optical modes to be measured. Due to the much higher energy of an X-ray photon, with respect to a neutron or electron, a wide range of energies and momenta can be transferred in an x photon-scattering event. When high-energy X-rays are employed, momenta are comparable to the inverse lattice spacing of typical condensed matter systems so that, unlike Raman scattering experiments with visible or infrared light, the full dispersion of low-energy excitations in solids can be probed by IXS [672]. This is however, a great experimental challenge because an extremely high-energy resolution is necessary, given that typical photons with an energy above 10 KeV are employed to study phonon excitations in the meV region, so that a relative energy resolution of at least $\Delta \mathrm{E}/\mathrm{E} = 10^{-7}$ should be provided. The great interest in using X-rays, derives from the much higher energy of the probing particle than the exchanged energy, so that the range of energy transfer is practically unlimited in the accessible Q transfer range (1–100 nm^{-1}). Specific advantages of the IXS technique, with respect to inelastic neutron scattering (INS), are related to the velocity of thermal neutrons (~1000 m/s), which limits the possibility of investigating the acoustic dynamics of disordered materials with a smaller speed of sound. In fact, in these cases the lack of translational periodicity, such as that found in crystalline materials, dictates that the acoustic excitations must be measured at small momentum transfers. Another important advantage, particularly relevant in high-pressure studies, regards the required sample volume, typically several mm^3 in INS whereas using IXS, measurements can also be performed with the typical volume contained in a DAC; in fact, microfocusing to a few microns approximately corresponds to the scattering volume in the hard X-ray region. Sample environments allowing IXS measurements on fluid and solid samples, up to pressures in the order of 100 GPa and temperatures as high as 1000 K by resistive heating, have been recently proposed [673]. Although high-pressure studies become increasingly more difficult for light elements ($Z \leq 20$) whose optimum thickness should be 1–2 orders of magnitude larger [670], many examples of studies in low-Z materials are reported [674, 675]. For these reasons, the main application of the IXS technique has been, so far, the study of relaxation dynamics, sound propagation, and transport properties in disordered systems such as glasses, fluids and polymers. The collective dynamics of biological and confined systems are presently great challenges.

A particular case of non-resonant IXS, X-ray Raman scattering (XRS), offers the possibility to access the core electronic level through a high-energy scattering process. The measure of K edges of light elements, under high-pressure conditions, is indeed not possible by direct absorption because their binding energy falls in the soft X-ray region. Observation of carbon hybridization changes with pressure in graphite [676], benzene [677] and fullerene [678], identification of the molecular orbital responsible for the intermolecular association in the ϵ phase of oxygen [679], and the change from the 3- to the 4-coordination of boron in B_2O_3 glass under pressure [680], are some representative examples of the applications of the XRS technique to model systems.

In resonant IXS (RIXS), the energy of the incident photon corresponds to that of an atomic X-ray absorption edge of the system. This resonance can greatly enhance the inelastic scattering cross section ensuring that the electronic properties of the element under scrutiny can be selectively observed. In principle, a differentiation of the chemical bondings, valencies and inequivalent crystallographic positions relative to the same chemical element is also possible. In addition, as already stated, a very large variation of the momentum transfer is attainable and the dispersion of the excitation can be studied over multiple Brillouin zones. RIXS spectroscopy, and conversely X-ray emission (XES), are sensitive to electron-electron interactions and provide complementary information to absorption spectroscopy. Orbital splitting, spin and oxidation states can therefore be probed through element-specific excitations, in addition to local symmetry and coordination. An extensive summary of different properties, such as magnetic and spin states, of transition metal compounds as obtained by XES studies are reported in ref. [671].

Besides the X-ray based techniques discussed so far, there are a plethora of new applications of synchrotron X-ray beams to the study of samples compressed inside a DAC. Nuclear resonant scattering of synchrotron radiation allows the observation of the phonon energy spectrum. Selective element excitation is possible and the scattering, due to the nuclear resonant excitation and non-resonant electronic scattering, can be discriminated by taking advantage of the longer lifetime of the nuclear excited state and the pulsed synchrotron radiation. Generally, the energy of the incident synchrotron radiation is tuned, with respect to the nuclear resonance, while the total yield of the delayed fluorescence photons is monitored, thus providing a direct measure of the phonon DOS [681, 682]. Using this technique the phonon DOS of Fe has been obtained up to 150 GPa [683], which is beyond the typical pressures of the core-mantle boundary of Earth. This is relevant for acquiring vibrational, elastic and thermodynamic properties of Fe, crucial in understanding core static and dynamic properties. X-ray magnetic circular dichroism, a powerful local probe of magnetism, has been used to study a magnetic transition in magnetite between 12 and 16 GPa, attributed to a spin transition of Fe^{2+} [684]. Another recent important advance is the possibility of performing X-ray tomography in samples also contained in the DAC, by recording and reconstructing 2D images collected with a fine scan of the rotational angle [685]. EOS of amorphous materials can be obtained by this method, through direct volume determination, whereas breakthrough applications regarding the characterization of heterogeneous samples under extreme pressure conditions can be envisaged.

4.7.3 *Neutron diffraction*

In X-ray diffraction experiments, the incident wavelengths are of the same order as the atomic radius and electrons, which are the scattering centres distributed over the entire atomic volume, the intensity of the scattered light depends on the scattering angle (θ) because of the phase difference between scattered photons from different points of the atomic volume. The intensity decrease with the scattering angle is described by the atomic form factor. In contrast, neutrons interact with the atomic core, so that the coherent scattering amplitude is not dependent on θ and λ, thus also allowing the attainment of diffraction data for high values of the scattering vector Q. Therefore, neutrons are efficiently scattered by hydrogen (deuterium) and other light atoms such as N, C and O. Using neutron diffraction studies, it is therefore possible to identify these atoms, and hydrogen in particular, which constitute simple systems and many important minerals, especially when heavy elements, which dominate the X-ray scattering, are also present. In addition, given that the nuclear cross section is independent of the number of electrons, iso- or near iso-electronic compounds can be distinguished. These aspects are added to the fact that neutron diffraction is the best technique to study magnetic structures, thus making neutrons more suitable than X-rays in many studies, even though the low flux of neutron sources requires large samples with typical dimensions of 10–100 mm^3. As a consequence, high-pressure neutron studies are very few and limited in pressure, compared with X-ray diffraction studies. An interesting comparative analysis of the potentiality and applications of the two techniques to different current topics in the framework of condensed matter science has recently been reported [686]. Due to limitations in the sample dimensions, the majority of neutron scattering high-pressure studies have been performed using the Paris-Edinburgh cell [312]. The phonon density of states, measured by inelastic neutron scattering as a function of pressure, provided valuable information about the lattice dynamics of semiconductors (Ge, GaSb) and bcc iron [687, 688], but also about the structure of hydrous minerals and disordered hydrous liquids and glasses [315]. However, water is by far the most extensively investigated model system using neutron diffraction, both in the fluid and in the crystal phases, taking advantage of the large sample volume attainable through the Paris-Edinburgh cell. The structure factor S(Q) of D_2O was obtained by studying the liquid as a function of P and T up to 6.5 GPa and 670 K, proceeding almost parallel to the melting line to prevent pressure-induced freezing [689]. The study was driven by the long-standing debate about the existence of a high-density form of liquid water (HDW), which above a few kilobars gradually replaces the low-density state (LDW). A local structure common to that of a simple liquid was reached with compression, although the H bonds still remain unchanged as attested by the slightly distorted tetrahedral coordination of the first neighbours. Several studies were also performed on low-temperature high-pressure phases of ice. The phonon dispersion of the I_h phase was characterized [690] as well as its transformation to the metastable high-density amorphous ice (HDA) [691, 692]. In addition, neutron diffraction studies were able to follow the transition between low-density (LDA) and HDA as a function of pressure and temperature, revealing that it occurs as a classical first-order transition [693]. Polyamorphism was also recently identified by this technique in a glassy phase, obtained by cooling an ionic salt aqueous solution, LiCl:6D_2O. A stepwise reversible transformation to a very high-density amorphous form occurs abruptly at 120 K and 2 GPa, likely due to a local reorganization of the water molecules around the Li^+ ion [694]. Recrystallization of the glassy solution at 4 GPa and 273 K gives rise to a high-density crystalline ice VII phase, hosting in its structure significant amounts of Li^+ ions [695]. The presence of salt modifies the structural properties, essentially due to an increase of the orientational disorder of the water molecules, which is likely responsible of the missed transition to phase VIII on cooling.

At the beginning of the 90s a new high-pressure cell, that could be equipped both with diamond or sapphire, was developed for neutron studies [696]. A careful description of the different anvil types, gasket materials and presses that can be employed in low-temperature neutron diffraction experiments, depending on materials under investigation and pressure range of interest, is provided in ref. [697], together with several examples demonstrating its application in the study of magnetic transitions, structural properties of polymers, carbon nanotubes and single crystals. Diamond small tips ranging from 1.5 to 0.5 mm allow maximum pressures of 20 and 50 GPa, respectively. Low absorption and scattering gasket materials (aluminium, niobium, copper and copper alloys) and sapphire, with semi-spherical central holes, are adopted to reduce contributions other than from the sample to the diffraction pattern and to increase the sample size. NaCl and a methanol-ethanol mixture are used as pressure-transmitting media in powder and single crystal diffraction experiments, respectively. Anvils made by cubic boron nitride have also been adopted but in this case, due to their opacity, pressure cannot be measured by the ruby fluorescence method but through internal reference materials. Besides the sample environment, the focusing, analysis and detection apparatus were also optimized for the study of magnetic structures by neutron diffraction. This cell was employed to study the magnetic ordering of crystalline oxygen in the high pressure α, δ and ϵ phases, exploiting the neutron sensitivity to both crystal and magnetic structures [698, 699]. Oxygen is indeed the only elemental molecule carrying an electronic magnetic moment and besides the low-temperature α phase, an antiferromagnetic order of the orthorhombic δ phase was proposed on the basis of the IR absorption spectra [700]. Antiferromagnetic order persists in this phase up to room temperature and, interestingly, three different magnetic structures can be identified. This occurrence has been intimately correlated to the dominant in-plane interaction of the oxygen sheets [701].

4.7.4 *Nuclear magnetic resonance*

Nuclear magnetic resonance (NMR) spectroscopy is one of the most powerful techniques in probing the structure, dynamics, reaction state and chemical environment of molecules. Megahertz radio frequencies (rf) are the characteristic frequencies, employed to probe the different spin states, created by the application of an external magnetic field whose strength determines the energy separation of these states. Once excited, the nuclei relax according to the structure and dynamics of the surrounding molecules. Diamond is transparent in the radio frequency region, thus in principle, permitting NMR studies in the DAC. However, the application of this technique to probe samples in a diamond anvil cell is extremely rare, due to intrinsic experimental difficulties such as the small sample dimensions, the reduced fraction of excited nuclei and the long relaxation time (typically between 0.1 and 10 seconds). In addition, the rf signal is efficiently shielded by metals so that the gasket must be composed of materials able to minimize the shielding of the rf signal (phosphor bronze), thus allowing an increase of the signal integration time. For the same reason the DAC should also be nonmagnetic. Beryllium-copper alloy has been adopted by several groups for the DAC realization [702–705], even though a titanium-based alloy has been employed to build a heatable cell for geoscience studies [706]. Split-pair rf coils are adjusted in series around the diamond tips and the gasket, the so-called split or saddle coil configuration, with the rf magnetic fields parallel to the conductive metallic gaskets, in order to minimize the shielding effects of the gasket and guarantee the best penetration into the sample region, which should therefore be as large as possible in diameter and relatively thin [703, 704]. A slightly different sample environment was adopted to measure the ortho-para conversion of hydrogen up to 12.8 GPa [707, 708]. A detachable rf hairpin-type resonator has been realized to avoid the need to remodel the saddle coil each time the gasket is changed and epoxy, or similar glue, is used to fix the coil to the diamond tips [709]. More recently, a single-solenoid rf probe, providing a higher signal-to-noise ratio has been set-up [710], allowing measurements in liquid methanol up to pressures exceeding 4 GPa [711]. A remarkable increase of the hydrogen-bonding interaction in liquid methanol and a decrease of the O-O distances with rising pressure are evidenced, testifying to a direct compression response of the first coordination shell. A smaller and further improved version of this latter rf coil, fixed around the sample by suitably laser drilling the diamonds, was employed to measure the collision and diffusion dynamics of molecular hydrogen in the hydrogen clathrate hydrate [712], and to compare the hydrate system with pure H_2 [713]. Interestingly, the fast diffusion of the H_2 molecules through the water hydrate cages is almost insensitive to the pressure increase. A new opposed anvil high-pressure cell able to generate pressures up to 9 GPa on remarkably larger samples, by approximately two orders of magnitude than in the DAC, has been recently presented [714]. In order to be employed in NMR measurements, both the clamp cell and the gasket are made of non-magnetic Ni-Cr-Al alloy, whereas non-magnetic tungsten carbide is used for the anvils. A great advantage of this cell is the possibility of using argon as a pressure-transmitting medium to obtain good hydrostaticity. Pressure can be measured *in situ* by the ruby fluorescence method employing an optical access, realized with a moissanite window. Anticonical-shaped gaskets have been realized to maximize the sample space, which also hosts a copper coil whose wires are placed in grooves on the gasket and are insulated by diamond powder and epoxy resin. Pressure and temperature dependences of the ^{63}Cu nuclear-quadrupole-resonance frequency of Cu_2O, and the in-plane Knight shift of metallic Sn, Pt and Cu have been measured up to 10 GPa using this cell.

Chapter 5

Principles of Chemical Reactivity Under Pressure

In this chapter the basic concepts regarding the effects of pressure on chemical reactivity will be presented, focusing attention on the aspects of interest for treating reactions occurring at pressures in the order of or greater than 1 GPa, whose description is the main purpose of this book. The peculiarities of these pressure conditions regard the physical state and the electronic properties of the molecular systems, which can greatly differ from ambient conditions. Condensed phases, both liquid and solid, characterize the pressure regime of interest, posing specific additional requirements in both the nucleation and in the propagation steps. Requirements for nucleation are mainly of steric nature and concern the successful realization of the proper relative conformation to match the conditions for achieving the transition state. In the propagation step, the dynamics of the system, mainly related to the transport of reactants to reactive sites, becomes dominant. In addition, pressure can strongly modify the electronic distribution leading to the formation of species capable directing the reactivity along pathways not accessible at ambient conditions. This is actually a relevant aspect that allows a precise distinction with the reactions occurring in the kbar range, which have been extensively reviewed by several authors [715–717] and will be not considered here.

Within this framework, we will first discuss the thermodynamic relations accounting for pressure effects on chemical equilibria and reaction rate. The environmental effects, with particular reference to intermolecular interactions, on the dynamics of high-pressure reactions will be considered. The importance of the environment is particularly relevant in condensed phases where geometrical constraints, determined by the relative distances and orientation among the molecules, are decisive in selecting the reaction pathway making a discussion of the connection between the system topology and the kind and mechanism of the reaction mandatory. In the fluid phase, the effect of pressure on the rotational and translational motions are driven by the viscosity changes that, beside selecting the approach geometry among the molecules, also regulate the energy barriers for molecular rearrangements needed in the various reaction steps. The same considerations also apply to reactions occurring in pure crystals where, however, the situation is complicated by the twofold role of the lattice which acts as solvent and reactant. In fact, due to the many-body nature of the solid state, collective motions transport the molecules to the reactive sites (diffusion) along with the perturbation, due to a chemical transformation from one to the other sites (propagation). This makes a chemical reaction in a perfect crystal a self-consistent and highly correlated process of the creation of perturbations which, in the case of a solid state reaction, are the product molecules. Since all these processes driving and propagating a chemical transformation are also strongly affected by temperature changes, it will be necessary to consider in many instances the pressure and temperature parameters simultaneously. Finally, despite the limited knowledge that we presently have of the pressure-induced changes of the electronic structure in molecular liquids and solids, we will introduce some general principles that can help the rationalization of both purely pressure-induced and high-pressure photo-induced reactions.

5.1 Pressure effects on chemical equilibria

A chemical reaction where reactants and products are kept to constant P and T conditions can be described by a balanced equation of the type,

$$aA + bB + \cdots \rightleftarrows yY + zZ + \cdots$$

The Gibbs energy for this system depends on P, T and on the number of moles of reactants and products (n_i). The total derivative dG is obtained by:

$$dG = \left(\frac{\partial G}{\partial T}\right)_{P,n_j} dT + \left(\frac{\partial G}{\partial P}\right)_{T,n_j} dP + \sum_i \left(\frac{\partial G}{\partial n_i}\right)_{T,P,n_{j\neq i}} dn_i, \qquad (5.1)$$

and by using $(\partial G/\partial T)_{P,n_j} = -S$ and $(\partial G/\partial P)_{T,n_j} = V$,

$$dG = -SdT + VdP + \sum_i \mu_i dn_i, \qquad (5.2)$$

where $\mu_i = (\partial G/\partial n_i)_{T,P,n_{j\neq i}}$ is the chemical potential of the species i. Therefore, for a reaction occurring at constant T and P values dG results,

$$dG = \sum_i \mu_i dn_i. \tag{5.3}$$

The chemical potential can be expressed in terms of activities (a), making this formulation valid for a general equilibrium system made of gases, liquids or solids,

$$\mu_i = \mu_i^0(T) + RTlna_i, \tag{5.4}$$

where μ_i^0 is the chemical potential of the species i in its standard state, which is defined as the pure state of the component at the T and P of the reaction. Taking into account the stoichiometry of the reaction we can write the change in the Gibbs energy for the reaction reported above as:

$$\Delta G = a\mu_a + b\mu_b - y\mu_y - z\mu_z \cdots, \tag{5.5}$$

and by using Eq. 5.4,

$$\Delta G = \Delta G^0 + RTlnK, \tag{5.6}$$

where K is the equilibrium constant given in terms of the activities a_i of the components in the system at the equilibrium:

$$K = \frac{(a_Y)^y(a_Z)^z \cdots}{(a_A)^a(a_B)^b \cdots}. \tag{5.7}$$

When a substance is in its standard state it follows from Eq. 5.4 that $a_i = 1$, therefore if all the species involved in the reaction are in their standard state $\Delta G = \Delta G^0$. When the reaction system is in equilibrium, the Gibbs energy must be a minimum with respect to any displacement from the equilibrium position ($\Delta G = 0$) and we obtain,

$$\Delta G^0 = -RTlnK. \tag{5.8}$$

The dependence on pressure of the chemical potential μ_i^0 is given by,

$$\left(\frac{\partial \mu_i^0}{\partial P}\right)_T = V_i^0, \tag{5.9}$$

where V_i^0 is the molar volume in the standard state. Therefore, the dependence of the equilibrium constant on pressure is expressed as,

$$RT\left(\frac{\partial \ln K}{\partial P}\right)_T = -\Delta V^0, \tag{5.10}$$

where ΔV^0 is the difference of the molar volumes, in the standard state, between the products and the reactants. Equation 5.10 shows that the equilibrium constant K decreases with increasing pressure at constant temperature when a volume expansion occurs, while, accordingly, reactions involving a volume contraction are characterized by equilibrium constants increasing exponentially with pressure. As an example, a pressure increase from ambient conditions to 1 GPa and a volume contraction of 10 cm^3/mol, lets the equilibrium constant increase by almost two orders of magnitude. What is therefore obvious is the role of pressure as a tool for tuning chemical equilibrium for reactions associated with a volume variation and not controlled by dynamic factors. These conclusions, obtained according to Eq. 5.10, assume molar volumes not changing with pressure (ΔV^0 constant). This assumption holds in very limited pressure ranges, whereas for larger pressure variations the use of Eq. 5.10 requires the knowledge of the compressibility data over the pressure, temperature and composition ranges of interest. Limiting attention to condensed phases, the treatment of liquid solutions is complex because of the necessity to know the activity coefficients of the different species. A large amount of activation and reaction volume data, accounting for the influence of pressure on chemical reactions in liquid solutions of organic and inorganic systems, is presently available [715, 716]. In contrast, much less is known about the behaviour of pure systems which are the main subject of this book.

The volume contraction of a reaction can be calculated from the difference between the van der Waals molar volumes of products and reactants. This approach implies the employment, in the computation of the intramolecular distribution function, of the molecule in the dilute gas. Different approaches have been proposed to overcome this approximation. Chandler assumed the repulsive forces among the molecular cores, shaped as hard spheres, to be dominant with increasing pressure, succeeding to explain equilibria constants not attainable by using standard van der Waals radii [718].

In spite of the severe approximation, Hamann attempted a classification of different types of reactions depending on characteristic average volume changes and on the number of additional chemical bonds in the products

Table 5.1 Volume change ΔV^0 referred to the formation of one mole of product, and the number of additional chemical bonds characterizing different reaction classes regarding liquid hydrocarbons [719]

Reactions	Net increase of covalent bonds	ΔV^0 (cm^3/mol)
dimerizations		
2(1-pentene) → 1-decene	1	-30
2(1-hexene) → 1-dodecene	1	-28
2(1-octene) → 1-hexadecene	1	-27
trimerizations		
3(1-hexene) → 1-octadecene	2	-36
cyclizations		
1-pentene → cyclopentane	1	-15
1-hexene → cyclohexane	1	-17
1-heptene → cycloheptane	1	-20
3(acetaldehyde) → paraldehyde	3	-36
isomerizations		
n-pentane → 2-methyl butane	0	+1.2
n-hexane → 2-methyl pentane	0	+1.2
n-hexane → 2:2-dimethyl butane	0	+2.0
cis-2-butene → *trans*-2-butene	0	+2.6

with respect to the reactants [67]. The volume changes for different types of reactions are summarized in Table 5.1. This approach, although qualitative, represents a useful method to identify the kind of reactive process under examination. Nevertheless, it is sufficiently reliable only when non-polar molecules react giving rise to non-polar products such as those reported in Table 5.1. Only in this case can the intermolecular interactions be considered of the same type, both in reactants and products, and environmental effects are negligible.

On the contrary, relevant volume variations can occur when the reactions involve polar species or ions. A reduction of the molar volume is generally observed in these cases as a consequence of the more efficient packing of the molecules. As will be discussed in the last section of this chapter, the effect of pressure on the electronic density distribution is quite relevant and a quite common effect, also on neutral molecules, is a strong polarity increase which can also determine the formation of radicals or ions, precursors of purely pressure-induced reactions [377–379].

All these general arguments also apply to equilibria in condensed phases even though the kinetics of the reactive process can be considerably slower. As a matter of fact, the molecular mobility, including both orientational and translational motions, is considerably reduced by the density increase. In the liquid phase these processes are regulated by the viscosity change with pressure, whereas in the solid phase, the mobility is further reduced because of the steep increase of the energy barriers height according to the structural environmental constraints (steric hindrance). From a thermodynamic point of view, this is a very important aspect because the system can explore secondary minima (metastable state) in which it can be indefinitely trapped. Equilibria of a mixture of two species in the solid phase can be described by extending a model developed by Drickamer to explain electronic transitions in iron compounds [12]. The Gibbs free energy of a mixture of two species at equilibrium, whose individual values are G_a and G_b, is formulated as,

$$G = N_0[(1-x)G_a + xG_b + x(1-x)\Gamma] - T\sigma_{mix}, \tag{5.11}$$

where N_0 is the total number of molecules, x is the fraction of component a converted to b, and Γ is an interaction term indicating that G, in condensed phases, cannot simply be derived by the free energies of the single components, representing a contribution from the environment weighted by the fraction converted. In an ordered single crystal, this term accounts for the cooperative effects typical of an ordered lattice which will be described later. All G_a, G_b and Γ terms depend on pressure and temperature. σ_{mix} is a term depending on the relative concentration of the two species,

$$\sigma_{mix} = k_B[N_0 lnN_0 - N_0 x lnN_0 x - N_0(1-x)lnN_0(1-x)]. \tag{5.12}$$

Finally, by applying the equilibrium condition $\left(\frac{\partial G}{\partial x}\right)_{P,T} = 0$, defining $K = \frac{x}{1-x}$ and $\Delta G = G_b - G_a$, one finds:

$$lnK = -\frac{\Delta G + (1-2x)\Gamma}{k_B T}. \tag{5.13}$$

The behaviour of K with pressure and temperature requires the knowledge, or a suitable formulation, of Γ. This choice drives the level of approximation introduced. Expansion of Γ as a function of pressure, has been proposed by Drickamer using coefficients depending on the relative volume and bulk modulus variation [720].

5.2 Pressure effects on reaction rates

The mechanism of a chemical reaction is commonly derived by the analysis of its kinetic behaviour through the employment of suitable models. Reaction rates and rate constants are obtained from the time dependence of different chemical and physical variables, according to the experimental technique selected to follow the reaction evolution. Analysis of the temperature effects on the rate of a chemical reaction are far more common with respect to those regarding the pressure effects, even though in the latter case, a clear distinction should be made depending on the pressure range of interest. Extensive studies have been reported both for organic and inorganic reactions up to 0.2 GPa [716], whereas kinetic data above 1 GPa are rather scarce.

With only a few exceptions, such as explosion and thermal deactivation reactions, rate constants depend on temperature and follow the empirical law,

$$\frac{dlnk}{dT} = \frac{E_a}{RT^2}, \tag{5.14}$$

where k, the rate constant, is a numerical term correlating the concentrations of the chemical species, involved in the reactive process, to the reaction rate. This relation can be integrated to give, if the term E_a is temperature independent, the Arrhenius equation,

$$k = Ae^{-E_a/RT}. \tag{5.15}$$

E_a, which has energy units, is the activation energy of the reaction and it is usually obtained by isobaric measurements of the rate constants at different temperatures, through the determination of the lnk vs $1/T$ slope.

A chemical reaction can be described according to the transition-state theory as a trajectory in which the maximum energy corresponds to the activated complex, the intermediate species between reactants and products. The Gibbs energy change between the reactants and the transition state, $\Delta G^{0\neq}$, is expressed through the equilibrium constant between the reactants and the activated complex $K^{\neq}$ as,

$$\Delta G^{0\neq} = -RTlnK^{\neq}, \tag{5.16}$$

and by using the Gibbs energy formulation in terms of enthalpy and entropy of activation,

$$\Delta G^{0\neq} = \Delta H^{0\neq} - T\Delta S^{0\neq}, \tag{5.17}$$

the activation energy E_a and the Arrhenius pre-exponential factor A can be expressed, respectively as:

$$E_a = \Delta H^{0\neq} + 2RT, \tag{5.18}$$

and,

$$A = \frac{e^2 k_B T}{hc^o} e^{\Delta S^{0\neq}/R}. \tag{5.19}$$

The two latter equations therefore relate the experimentally attainable parameters to the energetics of the reactive process, bringing the determination of the enthalpy and entropy of activation, respectively.

The effects of pressure on chemical reactivity can be intuitively related to the reduction of the molar volume. As a matter of fact, a reduction of the molar volume by 50% is observed in simple molecules (see for example N_2 [721]), both in the fluid and in the solid phases, for pressure increases of a few GPa. This remarkable reduction of the intermolecular distances can obviously dramatically slow down or even inhibit a reaction where a reorientation of the reacting species is required; on the other hand, if the relative orientation of the reacting species is favourable, the reaction rate will instead be accelerated. The clear involvement of the intermolecular interactions in the reactive process determines control of the reaction evolution by kinetic terms, and highlights the importance of performing kinetic studies as a function of pressure to obtain valuable information about the structural evolution of the system along the reaction trajectory.

A chemical reaction is often a succession of different elementary reactive steps which are characterized by their own rate laws, inferable from the stoichiometry. These considerations particularly apply to the reactions of interest in the present case, where the high density constraints generally require multistep processes. In most of the cases, the overall reaction rate coincides with that of the slowest elementary step, which is thus referred to as the rate determining step. As already stated, the macroscopic effect of the application of pressure consists of a reduction of the molar volume, by defining the reaction volume $\overline{\Delta V}$ as the volume change corresponding to the conversion of reactants into products, whose respective volumes are indicated by $\overline{V_R}$ and $\overline{V_P}$:

$$\overline{\Delta V} = \overline{V_P} - \overline{V_R}. \tag{5.20}$$

Analogously, the activation volume $\Delta V^{\neq}$ is given by the volume variation going from the reactants to the activated complex,

$$\Delta V^{\neq} = V^{\neq} - \overline{V_R}. \tag{5.21}$$

As already reported for the dependence of the equilibrium constant on pressure in Eq. 5.10, we can write the dependence on pressure of the reaction rate as:

$$\frac{\partial \ln k}{\partial P} = -\frac{\Delta V^{\neq}}{RT}. \tag{5.22}$$

Integration of Eq. 5.22 provides the basic relation employed to extract the pressure-independent activation volume $\Delta V^{\neq}$ using rate constant measurements as a function of pressure at fixed temperature values. Syntheses under high pressure are in principle always convenient for negative values of $\Delta V^{\neq}$, but the technical complications due to the high-pressure generation and sample confinement, limit the interest to those reactions presenting rather large values of $\Delta V^{\neq}$, typicallly less than -10 cm^3/mol. The change of some room temperature reaction rates, according to typical variations of $\Delta V^{\neq}$, are shown in Figure 5.1. However, the linear evolution of lnk with pressure is observed to hold only for small variations of pressure, revealing that $\Delta V^{\neq}$ can be considered independent of pressure only in these ranges. Generally, the pressure dependence of the activation volume is expressed through the compressibility coefficient of activation $\Delta\beta^{\neq}$ as,

$$\Delta\beta^{\neq} = -\left(\frac{\partial \Delta V^{\neq}}{\partial P}\right)_T. \tag{5.23}$$

Several empirical non-linear equations have been used to extract the relevant information from the pressure evolution of lnk. The most common formulation adopted is a quadratic evolution,

$$lnk = a + bP + cP^2, \tag{5.24}$$

which gives, at $P = 0$, $\Delta V_0^{\neq} = -bRT$ and $\Delta\beta_0^{\neq} = 2cRT$. Another expression employed is based on a fractional formulation,

$$lnk = a + \frac{bP}{c + P}, \tag{5.25}$$

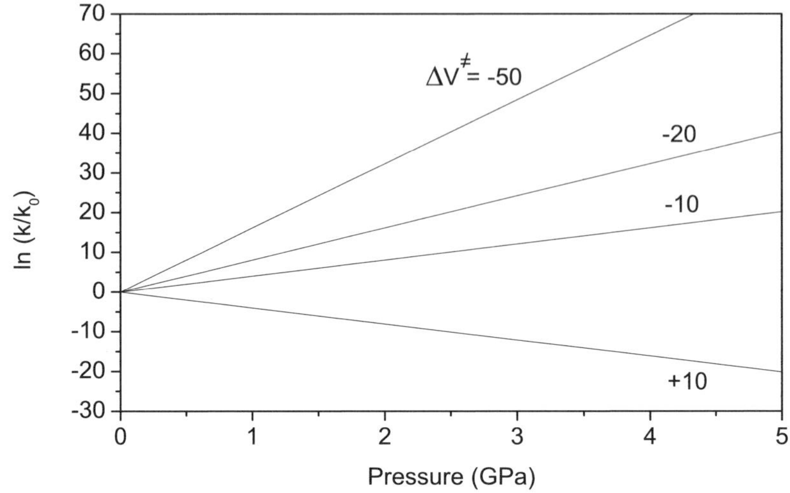

Fig. 5.1 Pressure evolution of the rate constant as obtained from Eq. 5.22 for different values of the activation volume. The evolution refers to ambient temperature conditions and assumes $\Delta V^{\neq}$ values constant over the entire pressure range.

which gives, at $P = 0$, $\Delta V_0^{\neq} = -bRT/c$ and $\Delta\beta_0^{\neq} = -2b/c^2$, but for pressures greater than $100c$ becomes independent of pressure, corresponding to a vanishing activation volume. For high-pressure values the exponential law,

$$lnk = a + b\left(1 + e^{-cP}\right), \tag{5.26}$$

gives also rise to a vanishing activation volume.

Several processes contribute to the overall reaction trajectory, from ground to transition state and within the transition state, and the activation volume, as obtained by the analytical laws reported above, contains the full complexity of these contributions. $\Delta V^{\neq}$ should therefore be expressed as a summation of all these contributions which are normally grouped in two main terms, the structural and environmental ones,

$$\Delta V^{\neq} = \Delta_s V^{\neq} + \Delta_m V^{\neq}. \tag{5.27}$$

The structural contribution $\Delta_s V^{\neq}$ reflects the volume variation of the system, due to the change of the nuclei positions in going from the reactant to the transition states. In other words, this contribution accounts for the changes in the bonding scheme due to bond cleavage and formation, and it is considered independent of the solvent and concentration in such a way, that it is also labelled as an intrinsic volume. $\Delta_m V^{\neq}$, the environmental contribution, results from the volume effects related to the interactions of the reactants and activated complex with the medium. This contribution therefore reflects the different packing of reactants and transition state complex with the neighbouring molecules and, for this reason, is expected to be important in high-pressure reactions in condensed phases. A quite detailed analysis of the possible contributions to the structural and environmental terms of the activation volumes have been discussed by Jenner, also taking advantage of several experimental results concerning the pressure effect on reactions in solutions [722].

Immediate information about the reaction trajectory can be gained for many reactions where the environmental contribution is negligible [717, 722]. $\overline{\Delta V}$ being the difference between the product and reactant volumes, we can define the quantity θ as,

$$\theta = \frac{\Delta V^{\neq}}{\overline{\Delta V}}, \tag{5.28}$$

which can provide a qualitative indication of the position of the transition state along the reaction coordinate depending on the θ value. For example, θ values close to 1, as observed for concerted Diels–Alder reactions in solution, unambiguously indicate that the structure of the transition state is close to that of the product molecule. Pericyclic reactions can occur through both concerted or stepwise diradical mechanisms. In the former case, the reaction is always accelerated by pressure because the volume of the transition state is smaller than that of the reactants. On the contrary, a stepwise mechanism may be characterized by a transition state also having a larger volume than the reactants, therefore being retarded by pressure. Therefore, in some cases the pressure behaviour of the reaction rates can be fundamental to identify the type of mechanism driving the reaction. These processes can also in principle be recognized by the θ value, being $\sim$1 in the concerted case, and significantly deviating from unity in stepwise mechanisms [723]. As also observed in many other competing reaction classes, the application of pressure can be very efficient in selecting the trajectory characterized by the more negative $\Delta V^{\neq}$. For example, selection of enantiomers having different activation volumes can be realized, as reported for Michael reactions of amines to chiral crotonates, where the enantiomeric excess increases from 10% at 0.1 MPa to 98% at 1.4 GPa [724]. Peculiar behaviour has been identified in some classes of reactions allowing the identification of specific contributions to the structural term. For example, concerted Diels–Alder reactions are generally characterized by $\Delta V^{\neq}$ and

$\overline{\Delta V}$ negative values and $\theta \sim 1$. Nevertheless, it has been reported that in some unhindered Diels–Alder reactions $|\Delta V^{\neq}| > |\overline{\Delta V}|$ [722]. This occurrence reveals an additional volume contraction of the transition state with respect to the products and which cannot be ascribed to environmental contributions, which are generally negligible in this class of reactions. A small polarity change of the transition state of intramolecular origin, was suggested for this contribution $\Delta_\mu V^{\neq}$ [725]. The steric volume of activation $\Delta_\sigma V^{\neq}$ is another contribution to the structural activation volume invoked to explain the pressure effect in competitive reactions regulated by steric hindrance (see references in [722]). Despite the lack of a full understanding of its physical meaning, the net experimental result is that pressure favours the most hindered process. Such an effect was interpreted as a pressure driven displacement along the reaction coordinate of the most hindered transition state [726]. This kind of contribution is particularly important for the reactions of interest in this book, where pressure exceeds 1 GPa and for this reason the molecular mobility is greatly reduced and steric contributions can even dominate the reaction kinetics.

The density increase in the condensed phases also reflects a growing importance of the volume changes due to the interaction with the environment surrounding the reactants and the transition state complex. These interactions should include dispersion and repulsive terms, hydrogen bonds and electrostatic contributions. Some of these terms can be negligible at ambient conditions but can rapidly build up at high pressure. This is, for example, the case of electrostatic contributions or interactions like hydrogen bonds which will be discussed later in this chapter for butadiene and pyridine, respectively. In general, if along the reaction coordinate the only intermolecular forces present are of the dispersive type (isopolar path), the only contribution to the activation volume is the structural one. However, when a change of polarity takes place along the reaction trajectory, a significant environmental contribution should be considered to account for the consequent changes in the interaction with the surrounding molecules. This term, indicated as the electrostriction volume, can be larger than the structural contribution, and it can also have an opposite sign, so that pressure driven polarity changes can be extremely important in opening or selecting reaction paths not possible at ambient conditions. A formal expression of this contribution can be formulated through the Drude–Nernst equation, an electrostatic derivation developed to quantify the volume contraction of a solvent around an ion. By assuming a spherical charge q, having radius

r, surrounded by a medium of dielectric constant ϵ, the electrostriction volume can be expressed as [727]:

$$\Delta_e V^{\neq} = -\frac{q^2}{2r\epsilon}\frac{\partial \ln \epsilon}{\partial P}. \tag{5.29}$$

The pressure evolution of this contribution can be derived for a liquid up to pressures of several kilobars by using the Owen–Brinkley equation to account for the dielectric constant changes with pressure [728],

$$1 - \frac{\epsilon^0}{\epsilon(P)} = A\epsilon(P)\ln\left(\frac{B+P}{B+1}\right), \tag{5.30}$$

where A and B are parameters not dependent on pressure. Volume effects ascribable to the environment, having a different origin from the electrostatic one, can also take place. These are generally due to a volume shrinkage because of solvophobic interactions, such as those involving neutral organic molecules in a polar solvent.

A summary of the key findings of this discussion can therefore be attempted. When the rate determining step involves the formation of a covalent bond (associative process), a negative structural contribution to $\Delta V^{\neq}$ is observed. On the other hand, when the rate determining step is characterized by a dissociative process, such as the breaking of a covalent bond, a positive structural contribution results. When a charged or polar transition state forms in the rate-limiting step from neutral reactants, a negative electrostriction contributes to $\Delta V^{\neq}$. In the same way, neutralization of charge in going from the reactants to the transition state leads to a positive environmental contribution to the activation volume. More complicated is the analysis of multi-step reactions, where the measured $\Delta V^{\neq}$ is the sum of the activation volume of the rate-determining step and the reaction volumes of all the pre-equilibrium steps eventually occurring prior to the rate-determining step. A nearly complete review of experimentally determined $\Delta V^{\neq}$ for organic reactions in solution are reported in refs. [715, 716, 729]. The $\Delta V^{\neq}$ values characterize different classes of reactions, therefore being useful for their identification. In Table 5.2 we report some examples which allow an immediate correlation between the reaction mechanism and the activation volume.

Experimentally, the determination of $\Delta V^{\neq}$ is performed through Eq. 5.22 by using the rate constant k measured at different pressures. The rate constant is obtained by a fit of the time evolution of the concentration, or of any other related quantity, of reactants or products with a suitable model. The choice of the kinetic model is definitely the basis of the analysis because

Table 5.2 Activation volume values (cm^3/mol) for various kinds of reactions [266]

Reaction	$\Delta V^{\neq}$
homolysis	5 to 20
polymerization (radical propagation)	~ -20
cycloaddition Diels–Alder	-25 to -40
cycloaddition intramolecular	-25 to -30
cycloaddition dipolar	-40 to -50
cycloaddition (2+2)	-40 to -55
ester hydrolysis	-10 to -15 (basic) >-10 (acid)
epoxide-ring opening	-15 to -20
Wittig reactions	-20 to -30

many different steps, also having comparable rates, may contribute to the kinetics. This is particularly true in solid state reactions where typically fast processes in the fluid phase such as diffusion, including both transport of matter to (nucleation) and from (propagation) the reactive site, are slowed down. In some cases, the choice of model is subordinated to the identification of the products, which restricts the possible reaction paths. In reactions characterized by a much slower step than the others, the overall reaction rate is determined by the rate law of the slowest process and simple models, such as those describing gas phase or dilute solution reactions, can be applied to denser environments. As an example, the mechanism of butadiene dimerization in the crystal phase could be successfully interpreted, taking advantage of the identification of a dimer (vinylcyclohexene) as the only product, thus limiting the possible simple kinetic laws accounting for the bimolecular nature of the process, and by the determination of the activation volume [730]. This is also a good example to evidence the possible differences in the kinetics of the same chemical reaction occurring in fluid and solid phases. This dimerization in fact follows a second-order kinetic law in solution or in the gas phase whereas, it is first order in the crystal, thus attesting to a completely different rate-determining step in the two cases and highlighting how reaction kinetics may be strongly related to the molecular rearrangement in the transition state. Another good example is provided by hydrogen transfer reactions, which exhibit a marked difference in the liquid and solid phases [731]. These reactions are generally accelerated by pressure in the liquid phase, therefore indicating a negative activation volume. On the contrary, the recombination of macroradicals in solid polymers, an example of solid state hydrogen transfer reactions where the H atom migration is the rate-limiting step, are strongly retarded by pressure. In all these cases, as also in the bimolecular recombination of polymer radicals generated by gamma radiolysis in single crystals, evidence is gained of a rate-determining step characterized by the need for extra volume because of orientational reorganization of the reacting molecules in the transition state. This is also the case of the solid state dimerization of butadiene, where the internal rearrangement (reorientational step) determines the first-order kinetic law and the positive activation volume [730].

The examples reported so far are all bimolecular reactions whose kinetics is dominated by the association process. Many reactions in the condensed phases are instead dominated by the transport of the reactants to the reactive sites. This is also true in compressed liquids where, due to the exponential increase of viscosity with pressure, diffusion controlled reactions are strongly retarded by pressure. This kind of reaction is quite common in the solid state and in many cases, such as polymerization or amorphization reactions, transport of the reactants to the reactive sites is important and can even be the rate-determining step. In these cases, the rate is controlled by the diffusion coefficient of the migrating species and in general, this contribution is revealed by a sigmoidal shape kinetic curve [732]. Several models have been proposed to describe diffusion-controlled processes and most of them are focused on single and two-phase polymerization reactions. Among them, a particularly useful model has been proposed by Hulbert for reactions occurring at the interface between two solid phases and described by two fundamental processes: the transport of the reactants to the reaction site, and the transformation itself, in terms of the breaking and re-forming of bonds [733]. This model is based only on the diffusion coefficient and it can therefore be applied to single and multi-component fluids and solids. In addition, although developed for planar interface reactions, this model can be applied to crystalline, polycrystalline and also powdered compacts [734], because no reference to the environment, but only to the diffusion coefficient, is made. Two distinct approaches are proposed to describe the nucleation and growth processes. In one case, which was originally developed by Jander for planar interface reactions, the reacting species is continuously coated by a product layer so that the reaction rate coincides with the diffusional growth of this layer. In the other approach, there are active sites where the products nucleate and subsequently grow.

In the first case, the unidirectional interfacial growth of the product layer is expressed by the following law,

$$y^2 = 2kDt, \tag{5.31}$$

where y is the thickness of the product layer formed at the interface which can be ≥ 0 only for $t \geq 0$, D is the time-independent diffusion coefficient of the migrating species and k is a constant. Under several assumptions regarding the particles and reacting interface geometries, and assuming the above parabolic growth law, the well-known Jander equation is obtained,

$$k_J t = \frac{2kD}{r_0^2} t = [1 - (1 - x)^{\frac{1}{3}}]^2, \tag{5.32}$$

where k_J is the rate constant, r_0 is the initial radius of the reacting particles and $x(t)$ is the fraction of reaction completed at time t. The main limits in reproducing solid state reactions through Eq. 5.32 are represented by the changes affecting most of the parameters during the reaction. As a matter of fact, the diffusion coefficient, the reactant activities and also the interface geometry can change during the reaction. There are several modifications of the Jander equation, where a dependence on time of the activity of the reacting species, or of the reaction surface, are taken into account. Details of these specific derivations are provided in ref. [733], including the Valensi–Carter equation [735] which has been modified, including an inverse proportionality over time of D because of the increase in defect concentration with the reaction progression, to give:

$$kt = \frac{Z - [1 + (Z - 1)x]^{\frac{2}{3}} - (Z - 1)(1 - x)^{\frac{2}{3}}}{Z - 1}, \tag{5.33}$$

where Z is a term representing the volume of the reaction product formed per unit volume of reactant consumed, therefore accounting for the change in volume during the reactive process. Knowledge of the equation of state of the materials involved in the reaction is therefore necessary to compute Z. Many spinel formation rates were nicely reproduced by this equation.

Models based on nuclei growth are definitely more versatile and many different formulations have been proposed to describe nucleation and growth rates, and to relate them to the kinetics of the transformations. These models account for the nature of most of the high-pressure solid state reactions, which are characterized by diffusion-controlled processes. Isotropic growth rates are commonly adopted, being a relevant issue in the choice of the nucleation rate law. This law should account for the variation, with the reaction time, of the initial number, N_0, of nucleation sites. In general, a first-order law $N(t) = N_0 e^{(-ft)}$ is assumed for describing the change with time of the number of active sites, f being a kind of nucleation frequency. The nucleation rate can therefore be described by a first-order kinetic law such as $I = fN_0 e^{(-ft)}$ and, depending on the ft value, we can span from an almost constant nucleation rate, $ft \ll 1$, to a negligible nucleation rate for large values of ft. A general form of the kinetic law describing the growth of the nuclei is,

$$\ln \frac{1}{1-x} = kt^n, \tag{5.34}$$

which summarizes all the rate laws proposed on the basis of different assumptions concerning the distribution and shape of the nuclei, and of the different diffusion laws adopted to describe the growth process. For instance, the information related to the reaction mechanism, the nucleation rate and the geometry of the nuclei are all contained in the n parameter. An analogous equation was also derived by Avrami for describing the growth of a crystal from a liquid phase [736]. An additional parameter t_0, can be inserted to account for the nucleation step and to interpret the kinetic data of several high-pressure polymerization reactions in the crystal phase,

$$\ln \frac{1}{1-x} = k(t - t_0)^n. \tag{5.35}$$

The n value is a fundamental parameter, which can be derived by kinetic analysis, to provide insight into the microscopic evolution of the reaction since its value is related to the growth geometry and to the nucleation rate. In Table 5.3, the n values corresponding to diffusion-controlled reactions, depending on different nucleation rates and growth geometries, are reported. As it can be seen from Table 5.3, n values smaller than 1 unambiguously indicate unidimensional growth processes. This is actually a case of model pressure-induced polymerization in the crystal phase, such as those reported for acetylene and ethylene [376, 737]. In acetylene, values ranging between 0.5 and 0.6 have been found in the fit of the kinetic curves, built by plotting the integrated absorption of infrared bands of polyacetylene as a function of time [376]. These data, once combined with the knowledge of the structural arrangement, identify the unidimensional diffusion-controlled growth of the polyacetylenic chains along the diagonal of the bc lattice plane as the preferential reaction path. An analogous procedure allowed to account for the different quality of crystalline polyethylene synthesized by crystalline ethylene at 3.6 and 5.4 GPa. The lower quality of the polymer

Table 5.3 Values of the n parameter in diffusion controlled reactions depending on different growth geometry and nucleation rates

Growth geometry	Nucleation rate	n
3D (spheres)	$I = 0$	2.5
	$I = k$	1.5
	$I = fN_0e^{(-ft)}$	1.5 - 2.5
2D (plates)	$I = 0$	2.0
	$I = k$	1.0
	$I = fN_0e^{(-ft)}$	1.0 - 2.0
1D (rods)	$I = 0$	1.5
	$I = k$	0.5
	$I = fN_0e^{(-ft)}$	0.5 - 1.5

obtained at the higher pressure, a topic that will be further discussed in the next section, should be addressed to the formation of gauche defects (bending of the polymeric chains) due to the opening of new reaction channels, as expected according to the anisotropic compression of the monoclinic cell. This occurrence is revealed by kinetic analysis, where the higher value of the n parameter indicates an increase in the dimensionality of the growth process with compression [737].

Employment of the Avrami law does not allow reproduction of the early reaction stages dominated by diffusion and reflecting, in a sigmoidal shape, where the slow initial step is followed by a steep increase of the slope (autocatalytic regime). This behaviour is rather common in single phase polymerization reactions, where due to the similar structures, the polymer grows as a solid state solution in the monomer crystal without phase separation, as actually observed in the ethylene crystal polymerization [737]. A kinetic model for this class of reactions has been proposed by Baughman [732]. The difference with the Avrami model consists in considering that the rate constant for nucleation depends on the fraction of converted monomer $x(t)$. The rate of the solid state polymerization is given by:

$$\frac{\partial x}{\partial t} = By(x)(1-x)\exp\left[-\frac{E_I(x)}{RT}\right], \tag{5.36}$$

where y is the average length of the polymer, expressed as a function of the conversion factor x, and B is a constant accounting for the monomer structure. $E_I(x)$ is the activation energy for nucleation and its dependenceon x is due to the structural changes occurring during polymerization. This model has been found to nicely reproduce the sigmoidal kinetic curves relative to the polymerization of substituted diacetylenes [732].

5.3 Environmental effects at high pressure

The discussion reported in the previous section clearly evidences the central role played in high-pressure reactions by the environment in which the reacting centres are located. This subject has partly been tackled in connection with the determination of the reaction and the activation volumes. Nevertheless, some other issues are also important to understand the mechanisms regulating the chemical reactions in fluids or crystal phases subjected to ultrahigh-pressure conditions. As previously pointed out, the reactivity is generally tuned by the mobility of the reactants, which is extremely reduced in these conditions. Diffusion and reorientational motions are the two relevant contributions and, whereas they can be straightforwardly connected to the increased viscosity in compressed fluids, an analogous simple description in the crystal phase is far more difficult. Geometrical and volume constraints could determine the feasibility of a certain chemical reaction or even select one specific reaction when multiple reactive paths could, in principle, be possible. As a matter of fact, reactions requiring a high mobility of the reacting molecules can be completely prevented unless thermally activated, but the confinement of the molecule in a favourable intermolecular arrangement can select a particular reaction which would not be allowed at ambient conditions. All these aspects are particularly important in the crystal phase, where knowledge of the structure appears mandatory to interpret the molecular mechanism of the reactions in terms of nearest neighbour interactions. For example, reactions occurring in different crystal polymorphs have been reported to be characterized by different reaction products [61, 732]. A high selectivity of the reaction path by the crystal arrangement is reported in hydrogen cyanide, where the linear alignment of the molecules in the crystal, because of the hydrogen bond arrangement, allows the selection of one of several possible polymers obtained, in solution, under different catalysis and solvent conditions [738]. Also, the formation of *trans*-transoid polyacetylene in the pressure-induced polymerization of acetylene, is ascribed to the preferential growth of the polymer along the diagonal of the *bc* plane of the orthorhombic cell, because of the favourable arrangement of the molecules along this direction [376]. In fact, the relative orientation among the molecules resembles the polymer structure. This is a quite general occurrence in high-pressure reactions in the crystal phase; in fact, in the polymerization of diacetylenes [732] and ethylene [737], the resulting crystalline phase is determined by the crystalline structure of the reactants, meaning that the polymer chains and the monomer form a solid solution throughout the reaction. A good example is provided by ethylene polymerization (see Figure 5.2) because here, a highly crystalline polymer is obtained at 3.6 GPa, where the polymer growth is suggested to occur involving the ethylene molecules located along the a axis, where the shortest C-C contacts are realized and the CCC angle (131°) is quite favourable for the carbon hybridization change in going from the monomer to the polymer. On increasing the pressure to 5.4 GPa, the greater compressibility along the b axis makes the distances and orientations between the molecules sitting on the vertex and at the centre of the monoclinic cell, similar to those characterizing the neighbouring molecules along the a axis, thus opening a competitive channel for the polymerization and decreasing the selectivity of the reaction process. This occurrence is clearly revealed by the formation of gauche-defected polyethylene. All these examples concern topochemically controlled reactions because they satisfy the *topochemical principle*, which identifies the preferential reactive path in a solid state reaction as that taking place with the minimum atomic and molecular movement [739, 740].

Besides the *topochemical principle* another useful concept to rationalize the reactive processes in the crystal phase is that of *reaction cavity*, introduced by Cohen [740]. The space occupied by the molecules involved in the reaction can be defined by size and shape in the starting crystal. This volume, possessing a well-defined shape, is the reaction cavity whose surface is intuitively defined by the force field exerted by the neighbouring molecules. An extension of the topochemical principle to the cavity environment suggests that those reactions which imply the minimum distortion of the reaction cavity will be favoured (see Figure 5.3), because the deformation of the cavity along the reactive path will correspond to a decrease of attractive forces or an increase of the repulsive ones, therefore being energetically unfavourable.

Flexible cavities along the reactive path allow accounting for the distortion produced by the reaction [61]. Reaction cavities have been distinguished depending on the interactions with the environment. In the case of non-specific and non-directional van der Waals interactions, only steric effects enter in the reaction evolution and the cavity is assumed to be passive. On the other hand, active cavities are those involving directional and

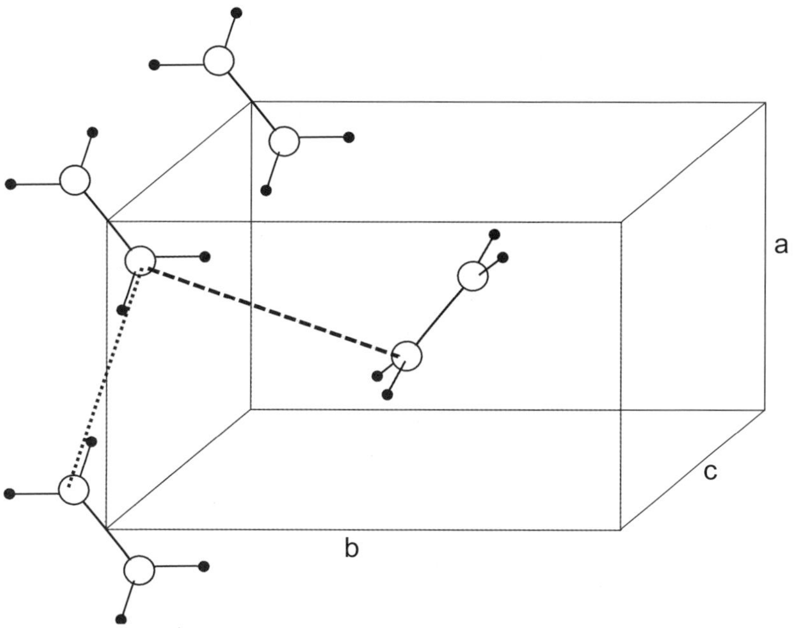

Fig. 5.2 Pressure-induced polymerization in crystalline ethylene. The shortest C-C contacts at 3.6 GPa regard the molecules aligned along the *a* axis (dotted line), whereas at higher pressure (5.4 GPa) this distance is comparable with that between the carbon atoms of the inequivalent molecules of the unit cell (dashed line).

stronger interactions, such as charge transfer or hydrogen bonds, with the reacting molecules. An enlightening example of the reaction cavity principle at work is represented by the high-pressure transformation of benzene to an amorphous hydrogenated carbon [741]. This reaction is triggered by pressure but it proceeds slowly above the reaction threshold pressure and only a limited amount of benzene reacts under these pressure conditions. On the contrary, once activated, the reaction strongly accelerates upon pressure release, coming to completion close to ambient pressure. The carbon hybridization change of a large amount of carbon atoms from sp^2 to sp^3 requires a strong deformation of the reaction cavity, so that the reaction propagation is prevented by large energy barriers originated by the steric hindrance of the neighbouring molecules to the molecular rearrangement. With releasing pressure, the cavity dimensions increase making the fast reaction propagation to the neighbouring sites possible.

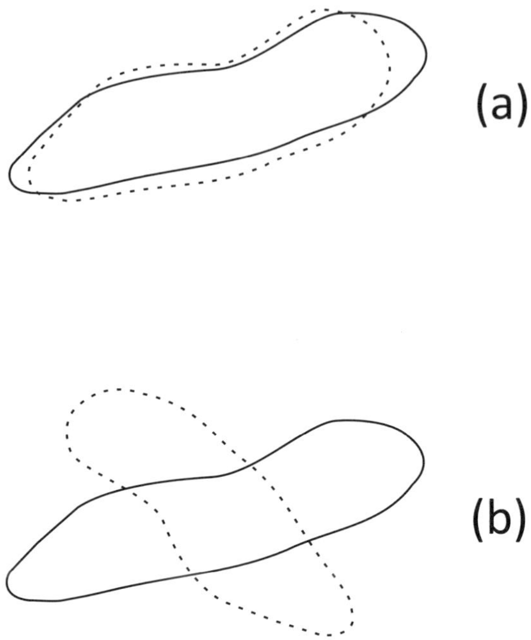

Fig. 5.3 Reaction cavity variation going from the reactant (full line) to the transition state (dashed line) for energetically favoured (a) and unfavoured (b) reaction paths.

Once the role of the environment in governing a solid state reaction has been defined, another fundamental issue that should be pointed out concerns the dynamic aspects entering the nucleation and propagation steps of the reaction. Nucleation is the primary reactive step in which the reactants are transported to the reactive sites, whereas propagation concerns the transmission of the perturbation to the crystalline environment due to the molecular transformation. The concept of mobility in the crystal phase is not immediate because molecules are indeed almost fixed in specific lattice positions and the possible degrees of freedom are limited to the amplitude of the thermal motion. Evidence of the role of temperature in high-pressure reactions is immediately gained by the slope of the instability boundary of several systems, so that a lower pressure is generally required to induce a chemical transformation in a crystal when the temperature is raised. Lattice motions, both translational and librational, represent the way in which favourable instantaneous distances and orientations are realized, allowing the occurrence of a chemical reaction. In addition, the perturbation due to the reactive event arising at one site is transmitted to the other sites by the lattice motions. The importance of cooperative effects in high-pressure reactions was suggested on the basis of Raman studies, where a softening of phonon modes, involved in the reaction coordinate, was detected [742]. The phonon softening results in a larger amplitude of both translations and librations bringing two molecules close enough to induce the reaction (nucleation). Thus, phonons in the crystal are analogous to collisions in the gas phase or solution reactions, playing the role of diffusion in crystals. Moreover, due to the collective nature of phonons, the processes are expected to be highly cooperative involving all the molecules of the crystal and not only those at the reactive sites (propagation). The active participation of lattice phonons in inducing a high-pressure reaction was clearly evidenced in the case of benzene [642]. The reaction threshold pressure was determined along several isothermal compression paths, whereas the nearest neighbouring contacts were obtained by the structural data at every P-T point where the reaction initiated. By taking into account the thermal motion, in particular zone boundary acoustic modes, which are most effective in reducing the nearest neighbour C-C contacts while maintaining the parallelism of the π electron densities, it could be demonstrated that the reaction always occurs once the same distance between nearest neighbouring C atoms (about 2.6 Å) is realized, independent of the P-T conditions (see Table 5.4). It is also extremely interesting that this distance, also supported by *ab initio* molecular dynamics simulation, is much shorter than the van der Waals separation (3.4 Å), often taken as a reference distance for the molecular instability [743].

Similar observations were also reported in other high-pressure reactions, and specifically in the pressure-induced polymerization of tetracyanoethylene [744] and trimerized thiazyl fluoride $(NSF)_3$ [745]. In tetracyanoethylene, the reaction takes place at 6.0 GPa involving CN atoms of the nearest neighbouring nitrile groups. In addition, in this case, the distance obtained from structural data is shorter than the van der Waals distance by 13%, analogously the computed inter-ring SN distance is much shorter than the van der Waals distance. Phonon assistance also applies to the already mentioned polymerization reactions in the crystal phase of ethylene and acetylene, where the nearest neighbour distances are much larger than the van der Waals contacts. For example, in the acetylene case the shortest distance among the nearest neighbouring carbon atoms is that involving the molecules aligned along the diagonal of the *bc* plane, which reduces from 3.46 to 3.05 Å in going from ambient pressure to 4 GPa (see references in ref. [376]). This distance obviously reduces with lowering temperature,

Table 5.4 Shortest C-C contacts of neighbouring molecules located along the *a* axis at the different P-T reaction thresholds. Structural distances are computed through diffraction data, whereas the thermal contribution, which accounts for the translational amplitude, is classically computed. The minimum instantaneous distance is finally obtained by subtracting the thermal contribution from the structural distance. Details of the calculations are reported in ref. [642]

Reaction conditions	Structural distance (Å)	Thermal contribution (Å)	Minimum contact (Å)
14.6 GPa, 643 K	3.017	0.432	2.585
23 GPa, 540 K	2.912	0.360	2.552
32.1 GPa, 423 K	2.838	0.294	2.544
41 GPa, 298 K	2.786	0.234	2.552

nevertheless the onset of the reaction is about three times higher (9 GPa) at 200 K. This observation can only be explained by taking into account the thermal motion; in fact, the translational mean square amplitude, which can be classically computed, decreases from 0.5 Å at 300 K to 0.2 Å at 200 K, a much larger reduction of the pure volume contraction due to the temperature effect. An indirect confirmation of the active role of lattice modes in triggering and propagating a reaction is provided by the pressure-induced reactions in propene [746] and pyridine [747]. The pressure threshold value for the oligomerization reaction in fluid propene exhibits a quite steep increase on lowering temperature (from 0.9 GPa at 370 K to 4.0 GPa at 270 K), but is not observed below 250 K or increasing the pressure above 20 GPa. Isobaric heating cycles have shown that the instability boundary inverts the slope sign below 270 K, following the melting line (see Figure 5.4). This is clear evidence that the reaction is not solely driven by density and, given the glassy nature of the low-temperature and high-pressure solid phase, the missed reactivity in the solid should be ascribed to the lack of collective motions able to trigger the reaction. Analogous observations have also been made for pyridine. Crystallization of the fluid at ambient temperature can result both in a glassy and in an ordered orthorhombic phase ($P2_12_12_1$, D_2^4), the reaction being observed by compressing the sample at about 17 GPa in the ordered phase, whereas it is not observed in the glassy phase up to pressures as high as 25 GPa.

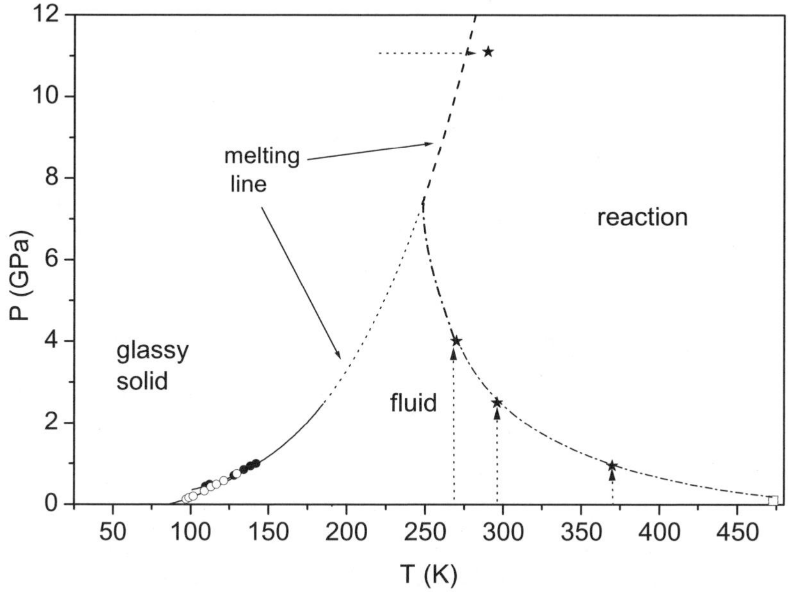

Fig. 5.4 Stability diagram of propene. The stars indicate the P-T values where the reaction was induced along isothermal compression or isobaric warming experiments, described by the arrows. Reactivity from the solid glassy phase is observed only when the sample is melted.

Luty and Eckhardt have developed a model treating the reactions in ordered crystals as the result of a highly cooperative process in which all the molecules are involved [748]. The model is formulated in such a way that the perturbation represented by the chemical reaction, and the relative crystal response, can be related to measurable quantities. Here, the local perturbation produced by pressure, or by the chemical transformation of a molecule at site n, has a mechanical nature and can be expressed by means of an elastic stress tensor. This mechanical local field couples to a set of coordinates $Q(n)$, such as translational and rotational coordinates, able to generate a non-local response. The dynamical variable Q can include any set of coordinates of interest for the process under consideration. The Hamiltonian system includes three different contributions: a single molecule term, the coupling between the molecular variables at different sites through a force constants matrix W, and the coupling to the dynamical variables by an operator of the local stress. The local effective field which couples to the variable Q is expressed as a mean field,

$$h^Q(n) = \sum_{n'} W(n,n')\langle Q(n)\rangle + V^Q(n), \tag{5.37}$$

where $V^Q(n)$ is the perturbing field. The response of the system to this local field is expressed, in the linear approximation, by a response function X such as,

$$\langle Q(n)\rangle = \sum_{n'} X(n,n')V^Q(n'). \tag{5.38}$$

According to the model, a perturbation at one site is transmitted to all the other sites as a collective process, the molecules at the different sites being related through the force constant matrix W. The response function X, which represents an effective mechanical susceptibility of a reaction cavity subjected to the mechanical perturbation produced by a chemical reaction, is the quantity that can be related to experimentally measurable properties such as elastic constants, phonon frequencies and Debye–Waller factors.

Once a reaction cavity is defined, we can consider that at every crystal site a force field, determined by the surrounding molecules and that can be described as a chemical pressure, acts. By introducing a perturbation of chemical nature, such as a reaction product or an excited or ionized molecule, an excess of energy is produced at site n that can be viewed as work against the chemical pressure and it can therefore be termed as deformation energy. Different contributions can be identified. First of all, the self energy of the perturbed site n, i.e., the energy ΔE_0 required to create distinct perturbations, will be proportional to the concentration of perturbed sites and will always be negative. Moreover, a contribution accounting for the collective nature of the perturbation, and ascribable to the interaction of the perturbation with the molecular degrees of freedom, should be considered. This term can be expressed in terms of an elastic multipole representation. The net force acting on the molecule n will be given by the sum of the different forces $\underline{\nu}(n\xi)$, each due to the atom ξ in the cavity, at distance $\underline{r}(n\xi)$ from the molecular centre of mass,

$$\underline{V}(n) = \sum_{\xi} \underline{\nu}(n\xi), \tag{5.39}$$

with the components of the elastic dipole moment expressed as,

$$P_{ij}(n) = \sum_{\xi} r_i(n\xi)\nu_j(n\xi). \tag{5.40}$$

The energy required to create the perturbation at the site n is formulated as,

$$H_n = \sigma(n)\left[\Delta E_0 + \sum_i Q_i(n)V_i(n) + \sum_{ij} \epsilon_{ij}P_{ji}\right], \qquad (5.41)$$

where ϵ represents the local strain which can be assumed to be homogeneous within the cavity. The great advantage of using this approach, formalized by Eq. 5.41, is represented by the fact that both the shape of the reaction cavity and the distribution of forces inside the cavity, the chemical pressure, are explicitly contained. Furthermore, since the force acting on each molecule results from the summation of the interactions with all the atoms of the different molecules, it is clear that all the collective motions are already included in the model.

Some examples regarding experimental determination of structural parameters by X-ray diffraction, IR absorption, and optical and electron microscopy measurements have been reported (see for example ref. [61] and references therein). Through these techniques it is possible to gain insight into the strains and stresses, determined by a distortion of the reaction cavity, by measuring the crystal modifications taking place during the reaction. The evolution of lattice parameters and molar volume can be monitored during the reaction by X-ray diffraction which, in some cases, could also provide refined data so that the changes could be followed at the atomic level. Optical microscopy is employed to evidence the stress field created by the reaction in the crystal or to monitor defect formation.

5.4 Effects of high pressure on the electronic structure

There is experimental evidence of the pressure effect on the electronic structure of molecules. The dielectric constants of hydrocarbons increase by more than 10% in a few kilobars [749], whereas vibrational frequencies generally increase between 1–10 cm^{-1}/GPa, both effects being related to a change in the charge distribution in the system. A more direct insight into the pressure tuning of electronic properties is provided by electronic absorption and emission spectra. As a matter of fact, many molecular systems exhibit a colour change with increasing pressure (piezochromism), an effect obviously related to a pressure shift of the absorption edge and representing a sharp evidence of the pressure effect on the orbital energies and therefore in the HOMO-LUMO separation. In general, but not necessarily, a red-shift is also observed in very simple systems such as diatomic molecules, and as also stated above, this is accompanied by other phenomena such as relevant changes in the vibrational properties. Observations of this type include, for example, oxygen, which becomes red above 20 GPa together with the appearance of new vibrational modes indicating a molecular association [483, 661]. In addition, nitrogen exhibits a gradual change of colour and a vibron softening prior to polymerization [750]. Despite the relevance of this subject, especially in relation to pressure-induced reactivity, the number and type of experimental and computational studies regarding the effects of pressure on the electronic states are quite few. A direct insight can be provided by studying the electronic spectra, both in absorption and emission, as a function of pressure. The two principal observations of phenomena consequent to a pressure increase are the already mentioned shift of the absorption edge, and the broadening of the absorption bands. A model attempting to rationalize these effects was proposed by Drickamer [12, 13] on the basis of two different contributions (see Figure 1.2). The first regards the relative vertical shift of the ground and excited states due to a different stabilization of the two states (structural changes), according to the density increase. This contribution primarily affects the energy of the electronic transition which, as already noted, generally decreases with pressure. This effect can have very important implications in reactive processes; in fact, due to the changes in the electronic distribution, the excited molecules can be extremely efficient in triggering a chemical reaction, and if the energy gap is sufficiently reduced, the thermal population of the excited state would not be negligible. The second contribution is related to the possible different compressibility of the two electronic states along specific configuration coordinates. This corresponds to a change of the relative position of the two minima and for this reason it is referred to as lateral shift. Macroscopic consequences of this effect should be searched for in the frequency shift of both absorption and emission bands, without a corresponding change of the energy gap between the two states necessarily taking place. In addition, since the two contributions may have opposing effects on the peak absorption energy, a small change of the optical absorption peak frequency may also be consistent with a significant change of the thermal energy threshold. Additional evidence of the pressure effect on the electronic transition is a broadening of the absorption bands. The model developed by Drickamer accounts for changes with pressure of the peak width, only if the force constants of the two states are different, providing an equation that relates the experimental observables, the peak maximum and the half width of the absorption band, to the thermal energy gap. A serious approximation involved in the model is that a single configuration coordinate is considered, whereas pressure, in general, couples to several vibrational modes so that a more complex configuration path should be considered. This model satisfactorily reproduced experimental data relative to different metal complexes providing evidence of thermal occupation of π^* orbitals by π electrons.

As previously mentioned, the implication of the occupancy of electronic excited states, in the reactivity of molecular systems, is particularly relevant because of the changes in the electronic properties in going from the ground to the excited states. Changes in the molecular polarity are indeed quite common, but dissociation can also occur in several systems, making these excited molecules active reactive centres. Among the different types of transitions ($\sigma \rightarrow \sigma^*$, $\sigma \rightarrow \pi^*$, $\pi \rightarrow \pi^*$, $n \rightarrow \sigma^*$, $n \rightarrow \pi^*$) the $\pi \rightarrow \pi^*$ transitions are generally the most sensitive to pressure, due to an appreciably lower overlap at ambient pressure than σ orbitals. These transitions obviously occur in unsaturated compounds which are therefore particularly reactive at high pressure. In general, a red shift with increasing density is expected for these transitions due to an excitation exchange between molecules. In the dipole approximation, the shift would be proportional to the density therefore providing, in case of a different density dependence, insight into other processes occurring upon excitation. For example, a difference in the dipole moment, or even in quadrupole interactions, between ground and excited states can be very important. If a significant increase of the dipole moment takes place upon excitation, a red shift of the corresponding transition energy with increasing the pressure results. The opposite occurs, i.e. a blue shift with increasing the pressure, when the configuration achieved upon excitation implies a dipole moment decrease. These arguments have been employed to explain the pressure dependence of the optical gap, related to the $\pi \rightarrow \pi^*$ transition, of several aromatic compounds such as anthracene [751, 752], tetracene, pentacene and azulene [751]. More recently, the study of the absorption spectra of benzene [631, 643] and some heteoaromatic molecules such as furane [753], indole [644] and pyridine [645], provided additional insight.

The case of benzene is particularly useful to understand the reliability of optical absorption experiments because of the possible comparison of absorption data obtained both in one-photon [643] and two-photon [631] absorption experiments. The density behaviour of the electronic origin relative to the $S_0 \rightarrow S_1$ transition, as determined by two-photon excitation profiles [631], is compared in Figure 5.5 with one-photon absorption data extracted by the saturated absorption band [643]. A linear shift with pressure results from the very precise determination of the 0-0 exciton line through two-photon excitation profiles, as expected for benzene because no dipole moment is acquired with the excitation to S_1. On the contrary, the strong non-linear red shift obtained by one-photon spectra clearly evidences that the simultaneous effect of pressure on the peak frequency and width prevents a reliable analysis of the saturated peak. The advantages of using two-photon absorption spectroscopy instead of conventional one-photon absorption techniques have already been described in Chapter 4, and can be summarized in the overcoming of the limitations posed by the diamond absorption edge, and by the small cross section which minimizes the production of excited species, potential nuclei of a chemical reaction. Two-photon absorption processes have been extremely useful, both through fluorescence and photo-induced reactivity studies, in providing an excellent insight on the pressure effects on the lowest electronic excited state of three different systems, enlightening the extreme variability of these effects and highlighting the importance of a direct characterization.

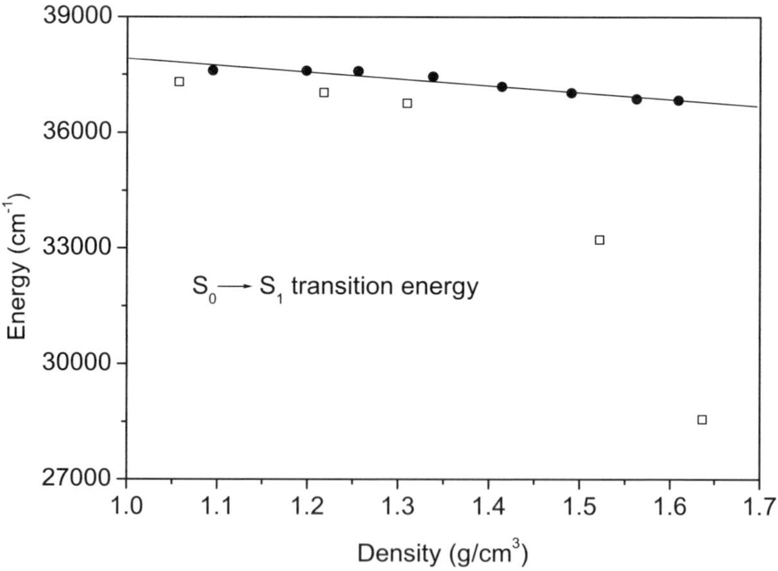

Fig. 5.5 Energy shift of the $S_0 \rightarrow S_1$ origin as a function of density as determined in two-photon [631] (black dots) and one-photon [643] (empty squares) absorption experiments. The linear shift with density of the two-photon data is evidenced by the full line.

In benzene, the simultaneous characterization with pressure, up to 18 GPa, of the two-photon excitation profile of the $S_0 \rightarrow S_1$ transition and of the $S_1 \rightarrow S_0$ emission spectra, allowed pressure-induced thermal mixing of S_1 and S_0 states to be ruled out as a possible origin of the high-pressure reaction. Conversely, these data show the progressive stabilization with pressure of excimers [631], as clearly evidenced by the intensity exchange between monomer and excimer emissions, tuned by pressure, and reaching a saturation (90% of excimer) above 10 GPa. The process is fully reversible indicating a pressure-induced modification of the S_1 surface as indicated in Figure 5.6. The excimer species can be related to that forming in the ground state by applying much larger pressures and is responsible for the reaction activation [642], so that both laser-assisted [643] and purely pressure-induced reactions can be related to the same triggering species.

Another interesting result regarding the pressure effect on the electronic 0 excited states has been obtained by two-photon induced fluorescence studies of pyridine [645]. Like most molecular systems where the lowest excited state has $n\pi^*$ character, pyridine is characterized by a low fluorescence quantum yield so that no fluorescence data were available in the condensed phases. On the contrary, the fluorescence intensity increases remarkably with pressure, being at 3 GPa in the solid phase II, six orders of magnitude larger than in the liquid. On the basis of the fluorescence yield increase with protonation, and knowledge of the crystal arrangement, the intensification has been explained as due to the inversion of the lowest $n\pi^*$ ($^1B_1$) and $\pi\pi^*$ ($^1B_2$) excited states, S_1 and S_2 respectively at ambient conditions, due to the increasing strength of the hydrogen bond network with rising pressure. A schematic representation of this occurrence is shown in Figure 5.6.

The last example reporting indirect evidence of the modification of the excited states derives from high-pressure photo-induced reactivity studies of ethanol [754, 755]. Here, a photochemical reaction was induced at high pressure by exploiting the dissociative character of the lowest electronic excited state, corresponding to a $n \rightarrow \sigma^*$ Rydberg transition, reached through a two-photon absorption process. The efficiency of this reaction, monitored through the consumption of the reactant, was found to depend on pressure. A comparative study of the photo-induced reaction in C_2H_5OH and in C_2H_5OD provided evidence, through the kinetic isotopic effect (KIE), that dissociation along the OH(D) coordinate becomes the rate-limiting step with rising pressure. This result has been interpreted on the basis of a decrease of the dissociative character with rising pressure, suggesting a relevant modification of the surface of the excited state with the creation of an energy minimum (see Figure 5.6).

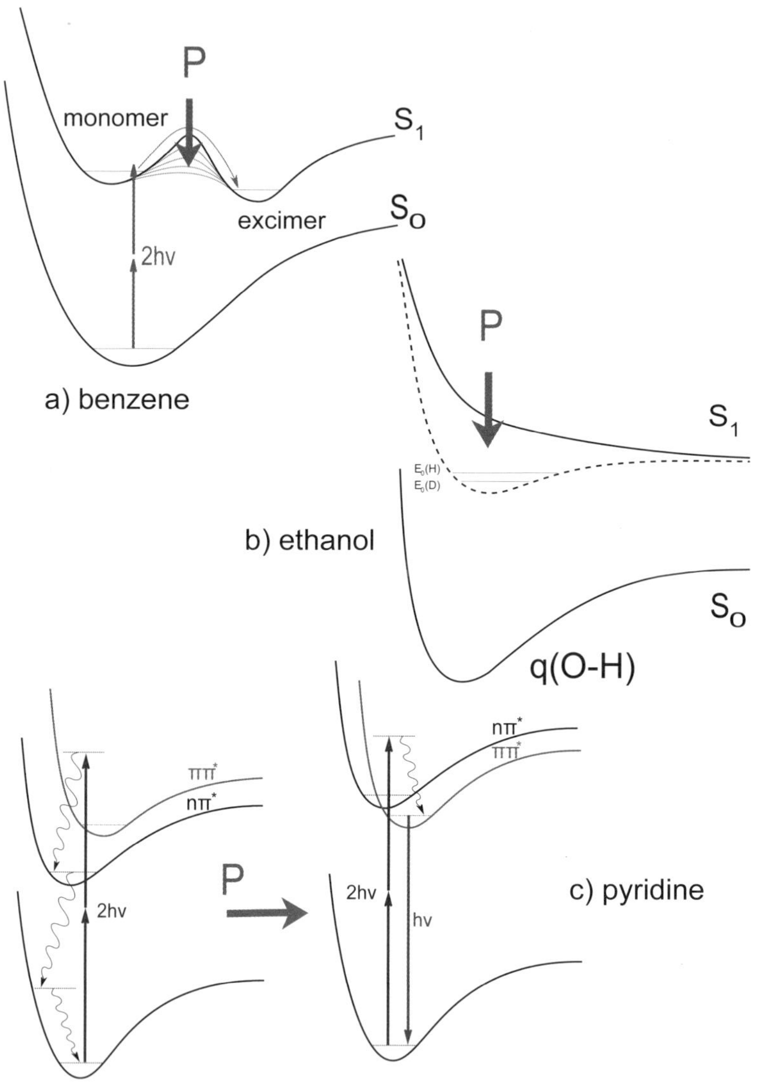

Fig. 5.6 Schematic representation of the pressure effects on the lowest excited state of (a) benzene [631], (b) ethanol (Reprinted with permission from S. Fanetti, et al. *J. Phys. Chem. B* **115** (2011) 15236–15240, Copyright (2011) American Chemical Society), and (c) pyridine (Reprinted with permission from S. Fanetti, et al. *J. Phys. Chem. B* **115** (2011) 12051–12058, Copyright (2011) American Chemical Society).

Computational methods represent another important source of information regarding the effects of pressure on electronic properties. Two examples are particularly useful for our purposes since they describe the molecular behaviour just before the reactive events. The density evolution of the molecular orbital profile of propene has been obtained by *ab initio* molecular dynamics simulation in the density functional approach [756]. The pressure-induced reactivity of this molecule had been experimentally studied in the fluid phase along different isothermal compression paths [746]. The calculation results indicate a rapid increase in the energy dispersion of the molecular orbitals as the pressure increases, whereas simultaneously the HOMO-LUMO energy gap decreases and new energy levels appear within the gap, suggesting the possibility of a degeneration of the density of states in a continuum and accounting for the general observation of a red shift of the electronic transition bands at high pressure. Interestingly, these results also present evidence of the role of the broadening of the energy levels which, together with the energy gap reduction, can give rise to thermal population of the excited states.

Other important features revealed by calculations regard the possibility to monitor the charge distribution within the molecule as a function of pressure or, from another point of view, as a function of intermolecular interactions. For instance, the pressure increase has been found to build up an electric dipole moment even in non-polar (centrosymmetric) molecules, such as *trans*–butadiene [757]. Before the reaction occurs, the charge redistribution gives rise to the formation of an extremely reactive zwitterion which triggers the oligomerization reaction. The formation of zwitterions was also found to be the primary reactive event in the study of the pressure-induced amorphization of benzene [642]. These species appeared because of the formation of a C-C bond between the nearest neighbouring equivalent molecules which determined a charge separation over two reacting molecules. Formation of zwitterions implies an ionic mechanism, and due to the fact that in the local-density approximation the unpairing of spins is precluded, a simulation in the local-spin-density approximation was also performed in the benzene reaction study to check if a radical mechanism was favoured. Consistent results with an ionic mechanism were found. It is important to recall that this triggering species can be directly related to that attained by electronic excitation and identified by emission spectra. A highly cooperative ionic mechanism was also found to be responsible for the reaction initiation in compressed nitromethane [758]. Experimental results regarding the polymerization of compressed trioxane by using ionizing radiation [759], which produces ionic centres, seem to support the effectiveness of ionic mechanisms in pressure-induced chemical reactions, and in particular, in polymerization reactions.

The results of the computational studies provide a considerable insight into the microscopic mechanisms of the reactivity. First of all, it is clearly evidenced that the reactive species are progressively constructed as the pressure is increased, through a complex mechanism of charge redistribution, which also modifies the molecular structure. This process begins well in advance with respect to the onset of the chemical reaction, as observed in the simulation of butadiene, but also indirectly evidenced in benzene where the excimer species produced under irradiation, and equivalent to the zwitterion found in the computation, is created in the crystal at very low pressure, with a concentration increasing with the density. The formation of these species can be viewed as the response of the system to contrast the high repulsive potential, originated by the modification of the interactions among neighbouring molecules. This consideration brings us directly to the second general output, regarding all the processes at work before and during the reaction. The entire reactive process has a highly cooperative character in the sense that all the molecular modifications, including the formation of polar species producing a favourable closer packing (electrostriction) in agreement with Le Chatelier's principle, are driven by the interactions with the surrounding molecules. In addition, the reaction does not generally occur through single and simple interactions of monomeric units, but more molecules are normally involved in each elementary chemical event, thus highlighting the collective nature of the reactive process.

Chapter 6

Chemical Reactions in Molecular Crystals

Chemical reactions at high pressure have been studied for a fairly large number of molecular crystals. The high-pressure reaction threshold ranges from approximately 1 GPa to several tens of GPa, depending on the structure and stability of the intramolecular bonding in the molecular units of the system. In several cases, molecular rearrangements or phase transitions occur before reaching the reaction threshold, and these processes are generally of primary preliminary importance to elucidate the reaction mechanisms. Here, attention will be confined to studies under static high pressures. These conditions have the advantage of a fine control of both temperature and pressure of the reaction, thus allowing for a more detailed study of the kinetics and the mechanism of the chemical reactions.

The variety of molecular crystals studied ranges from the simple diatomics of the second row elements of the periodic table to hydrocarbons, to more complex molecular systems like energetic materials or even biological molecules, to molecular multicomponent aggregates [27, 377, 378, 591]. A tentative classification within such a wide array is not easy beyond the obvious scheme of diatomics, triatomics and polyatomics, which can, in any case, be useful by itself. We may more profitably resort to thermodynamic or kinetic criteria [31, 760–762]. Let us indicate by R the initial low-pressure species of a transformation (e.g., a chemical react ion or a phase transition) and by P the final high-pressure species. The general free energy profile of the transformation can be represented as in the diagram of Figure 6.1, showing the stability region of R and P above and below some pressure threshold P_t, respectively. This profile depicts a substantially reversible transformation, where pressure helps overcome an energy barrier to reach a local minimum. The diagram is oversimplified due to the neglect of the full role of the energy barriers and of the possible change of the free en-

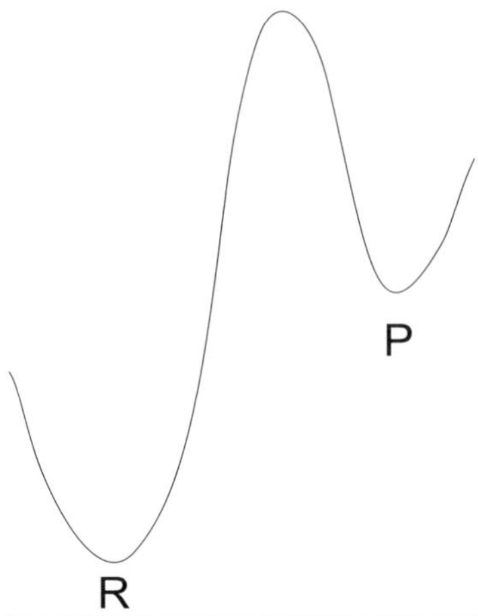

Fig. 6.1 Schematic free energy diagram for a high-pressure reaction. R and P indicate the reactant and product coordinates, respectively.

ergy profile upon an increase of pressure. More realistically, one can imagine a transformation of the free energy profile that is different at ambient and at high pressure, wherein R is the more stable species at ambient pressure, while P is more stable at high pressure according to the diagram of Figure 6.2. Both the forward R→P (at high pressure) and the backward P→R (at

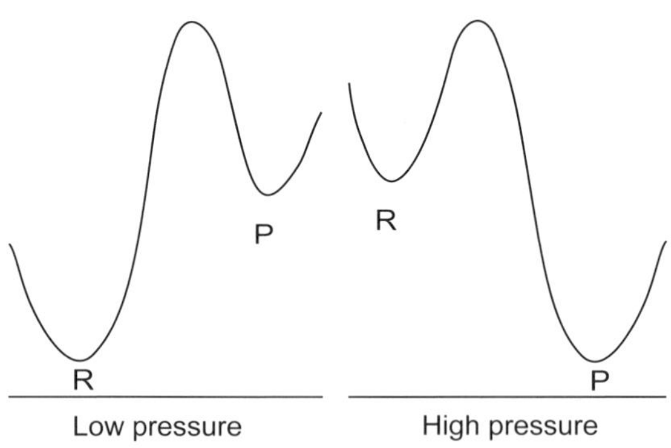

Fig. 6.2 Schematic free energy diagram for a reversible high-pressure reaction. R and P indicate the reactant and product coordinates, respectively.

low pressure) transformation require the overcoming of an energy barrier. If the energy barrier for the backward transformation is small (compared with kT), the process will actually be reversible both thermodynamically and kinetically, although a more or less pronounced hysteresis can be active. When, on the contrary, the energy barrier for the back transformation is sufficiently high, the recovery of the low-pressure material can exhibit a large hysteresis, as observed for non-molecular polymeric forms of low Z molecular crystals such as nitrogen and carbon dioxide, or the product can even be recovered in a metastable state at normal conditions. This latter, for instance, is the case with diamond, whose transformation at ambient conditions to the thermodynamically stable phase of carbon, graphite, is prevented by a large energy barrier. A quite different situation is encoun-

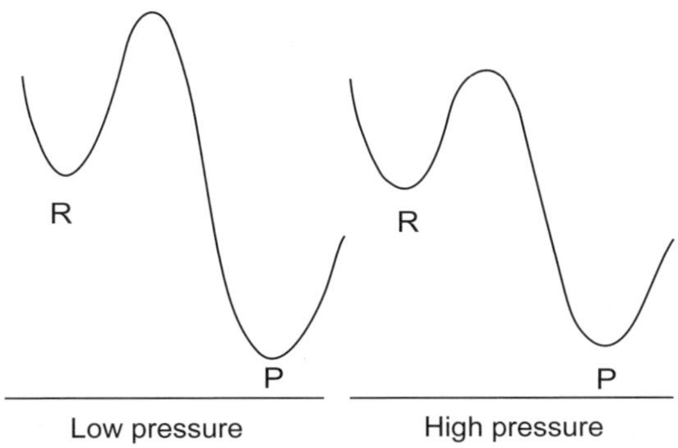

Fig. 6.3 Schematic free energy diagram for an irreversible high-pressure reaction. R and P indicate the reactant and product coordinates, respectively.

tered in cases when the product P is thermodynamically more stable than R, at both low and high pressures, as represented in Figure 6.3. When the low pressure transformation is hindered by energy barriers which are too high, increasing the pressure can act as a powerful activation tool, lowering the energy barriers of the reaction very much like temperature or catalysis in the more usual laboratory practice. In these cases, the reaction will occur irreversibly and the thermodynamically stable product will be recovered at normal pressure.

On the basis of the above considerations, one can distinguish between high-pressure reversible and irreversible transformations. However, the story is considerably more complicated. If pressurization is viewed as a chemical reaction activation tool, one may inquire on the combined effect of pressure and other activation tools such as temperature and photo-activation. Photo-activation has been found to be particularly useful and has been shown to be able to either lower the pressure threshold of the reaction, which is very important for practical applications of high-pressure reactions, or to trigger specific reaction pathways. In view of the peculiar and essential role of photo-activation, one could consider photo-activated high-pressure reactions as a specific category.

The formation of additional chemical bonds (as in condensations and polymerizations) is generally associated with negative reaction volumes and, in several instances, with negative activation volumes as well. Therefore, one expects that condensations and polymerizations should be the rule at high pressures. However, severe electron overlap and subsequent delocalization can give rise to atomization and metallization at high pressure. In addition, the tendency to achieve at high pressure the closest possible packing, can give rise to disproportionation and ionization. There are also several instances where the combined effect of pressure and photo-activation results in the formation of small molecules as is the case, for instance, in the chemical reactions observed in clathrate hydrates.

On the basis of these general considerations, we shall describe chemical reactions at high pressures in molecular crystals tentatively, distinguishing between reversible reactions, irreversible reactions, photo-activated reactions, ionization reactions and the formation of small molecules. This subdivision is only for the convenience of presentation, since the boundaries between the different types of reactions are not always well defined and, in fact, some systems will be discussed under different headings.

6.1 Reversible reactions

6.1.1 *Solid nitrogen at high pressure: the archetypal energetic material*

Interest in the behaviour of solid nitrogen at very high pressures arises from theoretical and practical motivations. The triple N-N bond is one of the strongest known chemical bonds and has a dissociation energy of 226 kcal mol^{-1}, which is significantly more than three times the dissociation energy of the single N-N bond (38 kcal mol^{-1}). Nevertheless, intense experimental and theoretical research activity has been carried out, particularly since the 90s, to ascertain if nitrogen clusters N_n with $n > 2$ (polynitrogen) could be stable or metastable in appropriate conditions or environments [763–767], an expectation supported by the existence and stability of the N_3^- azide ion. In fact, polynitrogen compounds are highly unstable for the dissociation into N_2 molecules and could act as high-energy density materials (HEDM) [768–770]. Several ionic (N_3^- [771], N_3^+ [772, 773], N_4^+ [774], N_5^+ and N_5^- [771, 775–778], N_6^- [779, 780], N_9^+ and N_9^- [781]) and neutral (N_4 [782–789], N_6 [786], N_8 [771, 783, 785, 786, 790–793], N_9 [781], N_{10} [771], N_{20} [794], N_{60} [767, 795]) polynitrogen clusters have been tentatively synthesized or predicted to be stable or metastable species on the basis of first principles calculations. In the polynitrogen clusters, nitrogen atoms are bound by single or by single and double chemical bonds. Since the energetic properties depend on the mentioned difference between the triple and single bond stabilities, it is evident that the ultimate high-energy density material will be an extended polymeric array of singly bonded nitrogen atoms, for which an energy storage capacity in the range 0.75-1.5 eV/atom has been predicted [796], considerably higher than for the best conventional high-energy materials. This polymeric form of nitrogen can be expected to form at high pressure.

Despite the simplicity of the molecular structure, the phase diagram of solid nitrogen is rather complex. This is a typical feature of molecular crystals, where small variations of the intermolecular interactions can produce changes of the crystal packing and phase transitions. At low pressure, quadrupolar interactions dominate in the nitrogen crystal giving rise to the classical Pa3 structure of linear molecules. At higher pressures, the importance of quadrupolar interactions decreases and new phases become stable, where the molecules are orientationally disordered in a spherical or disk-like fashion. The phase diagram of solid nitrogen has been discussed in detail in several places [45, 797–801]. Here, we shall only be interested in the transition from the ζ phase [799, 801–803] to the non-molecular polymeric phase, which is accompanied by a full reorganization of the chemical bonding. Dissociation of the N-N triple bond was first considered in shock-compressed fluid nitrogen above 30 GPa [804–808] and this opened opportunities to obtain non-molecular nitrogen in static experiments. It has been theoretically predicted [809–811] that at high pressure, non-molecular polymeric nitrogen is thermodynamically more stable than the molecular crystal and in particular, Mailhiot et al. [810] predicted a transformation pressure of $\approx$ 80 GPa to the stable cubic gauche form (cg-N), apart from possible kinetic hysteresis due to high-energy barriers. In fact, in early experiments nitrogen was found to be stable in the molecular form at pressures up to 130 GPa [750] or at 150 Gpa [812] at room temperature, despite some significant softening of the vibron, a probe of the weakening of the triple bond. Later experiments showed that at room temperature and above 150

GPa, the vibron band, a signature of the molecular phase, disappears and a transformation to a non-molecular phase occurs [813]. The new phase is semiconducting [814, 815], with an optical gap of 0.6 eV and remains semiconducting up to 240 GPa. The transition pressure is 140 GPa at room temperature and increases on lowering the temperature. The hysteresis is significant and this non-molecular phase was reported to be recoverable at zero pressure below 100 K. Evidence from the broadness of the Raman bands of the new phase (640 and 1450 cm^{-1}) implies that the sample is amorphous and therefore cannot be identified with the cubic gauche (cg-N) form, which should evidently be an insulator. The darkening of the sample [750, 812] can be explained assuming that the sample is an array of nitrogen atoms with single and double bonds.

The cg-N was finally obtained as a transparent material by Eremets et al. by the direct laser heating of molecular nitrogen above 2000 K at 120 GPa, whereas dark phases are obtained on heating at lower temperatures [512]. These results were later confirmed by Lipp et al. [816] who also identified the reddish phase, obtained by heating at 1400 K, as amorphous nitrogen, which finally becomes transparent at 2000 K, as expected for the polymeric nitrogen. The cg-N structure of polymeric nitrogen was definitely established by X-ray diffraction on the single crystal [801, 817]. The crystal unit cell and the Raman spectrum are shown in Figure 6.4.

The cubic gauche nitrogen is, as a whole, well characterized both experimentally and theoretically, although some problems are still unanswered. The crystal structure is a peculiar slightly distorted simple cubic, belonging to the space group $I2_13$ with unit cell parameter a=3.45 Å. From the equation of state measured in the 0–150 GPa range, a value of the bulk modulus B_0=290-340 GPa [817] has been obtained showing that cg-N is a superhard material. A value of the bulk modulus in the same range is obtained from the calculated elastic constants c_{11}=558 GPa, c_{12}=160 GPa and c_{44}=280 GPa [818–821]. Cg-N is an insulator with a wide band gap but estimates of the gap vary from 4 to 8 eV [811, 818]. The evolution of the band gap with pressure has also been discussed [818, 822, 823], and the band gap closure and atomization is predicted at 680 GPa. The vibrational spectra (see Figure 6.4), and its evolution with pressure, have been interpreted consistently with the space group and are in agreement with theory [824, 825], showing that cg-N should be dynamically stable even at very low pressure.

The metastability, stabilization and recovery of cg-N under normal conditions is a key issue for its possible use as a high-energy material. On

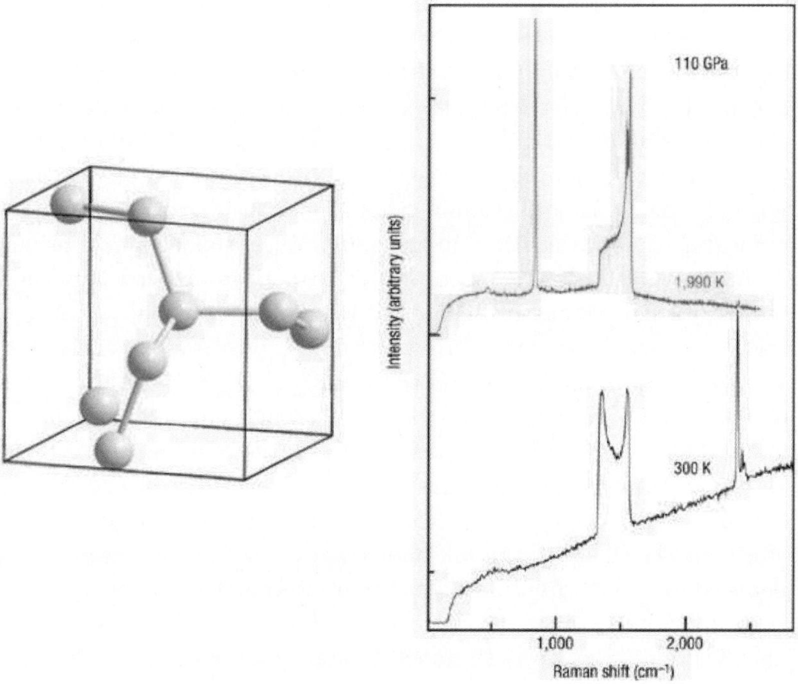

Fig. 6.4 Cubic gauche polymeric nitrogen: structure of the unit cell (left) and Raman spectrum of the stretching vibration (right) in the molecular (bottom trace) and polymeric (top trace) form (reprinted by permission from Macmillan Publishers Ltd: M.I. Eremets, et al. *Nat. Mater* **3** (2004) 558–563, Copyright (2004)).

the other hand [826], modelling of the shock-induced behaviour of cg-N casts some doubt on its practical use as a high-energy material in the pure state, since it predicts a slow transformation to molecular nitrogen through complex intermediate reactions. The role of defects, shear instabilities and surface stability and passivation [827] is thus evidenced. These problems stimulated extensive theoretical research to identify other polymeric phases of nitrogen that could be more convenient alternatives to cg-N for recovery under normal conditions [819, 822, 823, 825, 828–835].

From a perspective of finding less extreme conditions of formation of polymeric nitrogen, it would be important to clarify the atomic mechanism of the transformation. Experiments have shown that formation of cg-N occurs indirectly through an amorphous phase whose molecular structure is not known. In addition, the structure of the molecular phases (ς or κ) from which the cubic gauche phase is obtained is also not known. According to Erba et al. [836] the reaction mechanism does not depend on the structure of the starting molecular phase (either ς or ϵ), which may be surprising considering the collective character of solid state reactions and could imply the close similarity of the starting configurations. Possible pathways from molecular to polymeric nitrogen have been discussed on the basis of topological considerations, assuming that atomic displacements should only be minor during the transition, in the spirit of the topochemical principle. The transition has been discussed on the basis of minimized hypothetical structures of the starting molecular crystals [825, 837, 838]. Attempts to change the transition temperature with the use of common catalysts have not been successful. Of interest is the finding [839] that the transition to the non-molecular phase is accelerated by irradiation at 458 nm, while irradiation at 488 or 515 nm has no effect, but photo-activation of the reaction has not been further pursued. As a whole, knowledge of the transition mechanism seems so far incomplete, at least in the sense that the transition state for the molecular to non-molecular crystal is not known. Several problems connected to the synthesis of non-molecular nitrogen crystal have been discussed in considerable detail by Eremets et al. [840].

Chemical modification of the starting material is a possible route to the attainment of polymeric nitrogen at milder pressure and temperature conditions, than from pure molecular nitrogen. A natural choice in this direction is to attempt with azide salts containing the other known molecular form of nitrogen, N_3^-, where the nitrogen atoms are bound more weakly than in the nitrogen molecule. As a matter of fact, it has been reported [841] that at a pressure of 5 GPa, sodium azide can be photolyzed to give some kind of polynitrogen. Azide salts have a complex phase diagram [842]. The behaviour of sodium azide at high pressure has been investigated by Eremets et al. [843], obtaining results on the N_3^- polymerization later confirmed by Popov [839]. At room temperature and pressure above 50 GPa, the sample darkens and finally becomes completely dark at 120 GPa. The disappearance of the N_3^- vibron peak in the Raman spectrum is an indication of the formation of a non-molecular phase that, by comparison of the Raman spectra, can be identified as the amorphous non-molecular phase. Heating at 3000 K produces crystallization of the sample in the cubic-gauche structure. The reaction is reversible, and molecular sodium azide is obtained after releasing the pressure, although with a large hysteresis. However, the behaviour on decompression is rather complex. It has been reported that shear deformations applied by rotation of the diamond anvils accelerates the transformation which is completed at 80 GPa. An acceleration of the transformation is also realized by laser heating of the sample. Polymerization has also been foreseen in lithium azide at high pressure [844].

6.1.2 *Red oxygen*

Molecular oxygen has an open shell configuration with two unpaired electrons in the degenerate π^* orbitals and therefore has a magnetic moment. The intermolecular potential will therefore include magnetic interactions that give some peculiarities to condensed oxygen as compared with nitrogen. On the other hand, the possible association of oxygen molecules has already been detected in the gas phase, with the observation of $(O_2)_2$ dimers that are in a singlet state [845, 846]. Molecular aggregation should be easier in the solid at high pressure with frustration of magnetic interactions. It is therefore conceivable that high-pressure oxygen will bear only some resemblance with nitrogen. In addition, the phase diagram of molecular oxygen is rather complex. The high-pressure properties of solid molecular oxygen have been reviewed in great detail in a review article by Freiman and Jodl [847]. Here, we shall only discuss the transition to the ϵ phase, which at room temperature is stable above 9.9 GPa up to 96 GPa. A starting experimental observation is that in the ϵ phase, oxygen becomes red and the colour darkens on further increase of the pressure. The intense colouration implies a change of the electronic and, therefore, of the molecular structure. A second experimental finding is the significant volume reduction of ~10% at the $\delta - \epsilon$ transition, which is evidence for some kind of association between the O_2 units. Early X-ray diffraction studies established the monoclinic C2/m structure of the ϵ phase but were unable to resolve the structure completely. A first hypothesis on the association of the O_2 molecules in the ϵ phase was made by Gorelli et al. [483] from the analysis of the vibrational spectra [483, 848–850] and in particular, from the observation of the infrared activity of the vibron mode at ~ 1500 cm^{-1} (actually expected for any kind of cluster [851]), which is strongly enhanced by further increase of the pressure, and by the appearance of a second infrared mode at lower frequency (300 cm^{-1}). These results are consistent with the simplest associated unit: an $(O_2)_2$ dimer with a D_{2h} symmetry. Computational studies based on density functional theory suggested a non-magnetic phase, fully compatible with the reported vibrational spectra, where the O_2 molecules are arranged in linear herringbone-type chains in a Cmcm unit cell [851]. The crystal structure of the ϵ phase of oxygen was finally resolved using X-ray diffraction by Lundegaard et al. [661] and by Fujihisa et al. [852], and is shown in Figure 6.5. The ϵ phase is composed

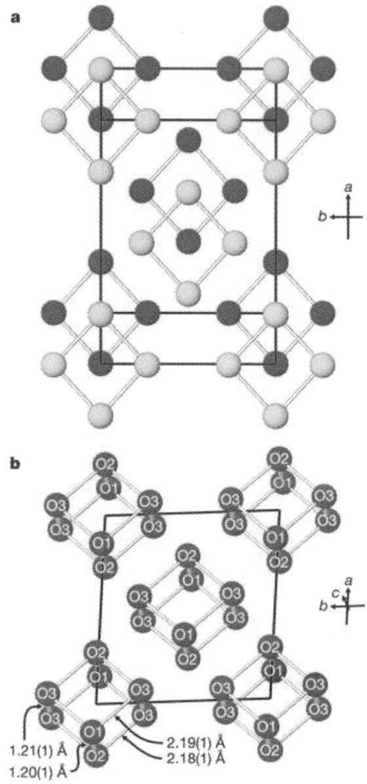

Fig. 6.5 Crystal structure and packing of the $(O_2)_4$ clusters (reprinted by permission from Macmillan Publishers Ltd: L.F. Lundegaard, et al. *Nature* **443** (2006) 201–204, Copyright (2006)).

by distinct $(O_2)_4$ clusters with O-O bond lengths of 1.20 and 2.20 Å. The cluster has a prismatic structure with the basal plane angles of 84° and 96° (17.6 GPa) and D_{2h} symmetry. The vibrational spectra are more complex with four Raman and three infrared active modes observed out of the nine Raman and seven infrared expected from isolated molecule (cluster) analysis. There is a continuity in the structural arrangement in the ϵ and the lower pressure phases of oxygen, in the sense that the prevailing feature is the parallel orientation of O_2 units in planar sheets parallel, in the case of the ϵ phase, to the *ab* crystal plane. The important interaction responsible for the formation of the $(O_2)_4$ clusters has been discussed in terms of π^* - π^* coupling by Steudel and Wong [853], who also report a molecular orbital energy diagram showing that the cluster is diamagnetic.

The formation of a new small molecule of an element at high pressure has been noted as a peculiarity of solid oxygen [853]. However, it will be shown in the following sections that there are other instances in more complex compounds where high-pressure reactions result in the formation of small molecules. It is worth noting the extraordinarily ample pressure range of stability of the ϵ phase.

6.1.3 *Carbon dioxide: a multiform solid*

Carbon dioxide is a chemical compound of widespread occurrence and is involved in many important chemical natural processes. It is a greenhouse gas of primary importance, participates in many biochemical processes and is formed in combustion reactions. Carbon dioxide also occurs in astrophysical ices and as a component of outer planetary atmospheres. As a solid at ambient pressure (dry ice), it is an important refrigerant and as a supercritical liquid, is an interesting solvent for green chemistry applications. The versatility of carbon dioxide is, however, particularly evident in its behaviour as a solid at high pressures. In the pressure range below 40 GPa, carbon dioxide crystallizes in a variety of structures (CO_2-I, CO_2-II, CO_2-III, CO_2-IV) very much like other molecular crystals [854]. In all these polymorphs, carbon dioxide retains the linear centrosymmetric structure. Claims have been discussed that in phase II, a strong dimeric association is at work [855] and that in phase IV, the molecule is bent [856, 857]. However, it has been demonstrated [484] that all the available experimental data can be explained with crystal structures composed by linear molecules. At high pressure (30–80 GPa) and temperatures above 2000 K, carbon dioxide decomposes into carbon (diamond) and molecular oxygen [858]. Above 40 GPa and temperatures lower than 2000 K, solid molecular carbon dioxide transforms into polymeric tridimensional arrays, recalling the structure of SiO_2. The phase diagram and stability boundaries are not well defined in this P-T region for two reasons, the metastability of the various polymorphs, the transformation thresholds depend on the conditions of the experiment (rate of pressurization and heating). By analogy with the many polymorphs of SiO_2, the transformation conditions will be different for the various polymeric structures that can be expected and have actually been obtained for carbon dioxide.

Four different polymeric forms of carbon dioxide at high pressure have been discovered and characterized to have a structure similar to the cristobalite, stishovite, coesite and amorphous silica analogues. A non-molecular phase of carbon dioxide CO_2-V was first synthesized by Iota et al. [859] upon heating the crystal at 40 GPa to 1800 K. Observation of a second harmonic generation and analysis of the Raman spectra in comparison with those of SiO_2 allowed identification of the product as an extended solid with tetrahedral coordination of the carbon atoms. Upon polymerization, a volume reduction of the order of 15% occurs. The new phase was later characterized by X-ray diffraction [860] as a corner-sharing structure of CO_4 tetrahedra and tentatively assigned as a tridymite-like crystal, a bulk modulus of 365 GPa was reported. Later experiments [526], with more accurate laser heating, showed that CO_2-V could actually be obtained at a much lower temperature (640 K) than originally found. The conclusion on the tridymite structure of CO_2-V has been questioned on the basis of several *ab initio* calculations [861–866, 835] that have rather consistently shown that the energetically favoured structure is of the cristobalite type. These calculations also reduced the estimate of the bulk modulus: CO_2-V is still a very hard material but with a bulk modulus in the order of 150 GPa. In more recent experiments [513], CO_2-V has been obtained at 40–50 GPa by direct laser heating to 1500 K and by synchrotron XRD, infrared and Raman spectroscopy and DFT molecular dynamics simulations, it has been demonstrated that CO_2-V actually has a distorted β-cristobalite structure which is shown in Figure 6.6. The tetrahedral angle in the CO_4 units (109.5°) and the tilting angle of the tetrahedra with respect to the c axis (38.4°) have been determined.

An amorphous silica-like structure of polymeric carbon dioxide has been obtained [867] by pressurizing the molecular CO_2-III crystal, at pressures of 40–48 GPa and heating at the boundary of the CO_2-II - CO_2-V transformation (300–680 K), and successive quenching. X-ray diffraction shows that the product is a glass, and infrared and Raman spectroscopy show a structure very similar to those of glassy SiO_2 and GeO_2 [868], as is shown by the Raman spectra reported in Figure 6.7. It was also concluded that this extended amorphous phase is the disordered counterpart of crystalline phase V. From the analysis of the static structure factor, it has been suggested that this amorphous form of polymeric carbon dioxide (called a:CO_2), which is a very hard glassy material if compared with a-SiO_2, could be an admixture of tetrahedral and higher coordination sites for the carbon atoms. A more accurate study of the vibrational spectra [869] and the identification

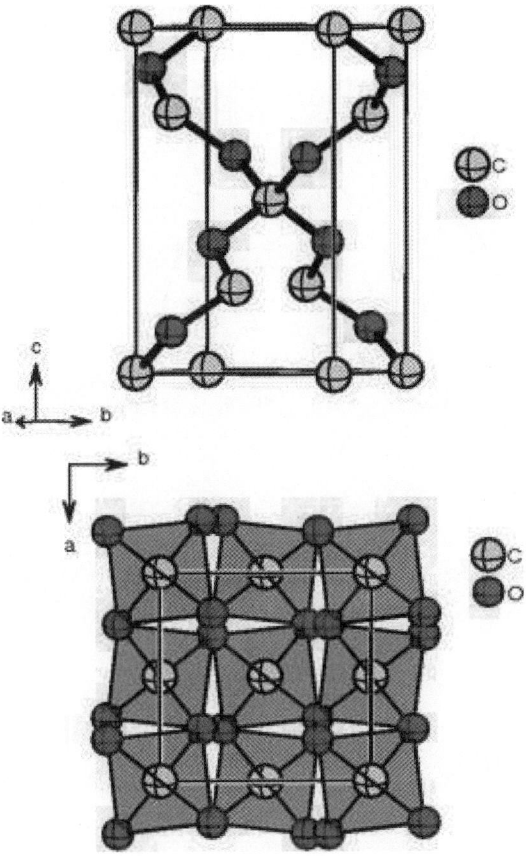

Fig. 6.6 The cristobalite-like crystal structure of polymeric carbon dioxide (CO_2-V) at 43 GPa. The lower drawing shows the structure of the unit cell along the c-axis where the partially collapsed arrangement of the CO_4 tetrahedra can be appreciated (reprinted with permission from M. Santoro, et al. *Proc. Natl. Acad. Sci. USA* **109** (2012) 5176).

of novel bands to be assigned as C=O stretching modes has shown that a:CO_2 contains a mixture of sp^3 and sp^2 hybridized carbon sites.

By isothermal compression of CO_2-II above 50 GPa and temperatures between 530 and 650 K, or by isobaric heating of CO_2-II to 1200 K at 50 GPa, a stishovite-like new form of polymeric CO_2 has been obtained (CO_2-VI) [870]. The partially amorphous solid has been characterized by angle dispersive X-ray diffraction and by Raman spectroscopy, which suggest that the carbon atoms in sp^3 hybridization sit in octahedral sites with a

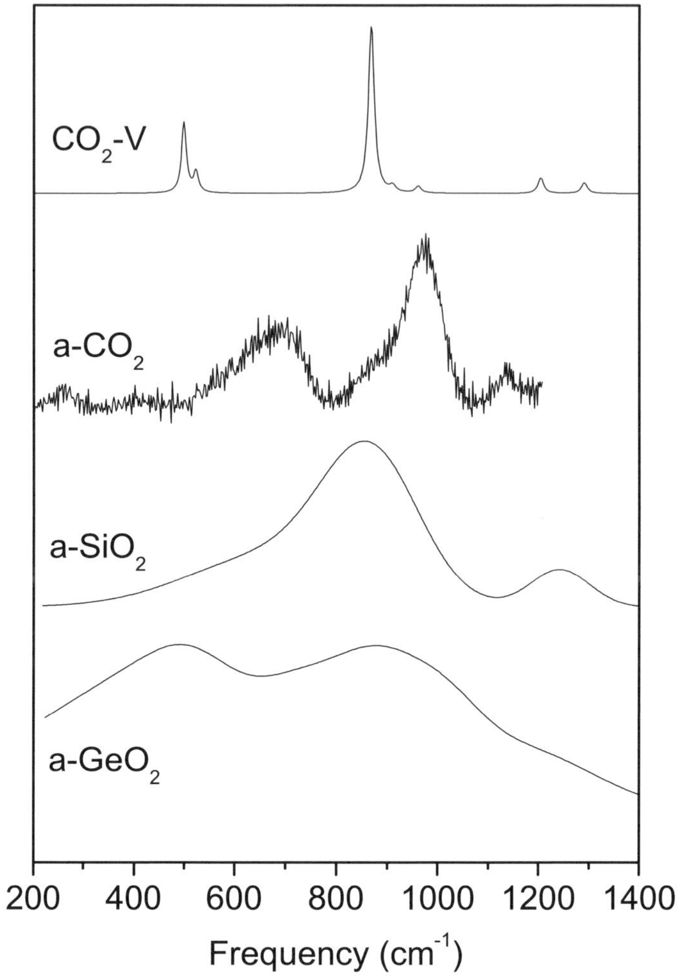

Fig. 6.7 Comparison of the Raman spectra of extended crystalline (phase V) and glassy carbon dioxide at 76 GPa with those of mass and pressure-shifted extended crystalline (phase V), glassy SiO_2 and GeO_2 measured at 66, 27.3 and 56 GPa, respectively, as described in ref. [867].

sixfold coordination. A close structural analogy exists between the atomic arrangements in CO_2-II and CO_2-VI. These results partly disagree with metadynamics simulations results which have identified the sixfold coordinated stishovite-like structure of CO_2-VI, as one of the possible partially polymeric structures intermediate to the attainment of a fully tetrahedral-layered structure (P-4m2) [871]. The calculated X-ray diffraction pattern and Raman spectra of the intermediate states are indeed in good agreement with experimental data reported for phase VI [870].

By laser heating the hexa-coordinated CO_2-VI above 50 GPa to 1300 K, a new polymeric form of carbon dioxide is obtained [872]. From the analysis of the vibrational spectra and correlation with SiO_2, the new form is identified as a coesite-like polymorph denoted as cI-CO_2. Under compression above 65 GPa, the Raman peaks split into two components or broadens. This is taken as evidence of a reversible phase transition to a different coesite-like structure denoted as cII-CO_2. Upon further pressurization above 90 GPa, the Raman signal disappears as an indication of the amorphization of the sample. These new phases of carbon dioxide, like all the other non-molecular phases, finally revert to the molecular phase although with quite large hysteresis. The Raman spectra of these polymeric phases are compared in Figure 6.8 with that of phase V.

The general rule, that at high pressures light elements behave like the heavier elements of the same group of the periodic table, applies with particular evidence to carbon dioxide, which has been found to exist in four different polymorphs with close analogues to silica. The major differences derive from the strength of the C-O bond and the stiffness of the O-C-O bond angle in comparison with SiO_2, thus explaining the low compressibility of the non-molecular polymorphs of carbon dioxide.

Despite the number of available first principles studies, an analysis of the transformation mechanisms from the molecular to the non-molecular phases of carbon dioxide has been attempted only recently [871]. Neglecting in the first instance if the transformation reported experimentally from CO_2-II to CO_2-VI [870] is complete or not, the topology of the crystal arrangement in the two phases clearly shows that small clear-cut rearrangements correlate the two phases. Similar topological correlations between molecular and non-molecular crystal phases have been discussed by Togo [866]. Also of some interest is the comparison reported by Holm [863] of the distribution of the valence electrons' charge density in the isolated molecule, in the dimer and in several crystal phases with that of the SiO_2 quartz.

Besides the fundamental interest of carbon dioxide behaviour at high pressures to understand the bonding properties of atoms and for the search of potentially novel materials, the stability of the non-molecular phases at high pressures and temperatures is of interest in planetary sciences for

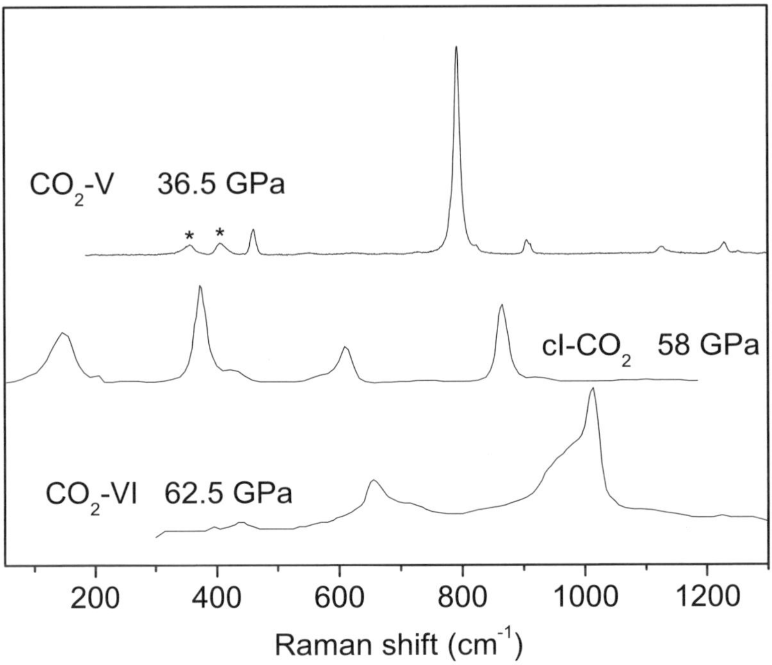

Fig. 6.8 Raman spectra of some extended phases of carbon dioxide. Bands marked by asterisks in the spectrum of CO_2-V are due to remnants of the molecular phases.

the possible stability of carbon dioxide in the Earth's lower mantle or in solution in silica analogues [835, 873].

6.1.4 *Formic acid*

Formic acid, the simplest carboxylic acid, is a case study of considerable interest for high-pressure experiments. At moderate pressures and high temperature, formic acid decomposes, by decarboxylation, to give hydrogen and carbon dioxide or by dehydration to give carbon monoxide and water [874]. It is of interest as a prebiotic material, and for its presence in astrochemical and planetary environments [875]. From a structural point of view, the major intermolecular interaction in formic acid is hydrogen bonding, an interaction that is particularly sensitive to pressurization. In addition, monomeric formic acid can exist in both the *cis* and *trans* conformation of the O-H bond, relative to the carbonyl group [876]. At variance with most carboxylic acids, formic acid does not form dimers in liquid or solid phases. At ambient pressure and low temperature, the crystal structure is dominated by planar chains of hydrogen-bonded molecules with the molecular units in the *cis* conformation [877, 878]. Experiments at high pressure on the solid have suggested that the high-pressure high-temperature crystal structure could be different from that at low temperature, involving a possible change to a *trans* or to a mixed *cis-trans* conformation [876]. However, more recent XRD experiments with synchrotron radiation [879] have shown that the high-pressure and low-temperature ambient pressure crystal structures are the same (orthorhombic $Pna2_1$). The important finding is the observed softening of the O-H stretching mode in the infrared spectrum, with an average variation of about 100 cm^{-1}/GPa, indicating the strengthening of the hydrogen bonds which become symmetric above 15 GPa. Above this pressure, therefore, the distinction between the *cis* and *trans* conformation disappears. At pressures above 40 GPa, the Raman spectrum broadens considerably and this is taken as evidence of a cross linking of the linear chains. The formation of intrachain C-O-C bonds leads to the formation of a three-dimensional polymer. The reaction is reversible although with a considerable hysteresis. In fact, a complete recovery of the original crystal structure of planar chains only occurs at 5 GPa.

6.1.5 *Sulphur. Polymeric and molecular phases*

In all the cases considered so far a tendency to increase the atomic coordination develops as a response to high pressurization. This is quite evident in the polymerization reactions of nitrogen and carbon dioxide, and also in the case of oxygen. In this respect, sulphur at high pressure exhibits some peculiarities, since closer packing is attained preserving a twofold coordination, even though a significant shortening of the intermolecular S-S contact, with increasing pressure, is observed.

At ambient conditions, sulphur crystallizes in an orthorhombic structure (S-I) with the classical crown-shaped S_8 molecular units [880]. Increasing the pressure above 36 GPa, the S_8 ring opens and a polymeric chain with tetragonal symmetry becomes stable (S-III) [881–885]. While the intrachain S-S distance remains similar to that in the isolated molecule, the interchain S-S distance shortens to 2.09–3.02 Å from a value in the range of 3.37–3.50 Å in S-I. If the S-I phase at pressures in the 1.5–3.0 GPa range is heated above 600 K, a new polymeric chain phase with trigonal symmetry (S-II) is obtained which can be quenched at room temperature. By heating above 650 K at 11 GPa, a new phase has been observed (S-VI) which from

X-ray diffraction, has been identified as a rhombohedral structure with S_6 molecular units [884]. The high pressure S-II, S-III and S-VI phases can be quenched at room temperature and are stable (or metastable) over a large pressure range from 83 GPa (or 30 for S-VI) down to about 3 GPa [885]. At pressures higher than 83 GPa, a new metallic and superconducting body-centred orthorhombic phase (S-IV) forms with sulphur atoms arranged in puckered layers [886]. The structures of the various phases of sulphur are shown in Figure 6.9.

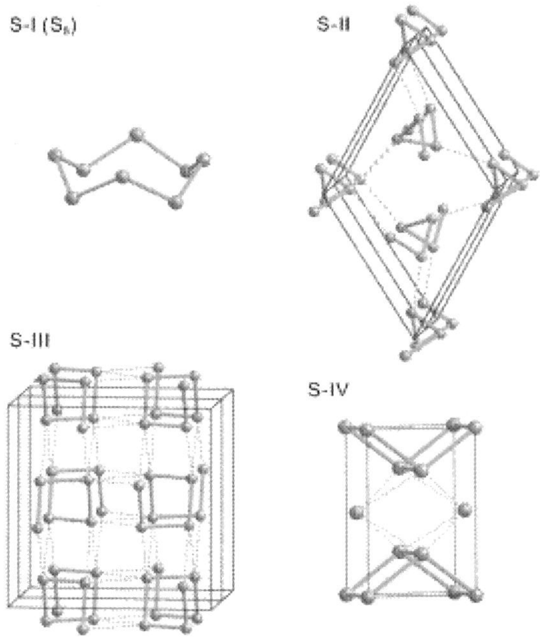

Fig. 6.9 Crystal structure of S-I, S-II, S-III and S-IV crystal phases of sulphur (reprinted by permission from John Wiley and Sons: W. Grochala, et al. *Angewandte Chemie Intl. Ed.* **46** (2007) 3620, Copyright (2007)).

A peculiarity of high-pressure sulphur is the alternating sequence from polymeric to molecular phases depending on the variation of the pressure-temperature conditions [884, 885]. It can be noted that the six-membered ring and the trigonal and tetragonal chain structure, are common features of the heavier elements of the group, Se and Te. On the other hand, it has already been mentioned that helical structures have been found among the theoretically possible structures of oxygen and nitrogen as well.

The complexity of the phase diagram implies that the phase stability boundaries obtained so far are kinetic, rather than thermodynamic boundaries. It is remarkable that the X-ray diffraction sequence, of phases outlined above, apparently disagrees with the sequence obtained from Raman studies when the excitation line is in the blue or green region. This must be interpreted considering that photo-activation at high pressures and temperatures, lowers the pressure thresholds for the structural changes; in fact, it has been shown that the crystal phases obtained from Raman experiments coincide with those obtained in X-ray diffraction experiments, only when using extreme care in selecting the incident wavelength and keeping the laser power as low as a few mW [887]. Photo-activation of chemical reactions at high pressures will be discussed in more detail in a following section of this book.

6.1.6 *Symmetry breaking and ionization at high pressures*

A possible response of a molecular system to pressurization is polarization or ionization, since ionic units can pack more closely than van der Waals aggregates. As a matter of fact, a charge transfer and ionization has also been observed in dense hydrogen at 150 GPa, with the formation of H_2^+ H_2^- pairs [888, 889]. As will be discussed in a following section, first principles calculations of the chemical reactions of unsaturated centrosymmetric hydrocarbons such as ethylene and t-butadiene have shown [207, 757], that before the onset of the reaction a charge separation occurs at high pressure as monitored by the presence of a transient electric dipole moment.

Symmetry breaking ionization has been observed in particular, in nitrogen oxides. Nitrogen dioxide NO_2 at ambient conditions is in equilibrium with the dimer N_2O_4. Another dimer with a nitrite structure, ON-O-NO_2, has been observed in the fluid [890]. On the other hand, nitrogen dioxide is a rather labile species and in rare gas matrices, at low temperature, a spontaneous ionization into nitrosonium nitrate $NO^+NO_3^-$ (which is a further isomer) has been observed [891]. At zero pressure, it has been reported that deposition of N_2O_4 on a cold window at 80 K gives rise to $NO^+NO_3^-$, while deposition at 14 K gives a mixture of isomers, which on heating, convert to $NO^+NO_3^-$ [892]. It has also been reported that temperature treatments and appropriate irradiation at ambient pressure can produce $NO^+NO_3^-$ [893]. There is also evidence that $NO^+NO_3^-$ is the stable structure in oxygen and nitrogen mixtures [894, 895], and therefore it is expected that pressurization could lead to its formation as well.

The first high-pressure (up to 7.6 GPa) experiment on N_2O_4 was reported by Agnew et al. [896]. Two crystalline phases of N_2O_4 have been identified. The α phase is obtained by crystallization under pressure and its structure is the same as the low temperature ambient pressure phase. The β phase is obtained by irradiation of the fluid. While the α phase is stable up to the highest pressure of the experiment, the β phase reacts at 2 GPa to give $NO^+NO_3^-$, which is then stable up to 7.6 GPa. The remarkable point is that in the explored pressure range, a reaction only occurs from the β phase. This has been explained assuming that the reaction occurs topochemically and, probably, through the nitrite isomer as an intermediate. A second remarkable point is that this reaction appears as a unique direct transformation from a molecular to an ionic crystal. It was later demonstrated [897] that at higher pressures, the α phase transforms into $NO^+NO_3^-$, probably through another intermediate γ phase. The crystal structure of $NO^+NO_3^-$, obtained by direct laser heating of nitrous oxide N_2O, has been determined by X-ray diffraction and found to be orthorhombic [898]. The structure is shown in Figure 6.10. The orthorhombic structure of $NO^+NO_3^-$ has been

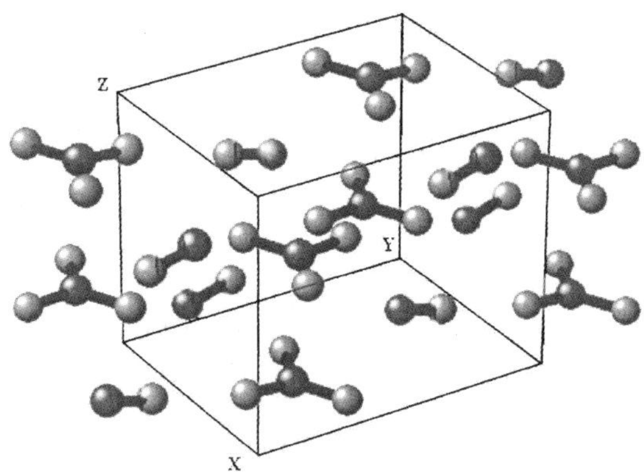

Fig. 6.10 The crystal structure of nitrosonium nitrate (reprinted with permission from Y. Meng, et al. *Phys. Rev. B* **74** (2006) 214107, Copyright (2006), by the American Physical Society).

confirmed in later experiments [899, 900], although there is no consensus on the exact space group.

As noted above, $NO^+NO_3^-$ has been prepared in two other interesting high-pressure experiments. At 10 GPa, under irradiation with a Nd-YAG laser, the ionic compound has been obtained from a N_2:O_2 mixture [894].

The oxygen excess of the mixture can be trapped in the $NO^+NO_3^-$ lattice. In a second experiment [895], irradiation by hard X-rays (10.2 keV) of a N_2:O_2 mixture at 0.5–1.5 GPa, produces the photodissociation of the molecules and the initial formation of the $NO_2^+NO_3^-$ ionic compound, which crystallizes in the $P6_3$/mmc space group. Increasing the pressure above 2 GPa, $NO^+NO_3^-$ is finally obtained.

Nitrous oxide N_2O is isoelectronic with carbon dioxide and some similarities can be expected, although N_2O on the whole, is more molecular and more ionic and at low pressure, the bulk modulus of carbon dioxide (around 90 GPa in phase III) is one order of magnitude larger than for N_2O (see appendix). The phase diagram of N_2O has not been studied in great detail, but similarities with the phase diagram of carbon dioxide have been discussed, at least in the low-pressure low-temperature regime. It has been reported that laser heating at high pressures gives nitrosonium nitrate. The reaction has been discussed in detail by Somayazulu et al. [900] and by Yoo et al. [899], who have obtained consistent results. According to [900], by heating to 1000 K, at pressures below 10 GPa or above 40 GPa, a dissociation into the elements is observed. In the pressure range 10–30 GPa, by heating above 2000 K, the dissociation is again observed, while heating below 2000 K, a new phase is obtained which is stable when the pressure is increased. From the vibrational spectra, it is apparent that the ionic character of the new phase increases with pressure and the new phase has been finally identified as a crystal of $NO^+NO_3^-$ with an orthorhombic structure, as already discussed. The formation of $NO^+NO_3^-$ occurs by the reaction 4 $N_2O \rightarrow NONO_3 + 3N_2$, as is evident from the presence of the nitrogen vibron in the infrared and Raman spectra of the reaction product.

Disproportion into the ionic compound, $CO^{2+}CO_3^{2-}$, has been reported by laser heating a mixture of carbon and oxygen at high pressure [901]. This reaction product is considered as an ionic form of $(CO_2)_2$, a kind of analogue of N_2O_4.

Disproportion and the formation of the ionic crystal of ammonium amide has been predicted, by first principles density functional theory calculations, to occur in solid ammonia at 90 GPa [902]. The ionic salt is predicted to be stable up to $\sim$250 GPa. In the present case, ionization occurs by proton transfer which becomes energetically favourable in the solid at high pressure. A similar transformation to the ionic solid $NH_4^+OH^-$ has been predicted for the ammonia monohydrate crystal at 5 GPa [903].

6.1.7 *Metallization at high pressures*

At extreme high pressures, the response of a molecular crystal to the increased electron kinetic energy can be the full delocalization of the electron distribution and hence the transition to a metallic conductive state. This can be accompanied (but not necessarily) by atomization. In the simplest molecular crystal of hydrogen, metallization at high pressure has been predicted a long time ago [52], although the transition has not been observed in low-temperature ($T<100$ K) experiments pursued up to 300 GPa [904, 905]. A room temperature transformation of solid hydrogen to a conductive phase has recently been reported for pressures larger than 260 GPa [906], and a large hysteresis is reported to characterize the back transition to molecular hydrogen. These results triggered new experiments and computational works whose results do not support the conclusions of Eremets et al. [906]. An updated review of this debate is reported in ref. [907].

In crystals of simple elements [908] and in heavier molecular crystals, the transition to the high-pressure metallic state has actually been observed and a brief review of the results is appropriate here, since the transition involves some interesting changes of the chemical bonding. Evidence in general derives from optical studies, from the temperature dependence of the resistance and from estimates of the gap closure. Metallization has been observed in the rare gas crystal of xenon at room temperature and pressures above 130 GPa [51, 405]. It is remarkable, considering the similarity of the electronic configuration, that a parallel transition has been observed also in CsI at 220 GPa [909, 910].

A metallic state has been known for some time for fluid oxygen [911]. In solid oxygen, metallization has been observed at the $\epsilon \rightarrow \varsigma$ transition occurring at 95 GPa [46, 912, 913]. Experimental evidence [651, 679] and *ab initio* calculations [914–916] show that the $\epsilon \rightarrow \varsigma$ is a first-order isosymmetric transition and that metallic oxygen remains in the molecular state. Evaluation of the bonding character in the metallic state and increased overlap of O_8 clusters have been discussed by Meng et al. [679]. Metallic oxygen is stable at high pressures and *ab initio* calculations by Zhu et al. [917] show that at a very high pressure of 1.9 TPa, metallic oxygen could reconvert to an insulator with a polymeric structure consisting of spiral chains.

Solid sulphur has been reported to metallize at 95 GPa [881, 918–920] in a structure initially supposed to be base centred orthorhombic. The closest S-S distances are in the range 2.082–2.441 Å, to be compared with the intramolecular distance of 2.06 Å in S_8 and the intrachain distance of 1.96 Å at 84 GPa. The actual structure of metallic sulphur was later determined to be incommensurate monoclinic [921, 922] in analogy with corresponding phases of Se and Te.

Metallization at high pressure has been reported in the molecular crystal of hydrogen sulphide, an analogue of water but with much weaker hydrogen bonding determining the crystal structure and packing. Although much simpler than for water, the phase diagram of hydrogen sulphide is still rather complex [923, 924]. It has been reported [925, 926] that above 30 GPa, the sample darkens and finally becomes black, but the transition is reversible. The signature of the molecular phase (the vibrational modes of H_2S) disappear above 42 GPa [925] and this is taken as evidence of the loss of the molecular character leading to a new phase V. The band closure and the transition to the metallic phase (phase VI) occurs at 95 GPa. According to these findings, hydrogen sulphide remains molecular up to 45 GPa, is dissociated between 45 and 95 GPa and becomes metallic at higher pressures. It has been supposed that in the metallic phase, the hydrogen atoms are located in the interstitial sites of a sulphur lattice. As a matter of fact, the formation of short S-S bonds accelerates with increasing pressure [927, 928]. *Ab initio* molecular dynamics simulations [927] suggest that the transition from the molecular to the atomic and to the metallic phase is actually more complex. According to the simulation, phase V in the 45–95 GPa range, is still molecular with short S-S contacts forming dynamically and the disappearance of the molecular vibrational modes is associated with hopping of the hydrogen atoms and formation of the ionic species H_3S^+ and HS^-. The remarkable point is that metallization occurs at the same pressure (95 GPa) as in the sulphur crystal, thus proving that the metallic behaviour is associated with the sulphur lattice forming at high pressure, with the hydrogen atoms intercalated within and between the sheets of sulphur atoms.

Metallization of solid iodine [48, 929] has been reported to occur at 16 GPa. At higher pressure, atomization takes place. The structural transformation occurring during the metallization and atomization is rather complex. However, experiments [49, 930–933] and first principles calculations [195, 934] show that the transition occurs through an intermediate incommensurate phase with atomization at 20 GPa. Metallization initiates in the intermediate incommensurate phase, where the persistence of the softened intramolecular vibration shows that the molecular character is not fully lost.

Actually, in the intermediate phase a distribution of bond lengths between the values in the free molecule and in the atomic phase has been found. A similar high-pressure behaviour has been observed in solid bromine with transition thresholds shifted at higher pressures [852, 934–937]. Theoretical predictions are available for solid chlorine metallization and its atomization at higher pressures [938].

Metallization in halogens is induced by overlap of the electronic clouds and subsequent electron delocalization. Therefore, this overlap could facilitate metallization of halogen-rich, more complex, molecular crystals. Metallization has actually been observed in the molecular crystal BI_3 with onset above 16 GPa [939]. It is of considerable interest that the molecular crystals of hexaiodobenzene [940] and iodanil (tetraiodo-p-benzoquinone) [941] show an insulator-to-metal transition at pressures of 35 and 30 GPa, respectively. From a structural point of view, it has been noted that at 10.3 GPa, the I-I shortest intermolecular contact is already significantly shorter (20%) than the sum of the van der Waals radii. Such shortening of the interatomic separation can lead to a charge transfer which is important explaining the conductivity increase at high pressures. An insulator-to-metal transition has also been observed in the molecular crystals of SnI_4 and GeI_4, with transition pressures of 12 and 16 GPa, respectively [942]. The transition has been explained as arising from the formation of I-I intermolecular bonds leading to the formation of linear polymeric chains. This could also explain the hysteresis of the back transformation upon unloading.

In recent years, interest has been growing in circumventing the difficulties of forcing hydrogen into the metallic state by alloying hydrogen with heavier elements. This stems from suggestions that in hydrogen-rich systems, like the group IV hydrides, chemically bound hydrogen has already undergone a precompression and, therefore, a feasible low-pressure threshold for metallization can be expected [943]. The crystal structure of silane has been determined in the low-pressure regime (10–25 GPa) [607, 944], but is not known at higher pressures. Several first principles calculations of the high-pressure phase transitions of silane have appeared with estimates of the metallization threshold [607, 832, 945–947]. The transformation of silane from an insulating molecular phase to a metal has been demonstrated to occur at 50 GPa [607, 948]. The structure of the conducting phase is interpreted as an ordered interstitial alloy. The hydrogen sublattice is very compact with the volume per hydrogen pair comparable with that for molecular hydrogen at 500 GPa. It has been reported that at 120 GPa, the metallic phase further transforms into a transparent phase. For germane GeH_4 [949, 950], only theoretical estimates of the high-pressure structures and transition threshold (70 GPa) are available. A first principles treatment has also been discussed for stannane [951] and disilane [952].

6.2 Irreversible reactions

In this section, chemical reactions induced irreversibly by pressure with products that can be recovered at ambient conditions will be analysed. As discussed in the introduction to this chapter, focus will be on chemical reactions where the products are in the lower energy state already at ambient pressure, but are prevented from occurring by some high-energy barrier. A typical example is that of polymerization reactions of unsaturated hydrocarbons. Under normal conditions, the dissociation energy of a carbon-carbon triple, double and single bond is typically 198, 145 and 85 kcal/mol, respectively. It is thus evident that the saturation of carbon-carbon bonds is thermodynamically favoured. This will *a fortiori* be true at high pressures on consideration that condensation reactions are associated with negative reaction volumes [378]. At normal pressure conditions, reactions of this type can be activated catalytically or photochemically. High pressure can be seen as an alternative activation tool which delivers energy to the system by mechanical work and forces the energy barriers to be overcome. However, the action of high pressure as an activation tool is extremely complicated. As already discussed, at high pressure the multidimensional energy landscape is modified and new reaction channels, not possible at ambient pressure, may open. Therefore, the high-pressure reaction pathway can differ from those activated catalytically or photochemically at ambient pressure. In addition, geometrical and structural constraints, accompanying the volume contraction, can impose some limitation to the reaction initiation or to the accommodation of products in the high-pressure confined environment. The variance of the scenario is enhanced when pressure and photo-activation are simultaneously used to induce chemical reactions [591]. The interplay and synergy of the two activation tools turns out to sometimes produce unexpected effects, due to the unusual exploration of the energy surfaces.

The details of the high-pressure behaviour of several of the reactions discussed in this section have been summarized elsewhere (see [377, 378, 591] and references therein). The discussion here will be organized for homogeneous categories of chemical compounds in an attempt to sort out analogies and similarities. The focus will be on issues such as:

- Pressure reaction thresholds and reaction evolution with increasing pressure
- Reaction products and selectivity at high pressure
- Reaction mechanisms

The photochemistry at high pressures will be discussed separately, although several reactions have been studied under both high pressure and irradiation.

6.2.1 *Unsaturated hydrocarbons*

We shall first consider high-pressure reactions of prototype unsaturated hydrocarbons with an isolated triple C-C bond (acetylene) or double bond (ethylene), the simplest conjugate hydrocarbon () butadiene and a methyl derivative of the latter two (propene and isoprene). At room temperature their reaction thresholds are 4.2 GPa (acetylene [376]), 3.5 GPa (ethylene [737]), 0.6 GPa (butadiene [629]), 3.1 GPa (propene [746]) and 1.1 GPa (isoprene [953]). On the whole and by comparison with other systems, these reaction thresholds are relatively low. However, a direct correlation between these pressure threshold values and the steric properties cannot be pursued because of the different aggregation states of these systems under the pressure conditions in which the reaction is induced. In fact, while acetylene and ethylene react in the crystal phase, the others are in the fluid phase at the onset pressure of the reaction. In this sense, the only meaningful comparison regards butadiene and isoprene, where the higher pressure threshold of isoprene can be directly related to the presence of the methyl group, which prevents the overlapping of the π bonding densities.

The reactions to be discussed here have been studied with *in situ* monitoring by spectroscopic (infrared and Raman) techniques. Since both Raman spectroscopy and pressure measurement through the ruby fluorescence method require laser irradiation of the sample, some care should be taken when comparing reaction thresholds, since the reactions can be photosensitive. This occurrence has been clearly demonstrated in the case of carbon monoxide [593], which will be discussed in the following sections, where the instability boundary considerably lowers when visible laser light is shone on the sample, even at the very low power required for pressure calibration, by the ruby fluorescence method. Also, the pressure threshold for the reaction has been found to be lower (3.9 GPa) using Raman spectroscopy as a probe [954] than when using infrared spectroscopy [955] and different pressure monitoring [376]. It should also be considered that the experiments have been carried out on polycrystalline samples with a possible wealth of defects, impurities and local strains, which can lower the threshold with respect to a perfect crystal. In the following sections this will be clearly illustrated with reference to some specific experiments regarding the benzene amorphization reaction [642]. In addition, a lowering of the pressure threshold can be produced by the non-hydrostaticity of the applied pressure. The importance of shear strain in high-pressure transformations was first discussed by Bridgman [293], reporting that several systems could be made unstable by deliberately applying shear strains using the rotation of the anvils, one with respect to the other. Consequences of the application of this kind of shear strain in a number of high-pressure reactions of molecular systems have been discussed by Zharov [759] and the general concepts have been illustrated by Gilman [956]. In addition, the relevance of shear strains to induce the solid state reactivity in molecular crystals has been discussed by several authors, also with reference to photo-induced reactions [957–959, 956]. As discussed in Chapter 5, theoretical models have also been elaborated in particular, for the initiation and propagation of solid state reactions [748, 960, 961], on the basis of concepts such as chemical pressure, reaction cavity and mechanical instabilities. These concepts, and in particular those of chemical pressure and reaction cavity, fall within the frame of the topochemical principle [962, 963] and are based on a structural or static point of view of the structure-reactivity relationships. However, as remarked in Chapter 5, the collective character of solid state properties and dynamics are also very important in molecular crystals, despite the large difference between intra- and intermolecular interactions, which apparently preserves several single molecule properties. Collective and aggregate properties have been recovered in theoretical approaches with the introduction of mechanical (elastic) instabilities and phonon assistance concepts [748, 958, 959, 961], to explain the mechanism of nucleation and propagation of a solid state reaction, from the active sites.

High-pressure experiments, as will be discussed in the following sections, have enlarged the implications of the collective character of crystalline material for chemical reactivity. The structural rearrangement preluding the reaction initiation should go beyond the reaction cavity scenario to involve aggregate properties more extensively. As already discussed in Chapter 5, the phonons involvement is not limited to the reaction propagation, but plays the role of a diffusion process for the approach of the molecules to the reaction site and to determine the reaction initiation itself.

The effect of increasing the pressure above the threshold is rather different for the reactions considered here, but a rationale can be found within the topochemical principle.

In acetylene, a polymerization reaction giving polyacetylene, the prototype of conductive polymers, occurs above 3.5 GPa and increasing the pressure, the reaction accelerates and the amount of the transformed monomer increases. In the purely pressure-induced reaction, at each pressure, the reaction proceeds up to a saturation [376, 955]. According to Sakashita et al. [955] at 14 GPa, only traces of unreacted monomer are observed in the infrared spectrum. However, it should be mentioned that in their experiment, some laser irradiation was used for the pressure monitoring. The key feature of the effect of increasing the pressure in the present case is that no profound changes of the end product occur. In Sakashita's experiment, it was found that the relative abundance of the *cis* and *trans* polyacetylene polymer formed in the reaction, changes at high pressure. The *cis* to *trans* ratio (estimated from the integrated intensities of characteristic infrared bands) changes from 1.3 at 6.4 GPa to 1.9 at 14 GPa. In the pressurization of polyacetylene [964], it has been found that the *cis* polymer abundance decreases at high pressure. Therefore, the increase of the *cis* form in the high-pressure reaction should derive from further transformation of the unreacted monomer. However, in the Ceppatelli et al. experiment [376], where laser irradiation was completely avoided, the only product of the reaction is the *trans* polyacetylene polymer, with traces of saturated carbon atoms. Apart from the extent of the transformed monomer, the results at 4.2 and 7.4 GPa are identical. It appears that Sakashita's experiment could have been affected by the laser irradiation. The kinetic analysis of the reaction shows a linear growth of the polymer [376]. The selectivity of the reaction can be explained, as already anticipated in Chapter 5, by considering the known crystal structure of the Cmca phase of the monomer and the theoretical work of Le Sar [965], showing that the shortest C-C intermolecular contact is between non-equivalent molecules in the *bc* crystal plane, which are also arranged in an optimal relative orientation as shown in Figure 6.11. As already discussed in the previous chapter, although the shortest contact, 3.46 Å at 4.2 K and 0 GPa, is estimated to reduce to 3.05 Å at 4 GPa, it is still too large for a significant overlap of the electronic distributions. However, if the lattice motion is taken into account, it can be estimated [376] that at 300 K and 4 GPa, the mean square amplitude

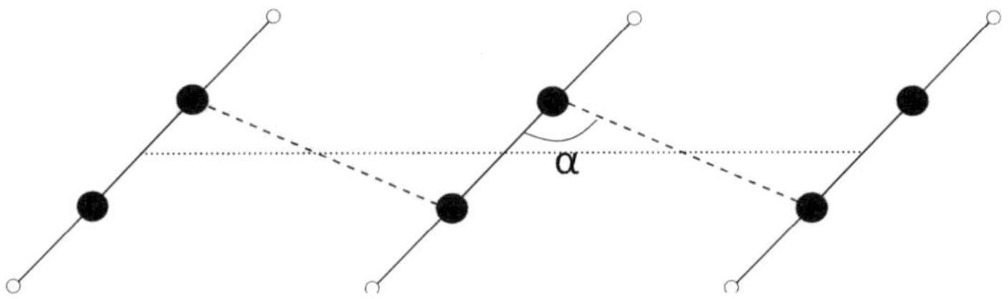

Fig. 6.11 Optimal orientation of the acetylene molecules along the diagonal of the *bc* crystal face for a topochemical addition to *trans* polyacetylene. The dashed lines indicate the shortest C-C contacts whereas the angle α is 109.1° at ambient pressure.

of motion associated with the translational phonons is of the order of 0.5 Å, which can bring the shortest C-C contact below 2.6 Å. Preliminary *ab initio* molecular dynamics calculations show that acetylene molecules in the crystal, readily react only when the C-C separation approaches this value. This is in essence, a dynamic interpretation of the topochemical principle. The high-pressure polymerization of acetylene has also been studied at low temperature (200 K) [376], where the reaction slows down and the transformation only starts at 9.0 GPa. The reduction of the translations' amplitude of motion at 200 K and 9 GPa to 0.2 Å, explains the effect of temperature on the reaction. Additional confirmation of the role of temperature is obtained by the study of the acetylene polymerization at 77 K [966], where the reaction threshold is found to raise to 12.5 GPa. It has also been reported that at this low temperature, the reaction proceeds further upon downloading, very much like has been found for the polymerization reactions to be described in the following section. The interpretation of the pressure-temperature thresholds are, however, carried out in simple terms of the intermolecular distances, without taking the vibrational amplitudes of motion into consideration. The advantage of the possible topochemical pathways of solid state and high-pressure chemical reactions, to obtain even single crystals, has been discussed [965–967].

The high-pressure reaction of acetylene has been studied using first principles [968]. The reaction occurs at a much higher pressure (25 GPa) than observed experimentally. This can be due to the poor representation of the intermolecular interactions or the inadequacy of the simulation duration. It has, however, been found that injection into the simulation cell of an excited bent molecule, corresponding to the geometry of the T_1 state, reduces the threshold to 9 GPa, confirming that the reaction is photosensitive. It is remarkable that the reaction products of the simulation are ordered polymeric chains, albeit with a mixture of *trans* and *cis* conformation and with additional cross linking. The intermediates of the reaction are bent monomers and dimeric aggregates.

Solid ethylene at high pressures reacts to give polyethylene. The reaction threshold has been known for a long time [969], but the reaction product and the reaction mechanism has only been elucidated recently [737]. The reaction has been studied at 3.6 and 5.4 GPa. At the lower pressure, the analysis of the infrared and Raman spectra shows that a highly crystalline polymer is obtained with only very weak bands of saturated carbons, due to the chain terminations. The result is confirmed by the kinetic analysis of the reaction, which shows a linear growth of the polymer. The result is remarkable by itself and, as already discussed in the previous chapter, can be interpreted on the basis of the ethylene crystal structure where, at low pressure, the shortest intermolecular C-C contact is along the a crystal axis which, consequently, is the direction of propagation of the reaction (see Figure 5.2). An ordered growth of the polymer along a unique crystal direction gives rise obviously, in a crystal with a low concentration of defects, to a perfect crystalline polymer with parallel-aligned polymeric chains. This interpretation is fully confirmed by a first principles molecular dynamics simulation [207]. The selectivity of the polymer growth is similar to that observed in acetylene. However, the behaviour at high pressure is distinctly different. In fact, while both acetylene and ethylene are rather anisotropic on compression, in acetylene the direction of the shortest C-C contact does not change, whereas in ethylene at higher pressure, the intermolecular contacts along the a axis (3.527 Å) and along the unit cell diagonal (3.635 Å) become comparable. Therefore, at higher pressure, growth of the polymer in ethylene is possible along different directions, so that a gauche-defected polymer is obtained. The kinetic analysis accordingly shows an increase of the dimensionality of the process at higher pressure. These results are confirmed by *ab initio* simulations [207]. The simulation shows a continuous broadening of the electronic states upon pressurization. The HOMO-LUMO gap at the initial volume (2097 Å^3) of the simulation cell is 3.5 eV and reduces at the cell volume of 727 Å^3 to less than 2 eV, with some novel states within the gap. The change of the electronic structure at high pressure is monitored by the build-up of a transient electric dipole moment that ranges up to 1.25 D, with an average value of about 0.4 D. This dipole moment arises from a transfer of the π electrons and suggests an ionic mechanism that obviously facilitates the linear growth of the polymer. In a qualitative view, the effect of the increase of the pressure is a way to modulate and engineer the crystal structure to change from a high-density to a low-density product polymer.

The high-pressure reaction in solid butadiene and its evolution with increasing pressure, is more complex than in acetylene and ethylene, with the formation of distinctly different products depending on the applied pressure [730, 970]. Although information on the crystal structure of butadiene is only partial [971], the results can reasonably be accounted in terms of structure-reactivity relationships. The reaction has been studied in the 2–7 GPa pressure range at room temperature. Since butadiene spontaneously dimerizes at room temperature once the fluid is compressed above 0.6 GPa [629], the crystals have been prepared at low temperature and pressure, compressed at low temperature to the desired pressure value, and then heated up to ambient temperature. Below 4 GPa, butadiene reacts to give the dimer vinylcyclohexene, as observed in the compression of the fluid at ambient temperature. The result is remarkable by itself for the selectivity of the reaction, considering that in the gas phase and in solution, a mixture of three different dimers is obtained. From the kinetic analysis of the reaction it is found that the reaction is first order and that the rate constant decreases with increasing pressure, therefore indicating a positive activation volume. These findings have been interpreted in terms of the three-step process shown in Figure 6.12. In the first step, a head-to-tail

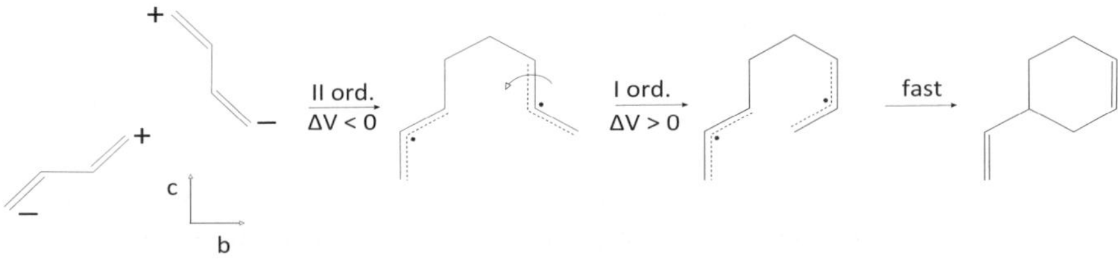

Fig. 6.12 Three-step reaction mechanism for the butadiene dimerization in the crystal at high pressure. Molecules are viewed in the *bc* crystal plane of the monoclinic cell.

attack, involving the two inequivalent molecules of the unit cell and driven by the lattice motions (phonons), occurs to give an open diradical. The rearrangement (internal rotation) of the diradical is the second step of the reaction, preluding the fast ring closure. It must be assumed that the diradical rearrangement, which is hindered by geometrical constraints in the crystal, is actually the rate-limiting step and this would explain the first order of the reaction and the positive activation volume. The dimerization rate suddenly drops above 3 GPa, to increase again for pressures higher than 4 GPa, where the formation of polybutadiene also starts to be observed. In these conditions, the dimerization rate increases almost linearly with pressure to again reach the rate value measured at low pressure, whereas the ratio of the polymer to dimer also increases with pressure. These results can be rationalized assuming that the linear polymer grows along some specific crystal direction, while dimerization occurs between non-equivalent molecules in a possible herringbone-type structure. The formation of the polymer releases in some way, the geometrical constraints, enlarging the reaction cavity for the dimerization and this could explain the increase in this pressure range of the dimerization rate. The butadiene reaction has been studied by first principles molecular dynamics in a disordered sample [757]. Since pressure in the simulation is very high, only the polymer formation is observed. An interesting finding is that a charge separation is induced by the high pressure with the formation of a transient dipole moment (with an average value of 0.7 D and a distribution up to 3 D). Therefore, polymerization occurs by an ionic mechanism which favours the growth of a linear regular polymer. It is also found in the simulation that at increasing pressure, a weakening of the double bonds occurs while the central single bond shortens, very much as is expected to occur in the first excited state.

The high-pressure reactivity of the methyl derivatives differs to a considerable extent from that of ethylene and butadiene. This could be due to the steric hindrance of the methyl group or to the different molecular packing in the condensed state. The reaction of propene has been studied at various temperatures in the liquid phase and the pressure threshold decreases with increasing the temperature from 11.3 GPa at 220 K to 1.5 GPa at 340 K [746]. No polymer is obtained by reacting propene in the experimental conditions considered and the product is a viscous material which, from the infrared spectra, can be characterized as a mixture of linear and branched oligomers. The transformation of propene is always incomplete. The kinetic analysis of the reaction shows a linear growth of the oligomers and a negative activation volume with a parabolic dependence of $\ln k$ on pressure. The activation volume has a value of -19 cm^3mol^{-1} at ambient pressure and of -5.1 cm^3mol^{-1} at 2 GPa, and becomes positive above 2.7 GPa, implying that the reaction is accelerated by compression at low pressure but is retarded at higher pressures. The formation of oligomers has been confirmed by a molecular dynamics simulation in a disordered propene sample [756]. The interesting result of the simulation, apart from the ionic character of the reaction, is the high collective character of the process. The reaction occurs through electron transfer and hydrogen migration and involves several molecules in the sample.

The high-pressure reaction of isoprene is of interest not only for a comparison with butadiene, but also because isoprene contains the basic structural motif of natural compounds like the terpenes and is the monomeric unit of important polymers. The high-pressure reaction of isoprene was first studied by Conant [106], while the effect of pressure on the isoprene dimerization has been studied by Walling and Peisach [972]. Recently, the reaction has been reconsidered by Citroni et al. [953] at four different pressures (1.2, 1.5, 2.0 and 2.6 GPa). The reaction product is a homogeneous transparent material consisting of a volatile component and a rubber-like solid. Analysis of the infrared spectra shows that the volatile component is a six-membered dimer, sylvestrene. The solid component is a mixture of *cis*-1,4- and 3-4-polyisoprene polymers. The activation volume of the dimerization reaction is negative at low pressures, but becomes positive above 3 GPa. This is evidence of an open chain diradical intermediate, very much like has been found for the butadiene dimerization. The similarity with the solid state reaction of butadiene is remarkable.

6.2.2 *Aromatics and heteroaromatics*

Aromatic systems, particularly in the crystalline state, are substantially more stable upon pressurization than the unsaturated hydrocarbons considered so far. For instance, it has been found that unreacted benzene is still present when the sample is pressurized to 50 GPa at room temperature [741]. This is one aspect of the general stability of aromatics under different thermodynamic or chemical environments, which makes aromatic, polyaromatic hydrocarbons (PAHs) and carbon aggregates, materials of widespread occurrence in the universe [973].

The high-pressure reaction of benzene is probably the reaction most carefully investigated under static high pressure, both experimentally and theoretically. The results obtained have been interpreted with concepts and models that are sufficiently general to be extended and applied not only to other aromatics, but also to many kinds of high-pressure reactions. Because of its relevance for the high-pressure reaction of benzene, it is appropriate to recapitulate the essentials of the benzene phase diagram, a story as old as a century, since the first determination of a polymorphism by Bridgman in 1914 [974]. In an isothermal compression at room temperature, a succession of crystalline phases (denoted as benzene I, II, III, III' and IV according to the nomenclature of Cansell et al. [975]) has been reported [23, 975] with transitions, in the order, at 1.4, 4, 11 and 24 GPa. The structures of phase I (orthorhombic space group Pbca) and of phase III (monoclinic space group $P2_1/c$) have been determined [976] from X-ray diffraction. The other phases have been identified, on the basis of the infrared and Raman spectra, from discontinuities in the slope of the vibrational frequencies and intensities as a function of pressure [975, 23], but their structure has not been determined experimentally and only scarce theoretical estimates are available [977–979]. Such a multitude of crystalline phases is a common feature in the weakly bound molecular crystals, as has been seen already for the diatomics and triatomics discussed in a previous section. The possible stability of several crystalline phases in solid benzene has also been discussed more recently on the basis of molecular dynamics simulations [980], using a model potential and of *ab initio* energy minimization [981]. However, the identification of these predicted phases with the experimental ones is not certain. In all the high-pressure experiments on benzene, it has been noted that the transitions are very sluggish. This led to an attempt to improve the quality of the crystalline sample in the diamond anvil cell by isobaric annealing at high temperature, within the stability range of molecular benzene [208]. As a result, it has been found that upon annealing, the infrared spectra simplify and no traces of phases III and III' are actually found. The monoclinic phase II appears to be stable up to the reaction threshold. Its structure is monoclinic and the space group, $P2_1/c$, is that indicated for phase III by Thiery et al. [976], which in their work assigned, as belonging to phase II, a stressed crystal where the I and II ($P2_1/c$) phases coexist up to very high pressure if thermal annealing is not performed [208]. The simplified phase diagram of benzene shown in Figure 6.13 resembles, at moderate pressures, that originally suggested by Block and Piermarini [982, 983]. This conclusion is supported by recent X-ray studies by Katriusiak et al. [984] of the intermolecular interactions in crystalline benzene, and by Zhou et al. [985] on the variation of Raman intensities with pressure.

The high-pressure chemical reaction of benzene has been studied at room temperature [741, 986] and at various temperatures by Cansell et al. [975] and by Ciabini et al. [642]. According to Cansell, different products are obtained at different pressure-temperature conditions, a result that has not been confirmed by the later experiments of Ciabini. The product of the reaction is an amorphous hydrogenated carbon. No loss of hydrogen has been observed in the reaction and, from the infrared spectra, an sp^3

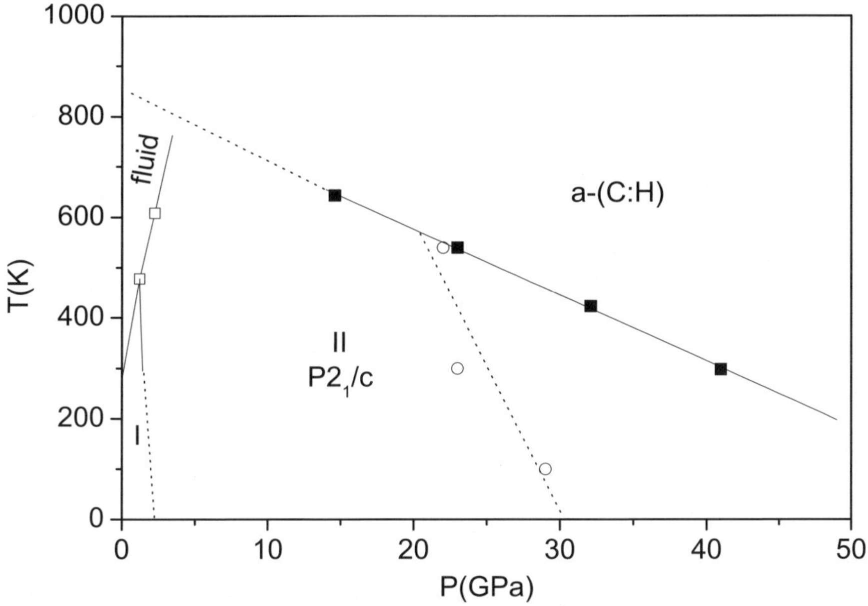

Fig. 6.13 Phase and instability diagram of benzene. P-T points where the reaction threshold has been identified in annealed (full squares) and not annealed samples (empty dots) are reported. Lines indicating the boundaries are reference guides.

to sp^2 carbon atom ratio has been estimated to be 3:1. An amorphous carbon obtained by chemical vapour deposition, with an sp^3 to sp^2 carbon atom ratio of 3:2, has an infrared spectrum quite similar to that obtained at high pressure [741]. The main properties of the product are on average: refractive index n=1.75, density ρ=1.39 g/cm^3 , bulk modulus B_0=80 GPa and an optical gap of 2.5 eV. These properties clearly indicate the non-molecular character of the product. Experiments with accurate annealing of the monoclinic crystal phase resulted in important consequences on the reactivity of benzene. In previous reports concerning not annealed samples, the room temperature reaction threshold was located at 25 GPa [741, 975, 982, 986], while in the annealed sample the threshold rises to 40 GPa [642]. The experiment thus exposes once again, the already mentioned active role of crystal defects or impurities on chemical reactivity in the solid state [987].

Concerning the mechanism of the chemical reaction, the first important observation is that in all the conditions the reaction proceeds up to a saturation, with detectable amounts of unreacted benzene even at 50 GPa. The incompleteness of a high-pressure reaction is generally ascribed to high geometrical constraints against the accommodation of the product in the confined crystal environment. In Figure 6.14, the IR spectrum of benzene is reported as a function of pressure, along the compression and decompression cycles of the sample. As can be deduced by the intensification of the C-H stretching band due to saturated carbon atoms, the reaction is accelerated on releasing the pressure and proceeds to the full consumption of benzene, with the product being recovered at ambient conditions. The enlargement of the reaction cavity with releasing pressure, in order to permit the hybridization change from sp^2 to sp^3 of the carbon atoms, is actually the issue explaining the reaction evolution.

Speculation about the possible involvement of some electronic excited state in the reactive process [979] was advanced on the basis of the pressure behaviour of the Raman frequencies in the ground and excited state. The role of the excited states, in general and in the specific case of benzene, will be discussed in detail later, in the section on photochemistry.

The onset of the reaction was determined through the infrared spectra of isothermal compression experiments by Ciabini et al. [642]. The reaction was located at: 41 GPa, 298 K; 32.1 GPa, 423 K; 23 GPa, 540 K and 14.1 GPa, 643 K (see Table 5.4). These data have been decisive in elucidating the reaction mechanism. Using the known equation of state and the variation with pressure of the lattice parameters, the closest C-C intermolecular contact could be evaluated. As already discussed in Chapter 5, it turned out that the closest contact was always between molecules along the *a* crystal axis ranging from 3.02 Å at the highest temperature, to 2.73 Å at room temperature. It should then be considered that the molecules vibrate in the lattice by the translation and libration phonons, with an amplitude that is temperature dependent. When the closest intermolecular contacts are corrected for the mean amplitude of motion, as has been discussed in Chapter 5, it is found that at all the temperature-pressure conditions investigated, the reaction occurs when the intermolecular contact lowers below 2.6 Å. This is a nice result in the spirit of the topochemical principle and shows that the collective motions of the crystal (phonons) play a key role in approaching the molecules and give rise to initiation centres. A further insight in the reaction process has been gained from *ab initio* molecular dynamics simulation and the trajectories of the simulation are particularly revealing. A typical trajectory of the molecular dynamics simulation is shown in Figure 6.15. Adjusting the simulation conditions such that the closest

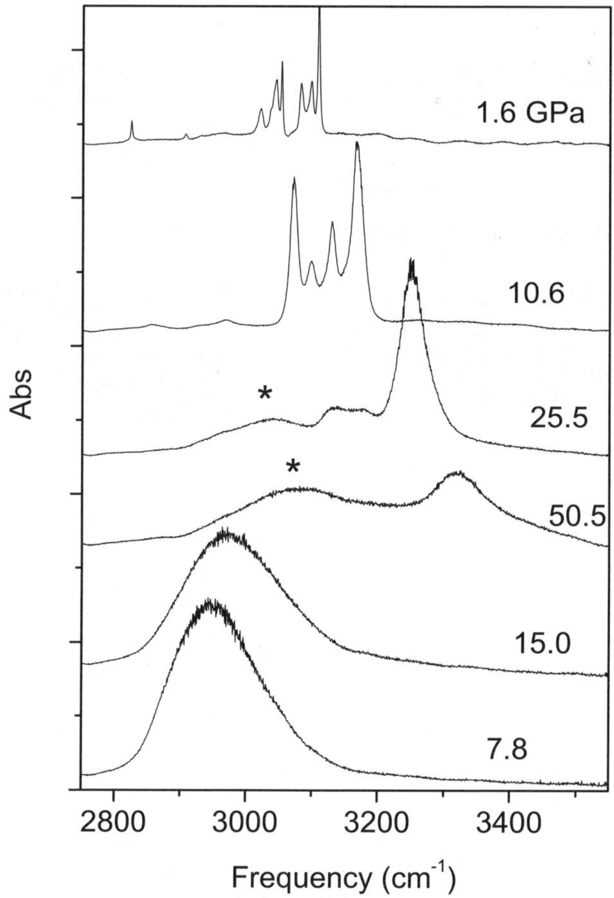

Fig. 6.14 Evolution of the benzene amorphization reaction on increasing and releasing the pressure from the infrared spectrum in the C-H stretching region. The absorption band marked by the asterisk is related to the formation of the amorphous network.

carbon-carbon contact is on average 3.0 Å, the contact oscillates around this mean value and the overlap of the electronic clouds is negligible for 400 ps. As soon as an instantaneous fluctuation lowers the contact below 2.6 Å, the electronic clouds overlap significantly and instantaneously the C-C contact drops to 1.5 Å, and a benzene dimer forms with a zwitterion charge. From this point, the reaction rapidly propagates in all the sample. From the simulation, the reaction threshold and the C-C contact acquire a real molecular meaning. These results show that the reaction proceeds

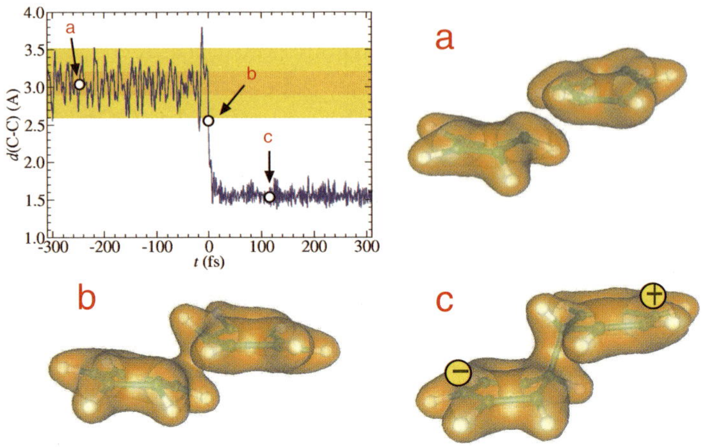

Fig. 6.15 A trajectory of the molecular dynamics simulation of the benzene amorphization showing the formation of dimers as initiators of the reaction (reprinted by permission from Macmillan Publishers Ltd: L. Ciabini, et al. *Nat. Mater.* **6** (2007) 39, Copyright (2007)).

through the formation of dimers as intermediates, which is not an unexpected finding in a high-pressure reaction. We have already discussed the formation of dimers in the high-pressure transformations of butadiene and isoprene. A variety of benzene dimers have been considered as possible van der Waals associates or stable compounds in appropriate conditions (see, for example, discussions in [984, 988–991]). The formation of dimers as intermediates has also been suggested for the high-pressure chemical reactions of dibromoethylene and tribromobenzene [992].

The high-pressure behaviour of linearly condensed polyaromatic hydrocarbons has been discussed by Drickamer [12]. Apart from the observed red shift of the electronic transitions, it has been found that in the case of pentacene, the electrical resistance greatly decreases with increasing pressure and that finally, metallic behaviour is reached. Although these conclusions have been recently refuted on the basis of spectroscopic data [993], association processes and electron delocalization have been confirmed. The structure of acenes at high pressure has not been fully characterized [751, 994, 993, 995]. With reference to the previous discussion on benzene, it is well known that dimerization of polyaromatics is easily photo-activated. However, dimerization is also induced by pressure, particularly in the presence of defects [996–999]. Dimerization in acenes generally occurs by a [4 + 4] cycloaddition reaction [1000]. In the case of benzene, according to the presently available evidence, the dimer has a stacked slipped conformation similar to the common van der Waals dimer [1001].

The features of the high-pressure phase diagram and chemical reactivity of benzene are to some extent shared by the single ring heteroaromatics pyridine, furane and thiophene. The lower symmetry compared with benzene and the subsequent different crystal packing can, however, produce differences that can help elucidate the mechanism of the ring opening. On the other hand, the internal strain due to the presence of the heteroatom, in particular in the five-membered rings, can lower the reaction threshold and this can be beneficial in studying the reaction mechanism.

The phase diagrams of these heteroaromatics [747, 1002–1007] have a complexity similar to that of benzene, particularly in the case of pyridine. A significant difference is that in pyridine, a metastable glassy phase exists and can persist up to high pressure. Room temperature compression of the fluid does not produce crystallization of phase I, which is the thermodynamically stable phase up to 1 GPa. In fact, the fluid crystallizes directly into phase II, which is the high-pressure thermodynamically stable phase, or can even solidify as a glass [747]. This is actually extremely important from the reactivity point of view because no reaction is observed, at least up to 30 GPa, when the glassy phase is compressed, whereas a reaction threshold pressure of 22 GPa is observed on pressing phase II [747, 1002]. This threshold is essentially the same as in benzene, in keeping with the similarity of structure, while it is significantly lower in furane (16 GPa) [1004] and in thiophene (10 GPa) [1006, 1007], as expected. As found in benzene, the high-pressure chemical reaction proceeds up to saturation. However, in pyridine the reaction is not accelerated on downloading [747, 1002]. The explanation is that while in benzene the reaction occurs as a structural feature of the crystalline array, in pyridine the reaction occurs at lattice defects. This is supported by the finding that the annealing of the sample decreases the reaction extent [747], an occurrence not observed in benzene. On the whole, a considerable amount of unreacted pyridine remains at the end of the reaction and this makes a full characterization of the product difficult. In any case, the reaction product is quite similar to the product of benzene amorphization. A remarkable observation is that upon pressurizing the glassy phase to 25 GPa, no reaction is observed and on downloading, the initial sample is completely recovered. The absence in the glassy phase of a chemical reaction has also been observed in propene [746] and, as discussed in Chapter 5, this has to be ascribed to the lack of diffusivity in the glass and to the lack of collective motions that approach the molecules to originate initiation sites. It can also be noted that in pyridine, the closest intermolecular contacts are not between parallel-stacked molecules [1003], as in benzene, and this is unfavourable for the formation of dimers or excimers. In addition, each pyridine molecule is hydrogen bonded to four other different molecules, and the constraints posed to the molecular motion by the tight H-bond network makes the formation of reaction nuclei extremely difficult.

In furane and in thiophene, the reactions reach a saturation depending on the pressure and then accelerate on downloading. In furane, the reaction product is rather similar to that found in benzene, but in addition, bands due to O-H stretching, to C=O stretching and to polyalkyether vibrations are observed. These bands and the formation of saturated carbon group vibrations below 3000 cm^{-1} give clear evidence of the ring opening. In thiophene the reaction product is a composite. The recovered product, besides the presence of unreacted liquid thiophene, has a fibrous appearance and has some vibrational features which resemble those of polythiophene and others that can be assigned to a polymerized non-cyclic product [1006, 1007].

It can be of interest in the present section to mention the results on high-pressure behaviour of the indole crystal which bears similarities with aromatic hydrocarbons. It has been reported [258] that crystalline indole has a remarkable chemical stability at pressures up to 25 GPa. Only traces of an amorphous product have actually been observed in the recovered sample upon downloading. The stability of indole has been associated with its particular crystal packing, which is dominated by C-H$\cdots\pi$ hydrogen bonding which prevents, as previously discussed in the pyridine case, the formation of appropriate intermediates. This result confirms the conclusion that high-pressure induced reactions in condensed phases of aromatics occur when the formation of precursors, from the superposition of π electron densities, is possible.

6.2.3 *Miscellanea*

The high-pressure reactions discussed in the present section are concerned with molecular systems that do not constitute a homogeneous class of compounds. However, in many cases, a comparison or correlation with the prototypical systems already discussed can be found on the basis of various structural features. In some instances we shall deal with isoelectronic systems, such as the dyads carbon monoxide-nitrogen and hydrogen cyanide-acetylene. In other instances we shall consider compounds with more than one of the reactive centres considered previously, as in the case of substituted acetylenes (cyanoacetylene, cyanogen, phenyl- and diphenylacetylene) and styrene. The discussion will focus in particular, on analogies and correlations concerning the reaction mechanisms.

Carbon monoxide is isoelectronic with nitrogen, but the intermolecular interactions and the crystal packing can be different because of the polarity of CO and the well-known head-to-tail disorder. Nevertheless, early determinations of the phase diagram by X-ray diffraction [721, 1008], showed that the phase diagram of carbon monoxide closely resembles that of nitrogen below 4 GPa, whereas above this pressure carbon monoxide reacts at ambient temperature, in contrast with the extreme stability of the different N_2 crystal phases. Also, in carbon monoxide, an extended phase which is sometimes indicated as polymeric, forms at high pressures [1008, 1009] and can also be recovered at ambient pressure, but easily reacts with atmospheric water when the high-pressure cell is opened. One major problem in high-pressure studies of carbon monoxide is that the crystal is highly photosensitive and, therefore, determination of thresholds or of the reaction products can be affected by the mild irradiation unavoidable in X-ray experiments, in Raman spectroscopy and in pressure monitoring by the ruby fluorescence method. Recently, the phase diagram and the chemical reaction, have been revisited with infrared spectroscopy using an infrared sensor for the pressure monitoring, in order to completely avoid any photophysical effects [593]. As shown in Figure 6.16, the known phase diagram has been substantially confirmed, but with all the stability boundaries slightly shifted to higher pressures. For instance, the room temperature reaction threshold is located at 6.0 GPa (instead of 4.2 GPa). In these more recent experiments, the high-pressure reaction has been studied at temperatures of 100, 200, 300 and 400 K, thus accessing the extended phase from different molecular crystal phases. Polymeric CO (p-CO) is an extended amorphous solid which can be recovered at ambient pressure in the high-pressure cell. p-CO easily decomposes upon irradiation to carbon dioxide and glassy carbon, with an energy content comparable with that of common explosives, and it has been shown that p-CO is eligible as an advanced energetic material [1010]. Once the high-pressure cell is opened, the product readily reacts with atmospheric water to give carboxylic acids. Information on the structure of p-CO has been obtained from spectroscopic (infrared and Raman) data. The end product of the reaction depends on the temperature-pressure conditions. Qualitatively, one can expect that

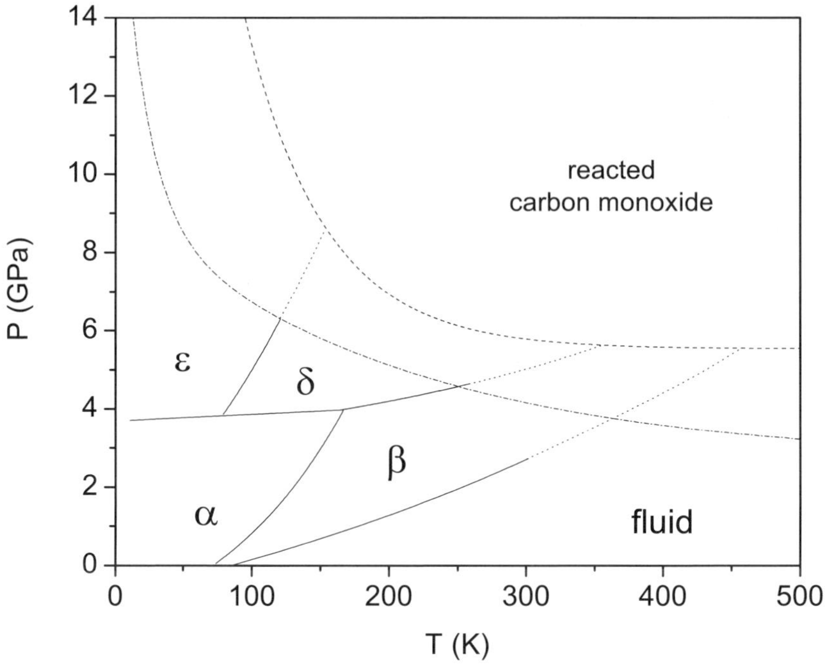

Fig. 6.16 Phase and stability diagram of carbon monoxide (Readapted with permission from M. Ceppatelli et al. *J. Phys. Chem. B* **113** (2009) 6652–6660, Copyright (2009) American Chemical Society). The dashed line indicates the stability boundary when irradiation of the sample with visible light is completely avoided, the dash-dot line is the instability boundary obtained when the sample pressure is measured through the ruby fluorescence method.

while increasing pressure should favour condensation (polymerization) processes, an increase of temperature should favour dissociation. Findings for the carbon monoxide reaction seem in agreement with this trend, as if the major effect of changing the conditions were due to temperature [593]. At temperatures below 300 K, the product is formed by polycarbonyl chains containing anhydride groups whose disorder is favoured by the intrinsic head-to-tail disorder in the CO crystal. The nature of the product is confirmed by the results of *ab initio* molecular dynamics simulations [1011], showing that the product is made of polycarbonyl-interconnected chains with attached five-membered cycles, involving a C-O-C feature and two side carbonyl groups. At temperatures above 300 K, the formation of carbon dioxide and of epoxy rings is observed. An open issue is whether these new products result from a different reaction mechanism, which is activated at high temperature or from a decarboxylation of the same p-CO formed at lower temperatures, although the latter alternative seems more likely.

Nitriles are characterized by the -C≡N group which, from the point of view of the electronic structure, bears resemblance with systems that have already been dealt with. The simplest molecule containing this group is hydrogen cyanide, H-C≡N, which is one of the ten electron molecules (considering only the outer shell) and is isoelectronic with nitrogen, carbon monoxide and acetylene. With the exclusion of cyanogen N≡C-C≡N, and other centrosymmetric molecules with two or more nitrile groups, nitriles have a high electric dipole moment (2.984 D for HCN) and the consequent electrostatic interactions and hydrogen bonding, whenever possible, heavily affect the crystal packing and the high-pressure behaviour. This is quite evident in hydrogen cyanide, which at ambient pressure crystallizes in a tetragonal structure with a head-to-tail alignment of the molecules along the unique axis [738]. Light distortion into an orthorhombic structure occurs at low temperature. The same behaviour is observed at room temperature, as a function of pressure, with crystallization in the orthorhombic phase at 0.2 GPa and a transition at 0.8 GPa to the tetragonal phase stable up to 1.3 GPa. Above 1.3 GPa, a reaction occurs with a colour change of the sample to red and black [1012, 1013]. The structure of the polymer is not known exactly. However, in the infrared spectrum of the product, the C≡N stretching mode is still present, together with absorption bands relative to N-H stretching and bending modes. The infrared spectrum is quite similar to those of some azulmic acids and in particular, to azulmic acid 5 of Volker [1014], which is a polymer with repeating unit -(H_2N)C(C≡N)-. The high-pressure reaction of HCN shows similarities with the reaction in solution or in liquid [1015]. The reaction has been studied by first principles simulations [1016, 1017]. It has been found that polymerization occurs at a much higher pressure than observed experimentally and various explanations have been found for this discrepancy. From the simulation, several possible structures of polymerized HCN have been found, but the absence of -C≡N groups rule them out as candidates for the product obtained in the high-pressure experiment.

Acetonitrile is known to trimerize into 2,4,6-trimethyl-1,3,5-triazine in liquid mixtures with methanol, at moderate pressures of 0.6–0.7 GPa [1018]. In the solid at room temperature, a polymerization reaction has been found [738, 1012, 1013] to occur at higher pressures (15 GPa), but the structure of the reaction product has not been reported. A discussion of the high-pressure conditions (including reaction thresholds) for the polymerization of various other organic nitriles has been reported by Zharov [759].

Cyanogen and cyanoacetylene are isoelectronic molecules and are isoelectronic with diacetylene as well. Their behaviour at high pressure is of interest because of their simple structure and crystal packing that, together with the nature of the polymerization product, allow some good inferences on the high-pressure reaction mechanism. At ambient temperature, solid cyanogen has two phase transitions at 0.5 GPa and 2.0 GPa. Above 3.5 GPa, the pale yellow colour of the sample intensifies with pressure and at 7 GPa, the lattice modes of the molecular crystal cannot be detected anymore [1019]. The transition is reversible with a 1 GPa hysteresis. From the persistence of the vibration of the nitrile group and appearance in the infrared spectrum of bands due to -N=C- bonds, the product is identified as poly(2,3-diiminosuccinonitrile) (p-DISN), a linear polymer with a dimeric [-N=C-C=N-] repeating unit and nitrile groups attached to the carbon atoms in a *trans* conformation. At pressures above 6 GPa, the colour of the sample further changes from brown to black at 10 GPa. The product can be recovered at ambient conditions and is also stable on heating above 600 K. The kinetics of this irreversible transformation has been studied [1020] and it has been shown that the reaction is diffusion controlled and has the overall activation volume $\Delta V^{\neq}$ = -3.3 cm^3/mol. The product is identified as paracyanogen and most likely as the ladder form of the polymer. The kinetic data are interpreted considering that in the first stage of the reaction, a reorientation of p-DISN occurs while in the second stage, a cycloaddition occurs between different chains of p-DISN. While the activation volume of the diffusion process is positive, the activation volume for a cycloaddition is typically in the range of - 10 cm^3/mol. The experimental value of -3.3 cm^3/mol is an average of the two contributions.

Cyanoacetylene is the first instance that we encounter where there are two different reacting groups whose high-pressure reactivity is not very different, or at least, exhibit comparable reaction thresholds. Cyanoacetylene is also of particular interest, since the high-pressure reaction has been studied in the single crystal [1021, 1022] thus avoiding, at least in part, the effect of defects. The packing in the crystal is dominated by linear chains of hydrogen-bonded molecules along the *a* crystal axis, with two molecules in the unit cell as is shown in Figure 6.17. A polymerization reaction occurs above 1.5 GPa. The molecules lying on the *ac* crystal plane react through the opening of the C-C triple bond, forming a polyacetylene backbone with the cyano groups attached in a *cis* configuration. There is

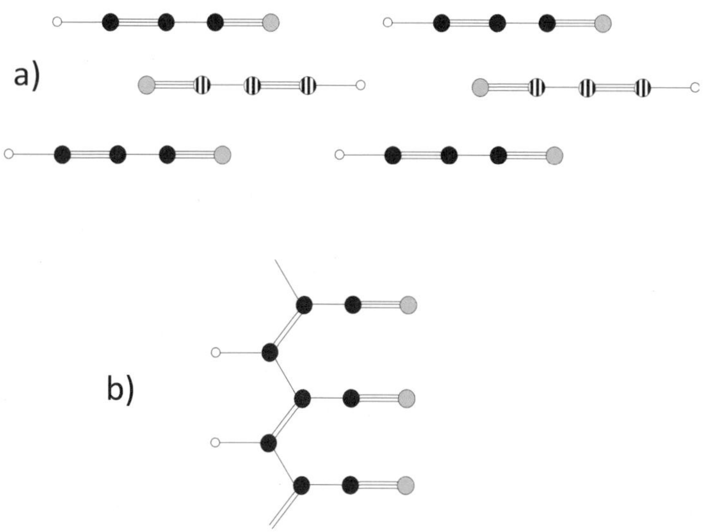

Fig. 6.17 Monoclinic crystal structure of cyanoacetylene projected along the b axis (a). The two inequivalent molecules of the unit cell are aligned antiparallel (molecules with filled and striped C atoms), the molecules are joined by hydrogen bonding between H (white) and N (grey) atoms forming infinite chains along the a axis. Each chain is surrounded by two parallel nearest neighbour chains which are involved in the polymer formation along the c crystal axis. The resulting all *cis*-polymer is shown in (b).

good evidence from the infrared spectra that the reaction product is indeed polycyanoacetylene (polyacrilonitrile). Looking at the crystal packing, this reaction appears to be a genuine topochemical reaction. This interpretation of the experimental results has been confirmed by recent *ab initio* molecular dynamics simulations [1023], which also show that at higher pressures, polyacrilonitrile chains will transform to polymers with fused pyridine rings which, at still higher pressures, will interconnect in a tridimensional array.

The reactive groups (C-C triple bond and phenyl group) present in phenylacetylene and in diphenylacetylene have, by themselves, quite different reactivities at high pressures, as has been seen from the reaction thresholds of acetylene and benzene. The bonding together of the two groups can change the scenario to some extent, either for the stabilizing effect of the phenyl substitution or for steric effects impeding or facilitating intermolecular approaches. Phenylacetylene has been reported to react above 8 GPa [1024]. The onset of the reaction is revealed by the bright red colour of the sample and by an observed pressure drop, although changes of the infrared spectra are small, particularly in the region of the H-C≡C vibrations. The reaction quickens above 10 GPa. Spectroscopic (infrared and Raman) evidence shows, as one would have expected, that the reaction involves the acetylenic group. On the basis of the infrared spectra, a *cis*-transoid conformation has been proposed for the structure of the polyphenylacetylene formed in the reaction and this is reasonable when also considering the packing of the monomers in the crystal. It can be noted how the phenyl substitution greatly increases the stability of the C-C triple bond, with the reaction threshold rising to above 8 GPa from the value of 3.5 GPa in acetylene. A quite different story is told by the reaction of diphenylacetylene. The onset of the reaction is revealed by the appearance in the infrared spectrum of sp^3 C-H stretching vibrations above 9 GPa [1025]. This points to the formation of chemical bonds between the aromatic rings of different molecules. As a matter of fact, in the crystal structure at ambient conditions, the shortest carbon-carbon contact is between atoms of the phenyl rings and these contacts will further decrease at high pressure because of the anisotropic compressibility of the material. Only at still higher pressure is the C≡C bond also involved in the reaction. On the whole, the recovered product appears as an amorphous hydrogenated carbon. The interesting point in the present case is that in diphenylacetylene, the reactivity of the two moieties is reversed compared with that of the individual units.

A polymerization reaction at high pressure has also been reported in dimethylacetylene [1026]. At 3.5 GPa, a loss of intensity of the C≡C stretching mode in the Raman spectrum is observed, with complete disappearance at 6.2 GPa. Correspondingly, an intense band develops at 1630 cm^{-1} with medium intensity structures in the 1900–2000 cm^{-1}. A characterization of the product has not been pursued. According to a model referring to the crystalline arrangement [1027] by *ab initio* methods, the first stages of the polymerization could be attributed to the formation of a dimer (tetramethylcyclobutadiene) which forms when molecules approach at distances close to 2 Å.

The high-pressure reaction of styrene allows a comparison of the relative stability of the vinyl and phenyl groups. The high-pressure polymerization of styrene in the liquid phase was first reported by Bridgman and Conant [107]. The high-pressure transformation in the solid state has been studied by Gourdain et al. [1028]. A definite threshold of the reaction could not be established precisely, because a slow partial transformation is already observed at low pressure. The fractional formation of a product increases with pressure, as a sigmoidal function, with a rapid increase at 15 GPa (which is taken as an effective threshold) and the transformation is almost complete (90%) at 32 GPa. Infrared analysis and comparison with spectra of commercial polystyrene, as a function of pressure, show that the polymeric product is indeed amorphous polystyrene. It is therefore seen that once again, the phenyl substitution has a stabilizing effect on unsaturated carbon-carbon bonds once this result is compared with the reactivity of ethylene. However, as said, the reaction starts very slowly at pressures as low as 3 GPa, much lower than the estimated threshold of 15 GPa. It has also been found that the benzene rings do not decompose at the highest pressure of the experiment (32 GPa). This can be ascribed to steric effects preventing the optimal reciprocal orientation of the phenyl groups required for the disruption of the aromatic rings. The styrene polymerization has also been discussed by Zharov under high pressure and shear deformation [759]. The reaction occurs at 10 GPa giving a regular, supposedly syndiotactic, polymer. It seems that in the present case, shear deformation is not very effective in reducing the reaction threshold. It is also reported that a further pressure increase, up to 40 GPa, results in a more irregular polymerization with chain branching.

The high-pressure reaction of carbon disulphide was first reported by Bridgman [294], who observed the transformation from the liquid at 4 GPa and 150 K, into a stable black solid. The material, subsequently called Bridgman's black carbon disulphide, was initially supposed by Bridgman to be a polymer similar to SiO_2. The structure of the polymer has later been clarified to be a linear polymer with -C(=S)-S- repeating units [1029–1031], although some cross linking of the chains is possible. The polymer, with density 1.92 g/cm^3, is the same as the polymer obtained by photopolymerization [1032]. As observed for styrene, the stability of carbon disulphide to pressurization is much greater in the solid. In fact, X-ray diffraction [1033] and Raman spectroscopy [1034] have shown that the orthorhombic structure, forming at 1.25 GPa at room temperature, is stable up to 9 GPa, the threshold pressure for the formation of the polymer. It is rather singular, by comparison with other molecular crystals, that no intermediate phase transitions have been reported for carbon disulphide.

Tetracyanoethylene is an interesting molecule for its ability as an electron acceptor to form charge transfer complexes with useful electrical and magnetic properties. At normal pressure there are two polymorphs: a cubic structure, that is supposed to be the thermodynamically stable phase at ambient conditions, which irreversibly transforms at 318 K in the monoclinic form and can be quenched to 4 K. Other not well-characterized forms have been reported (for a summary see [1035] and references therein). At room temperature, increasing the pressure, the monoclinic phase transforms to an intermediate amorphous phase between 2 and 6 GPa, and on releasing the pressure, the monoclinic or cubic phase is obtained depending on the rate of downloading [1036, 1037]. The phase behaviour is determined by the small energy difference between the phases, estimated to be only 0.04 eV at low temperature [1035]. At high temperature, the monoclinic phase stability arises from entropic contributions. A chemical transformation has been reported to occur between 6 and 14 GPa [1038–1041] with the formation of a polymer. The quite large pressure range is a consequence of the polymorph structure and depends on the irradiation conditions. The reaction seems to occur through the opening of the C-N triple bond, as revealed by the disappearance of the triple bond vibrational band. It has been reported that upon polymerization, a loss of nitrogen occurs and the C:N ratio raises from 3:2 to 7:1. On reducing the pressure, a cross linking of the polymer is observed. The reacted material has also been compressed up to 42 GPa and laser heated to approximately 2000°C, obtaining a mixture of sp^2-bonded carbon nitride and crystalline carbon [1041]. Carbon nitride presents a variable content of incorporated nitrogen which increases with pressure ranging from 24% at 18 GPa to 38% at 42 GPa.

There are several other reactions of molecular crystals that have been investigated under high pressure, but the reaction products and the reaction mechanism have not been fully characterized. Nevertheless, it is worth mentioning some of them briefly.

The high-pressure reaction of acrylic acid and methacrylic acid has been studied by Raman spectroscopy [1042]. The crystals exhibit phase transitions at 0.65 and 1.5 GPa, respectively, associated with a distinct change of the molecular packing arrangement of the hydrogen-bonded dimeric units. At higher pressures, a polymerization has been observed to give polycrylic acid and polymethacrylic acid, respectively. The reaction thresholds are not reported, but in a previous work Murli and Song [1043] reported a reaction threshold of 8 GPa for acrylic acid. The Raman spectra of polyacrylic acid obtained in the two experiments appear to be different. An interesting point is that in the Ostwald and Urquhart experiment, the polymers obtained from the two different polymorphs are identical. A high-pressure study has also been reported for methylmethacrylate [1044]. The methyl esterification of acrylic acid greatly increases the pressure stability and, apart from a crystalline phase transition at 10 GPa, the system is stable up to 30 GPa at 100°C where, according to the authors, a reaction occurs to form a dimer. The stabilization, induced by the crystal environment, is remarkable considering that liquid methyl methacrylate readily polymerizes at normal pressure. A high-pressure polymerization of a structurally related monomer, acrylamide, has been discussed by Zharov [759] under conditions of high pressure and shear deformation. The reaction is observed at pressures of tens of GPa and it is reported that the reaction occurs at the moment of deformation, by rotation of the anvils. The effect of pressure (up to 5000 atm) has also been studied on the photo-activated acrylamide polymerization in the 28–70°C range. It has been found that at a lower temperature, by increasing the pressure, the reaction is retarded while the opposite occurs at higher temperatures [1045].

Trioxane, a six-membered molecule with only single bonds, has been reported by Zharov [759] to resist polymerization at high pressures, even under shear deformation (however see the following). At pressures up to 3–4 GPa, the yield of polymer does not exceed 1%. However, it is found that if ionic active centres are introduced into the sample, the monomer readily crystallizes at 1 GPa to give polyoxymethylene. It is reported that the ionic centres are not active by themselves, but that the polymerization is actually induced by the increase of pressure under shear deformation.

The high-pressure behaviour of 65 chemical compounds, mostly molecular compounds, has been probed by Bradbury et al. [1046]. The highest pressure used in the experiment did not exceed 4.5 GPa. Under the conditions of the experiment only 14 of the compounds (acrylamide, p-phenylstyrene, potassium p-styrenesulphonate, itaconic anhydride, maleic anhydride, maleimide, 1,2,3,6-tetrahydrophthalic acid, 1,2,3,6-tetrahydrophthalic anhydride, acenaphthylene, p-benzoquinone, N,N'-phenylene dimaleimide, sulpholene, diphenylacetylene, s-trioxane) were found to polymerize and their reaction conditions are discussed. All the compounds considered contain a C=C double bond and the finding is rationalized considering that only compounds with a single substitution or with a cyclic 1-2 substitution, react at the pressure of the experiment. It is, however, evident that the high-pressure conditions used in the experiment could be too low for several of the compounds considered.

6.2.4 *Energetic materials*

Energetic materials and explosives have been studied extensively under shock loading, since these are obviously closer to the conditions of their practical use. Nevertheless, the study of energetic materials under static compression and at controlled temperature can be of considerable interest, since the fine control and tuning of pressure and temperature can furnish detailed information on the early stages of the detonation, in terms of the molecular mechanisms of the initiation. Static compression of energetic materials has been discussed in detail by Peiris and Piermarini [1047]. Here, attention will be limited to a few case studies, to enlighten the essentials of some reaction mechanisms that should supposedly depend on the distinct anisotropy of the compressibility of these systems.

A system of considerable interest for the simplicity of its molecular structure and for the peculiarities of the high-pressure reaction is nitromethane, CH_3NO_2. Nitromethane is a basic energetic material, a monopropellant which decomposes under shock compression and thermal initiation. Under static compression, nitromethane reacts slowly and study under these conditions can be useful to obtain information on the initiation process. From a more general point of view, nitromethane can also be of interest because of the possibility of internal rotation of the methyl group around the C-N bond, and because of the weak hydrogen bonding which will turn out to be quite important in regulating the crystal packing as a function of pressure and, finally, the reaction mechanism. There have been conflicting conclusions on the crystal structure as a function of pressure and on the equation of state of nitromethane (see [24] and references therein). CH_3NO_2 at room temperature, crystallizes in the orthorhombic system (space group $P2_12_12_1$ with four molecules per unit cell), the same structure as at low temperature. X-ray diffraction has given evidence of the stability of this structure up to 15 GPa. Infrared and Raman spectroscopy on the basis of slope changes in the frequencies versus pressure diagrams or of the appearance of new bands in the spectra, have suggested that some phase transitions may occur at high pressures, a hypothesis also supported by a number of models or *ab initio* calculations. These different interpretations of experimental data have been reconciled by a detailed study of the nitromethane crystal by X-ray diffraction and by infrared and Raman spectroscopy [24] up to 27.3 GPa, which is very close to the reaction threshold. The crystalline space group $P2_12_12_1$ remains unchanged over all this pressure range, where the crystal contracts rather isotropically with an overall volume contraction $V/V_0=0.69$. Observations can be rationalized in terms of a change of the intramolecular structure and subsequently of the intermolecular interactions. There are two limiting conformations of nitromethane, which correspond to an eclipsed and a staggered orientation of the two constituting groups. At low pressure there is in essence, a free rotation of the methyl group around the C-N bond and the relative population of the two limiting structures are equal. As a matter of fact, the rotational barrier of the methyl group has been estimated from neutron diffraction studies [1048] to be only 118 K (10 meV). Increasing the pressure, the relative population of the eclipsed conformation increases and above 11 GPa, this is the only conformation present in the crystal (see Figure 3.24). A constraint to the molecular conformation arises from the possibility for the eclipsed conformation to establish weak O$\cdots$H-C hydrogen bonds, which lock the structure in a network of nearly parallel chains of dimeric units (see Figure 3.25). The establishment of hydrogen bonds weakens the N-O and C-H bonds, giving rise to different vibrational frequencies that emerge in the spectrum with increasing pressure (see Figure 3.26). In conclusion, above 11 GPa a crystal form, denoted as phase II, is formed by hydrogen-bonded eclipsed molecules while below 4 GPa (phase I), the structure is made of freely rotating molecules. A mixture of the two structures is revealed in the intermediate pressure range by the vibrational spectra. The stability of phase II with hydrogen-bonded eclipsed molecules is demonstrated by the hysteresis found on releasing the pressure. The hydrogen-bonded network is stable to 3 GPa. Only on heating the sample above 50°C or lowering the pressure below 3 GPa, is the disordered structure recovered as documented by the infrared and Raman spectra.

The effect of high pressure on the thermally activated reaction of nitromethane has been studied by several authors [1049–1051]. It has been found that the results depend on temperature and that at high temperature, a decomposition occurs with the formation of carbon dioxide, and a recoverable and stable carbonaceous material, not characterized in detail. At room temperature, a chemical reaction occurs above 30 GPa [758, 1052]. From the shape of the infrared and Raman spectra of the product, characterized by partly structured broad absorption bands centred at 1200 and 3200 cm^{-1}, it can be seen that the product is amorphous and composite. Analysis of the spectra reveals features that can be assigned to vibrations of O-H, N-H and NO-H in the 3200 cm^{-1} region and to C-O, C-C, C-N and N-O groups in the lower frequency region where vibrations of unsaturated N=O and C=O groups are also present. These assignments are corroborated by results of *ab initio* molecular dynamics simulations [758] showing that the reaction product is an amorphous tridimensional array with terminal groups of various kind (COH, CNH, NOH, CNOH, CNO, NCO, OCO, ONO). The molecular dynamics simulation also shows that in the hydrogen-bonded high pressure crystal structure, the reaction is initiated by a hydrogen detachment and transfer to adjacent molecules forming the $CH_3NO_2H^+$ species and the aci ion $CH_2NO_2^-$, thus confirming the bimolecular nature of the initiation process. As a matter of fact, the kinetic analysis of the reaction gives evidence of the three-dimensional growth of the amorphous product and also shows, as expected, an increase of the reaction rate with pressure, as for a bimolecular-limiting step.

The most interesting aspect of the nitromethane reaction is the behaviour upon downloading. The product is unstable and on decreasing the pressure, a decomposition of the tridimensional array is observed. The strong 1200 cm^{-1} feature of the infrared spectrum weakens on downloading the pressure and finally disappears while a prominent peak develops at 1700 cm^{-1}. The changes in the 3200 cm^{-1} region are instead minor, showing that hydrogen-bonded associates are persistent on pressure lowering. It is notable that the development of a sharp absorption peak at 2350 cm^{-1}, due to carbon dioxide, can be taken as a probe of the formation of small molecules. In fact, on full downloading and opening of the cell, the sample disappears evidently being made of volatile molecules. The molecular dynamics simulation accordingly shows that decreasing the pressure, the amorphous solid decomposes with the formation of small molecules and among them carbon dioxide, water, formaldehyde, formic acid and hydroxylamine. Occasionally, when the high-pressure cell is opened, a sample remains on the cell windows which the infrared spectrum shows clearly to be N-methyl formamide. The abiotic formation of a molecule with a prototype amidic bond is quite remarkable but not surprising if one considers that the decomposition creates, in a confined environment, small molecules with the basic life elements H, C, N and O.

On the whole, it seems that in the static high-pressure nitromethane reaction, the basic process of decomposition of energetic materials into small gaseous or volatile molecules occurs in two stages: at high pressure, a condensation occurs to an amorphous intermediate solid which, when the high-pressure constraints are removed, decomposes into mostly volatile small molecules.

Studies of chemical reactions under the static loading of energetic materials have not been abundant, most likely because of the complex molecular structure of these materials. Rather, attention to static compression has been attracted by the determination or prediction [176, 1053, 1054] of properties (crystal structure, polymorphism, phase behaviour) and parameters (density, heat of formation) that are of primary importance for the performance, sensitivity, hazard and environmental impact of energetic materials.

RDX (see Figure 3.9) is an important energetic material. Recently [1055], the phase diagram has been reanalysed finding that in addition to the already known α, β and γ phases, a new phase, ϵ, is stable in a narrow P-T interval. It has been reported that a decomposition occurs in the ϵ phase by an autocatalytic reaction with a positive activation volume. On the contrary, the decomposition of the α phase occurs with a negative activation volume. This emphasizes the importance of the crystal structure in determining the high-pressure reaction mechanism. These static experiments also allowed the establishment that shock decomposition takes place from the γ phase [1056].

TAG-MNT is a nitrogen-rich energetic material made of two molecular ions (triaminoguanidinium and 1-methyl-5-nitriminotetrazole), which pack in the crystal with a network of hydrogen bonds. Flexibility of the molecular ions and strengthening of the hydrogen bonds lead to a stiffening of the structure at high pressure [1057]. Above 15 GPa, an irreversible chemical reaction occurs. Downloading and recovery of the product from pressures up to 25 GPa, show that the reaction has proceeded only partially. The Raman spectrum seems to only lack bands characteristic of the cation, which is evidently more unstable on compression. By compression to 60 GPa a full polymerization of the compound is obtained. Non-hydrostatic compression and shear strain exert effects on the intermolecular interactions and on structural features of the molecular ions, but do not influence the high-pressure chemistry. Irradiation with high power 457 and 488 nm laser lines induces photochemical reactions, resulting in the formation of molecular hydrogen.

6.2.5 *Photochemistry at high pressures*

Photo-activation of chemical reactions at high pressure is a practice that offers considerable advantages from several points of view. According to the concepts developed by Drickamer [12] already discussed in this book, the application of an external pressure induces changes in the electronic structure of the molecules and of their energy level diagrams, to such an extent, that the thermal population of some excited states can become possible even at ambient temperature. In this framework, pumping the system into an appropriate excited state is like supplementing an increase of pressure. This is an important, albeit simplified, scenario that primarily has the advantage of accelerating the reaction or, in other words, of lowering the high-pressure thresholds for the reaction. In addition, the identification of the relevant excited state opens the way to disclose the reaction mechanism. However, there are other attractive perspectives of high-pressure photochemistry. At high pressures, the free energy multidimensional surface of molecular crystals can change profoundly and it is quite possible that novel reaction pathways are opened by light absorption which are different from pure pressure-induced or ambient pressure-activated processes. This is an important source of selectivity in high-pressure photochemistry.

The primary effect of applying a high pressure is the change of volume, without simultaneously changing the temperature. Absorption of light can result in heating of the sample which can be inconvenient when full control of the external parameters is needed to study the kinetics of a process. In this respect, it has been found that a population at high pressure of the excited state in a limited extent as is realized by two-photon absorption, a process characterized by very low cross sections, avoids any heating of the sample. It has indeed been envisaged that excitation at an almost catalytic level [630] may suffice to photo-activate high-pressure reactions, exploiting the high-density conditions produced by compression. Two-photon absorption can also be useful to excite states in the far UV region by irradiating the system with visible radiation that is not absorbed directly by the sample. In this section, the several facets of high-pressure photochemistry will also be discussed with reference to several reactions induced purely by an increase of pressure that have already been mentioned.

The changes of the electronic energy surface with increasing pressure and the effect that these changes can have on the reaction mechanism is clearly illustrated by the chemical reactions induced in simple alcohols. The lowest absorption band of ethanol centred at 148 nm is due to the $n \rightarrow \sigma^*$ Rydberg transition which has a dissociative character. Experiments on liquid ethanol have been performed at pressures up to 1.5 GPa, irradiating the sample with the 350 nm laser line, which produces a two-photon transition to the Rydberg state [755]. The reaction evolution strongly depends on the pressure and requires irradiation powers above 100 mW. At pressures of 4.7 or 7.2 MPa, the main reaction product is molecular hydrogen, as revealed by the Raman analysis of the bubbles formed during the reaction, as shown in Figure 6.18. In addition, ethane and methane are identified as reaction products from weak Raman bands, the C-H stretching band of methane being used for the *in situ* monitoring of the pressure. The infrared spectrum confirms the formation of small amounts of methane and, in addition, carbon dioxide and water. The formation of hydrogen is remarkable by itself, even more so when the low pressure necessary to induce the reaction

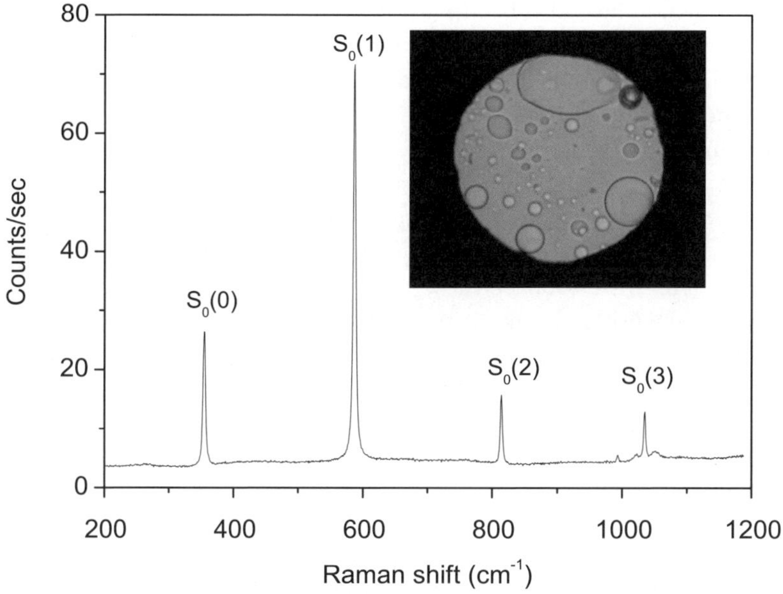

Fig. 6.18 Raman spectrum of bubbles of molecular hydrogen formed during the photo-induced reaction of ethanol at 7.2 MPa (reprinted from R. Bini, et al. *Chem. Phys.* **398** (2012) 262, Copyright (2012), with permission from Elsevier).

and the low radiation power are considered. It is also remarkable that the estimated rate of formation of hydrogen is in the range of 5.4–16 pmol/h^{-1} at the pressures of the experiment. The formation of hydrogen is also clear evidence of the dissociation along the O-H bond with the formation of hydrogen and methoxy radicals which trigger the reaction.

At higher pressures (0.5–1 GPa), the onset of more complicated reactions is revealed by the infrared and Raman spectra. The main products are ethane, 2-butanol, 2,3-butanediol and 1,2-diethoxyethane, together with some carbonyl compounds. The formation of these species is explained on the basis of two different dissociation paths involving, in order of importance, the splitting of the O-H and the C-O bonds, respectively. Whereas the amount of products ascribable to the latter dissociation path, ethane and 2-butanol, is almost unchanged with pressure, the relative amount of the compounds obtained through the O-H splitting channel reduces with increasing pressure, indicating a modification of the reaction efficiency along this path. For long irradiation times, when the amount of the reactant reduces, disproportionation of the products also occurs leading to the formation of methane, water and carbon dioxide.

Experiments have also been performed on methanol and its isotopomer CD_3OH, which also have a dissociative first excited state. Methanol has been studied at room temperature at pressures between 0.2 and 1.8 GPa, irradiating the sample at 350 nm with higher power that in ethanol [1058]. In methanol, the formation of hydrogen at low pressure has not been observed. At all pressures, the main reaction products are methoxymethanol and methylformate with small amounts of methane, water and unsaturated C=C species. At low pressure, the formation of ethylene glycol is also observed. The formation of these products can be explained, as already observed in the ethanol case, with reaction pathways that involve dissociation both along the O-H and C-O coordinates, whereas dissociation along the C-H coordinate seems not to be relevant. An interesting point raised by the methanol experiments is that the yield of the reaction products, deriving from the methoxy radical, increases with pressure up to 1 GPa and then decreases. The depletion of the reaction could imply that the dissociative character of the excited state changes with increasing pressure. To elucidate this point, some additional experiments have been carried out on ethanol and on deuterated ethanol EtOD [754]. The reaction of EtOD has been studied at 0.5 and 1.0 GPa. At 0.5 GPa, the reactivity of EtOD closely resembles that of the fully hydrogenated molecule. It can be noted that the formation of EtOH from EtOD demonstrated that the main dissociation involves the O-D coordinate. In all the reactions considered so far, it has been found that increasing the pressure, the reaction accelerates unless more favourable pathways are open. Instead in EtOD, it is found that at 1.0 GPa almost no reaction occurs. Actually, in EtOH the rate of consumption of the reactant also slowly decreases going from 0.5 to 1.0 GPa, while in EtOD the depletion is substantial. Since the reaction is definitely induced by the dissociative character of the excited state, the reasonable explanation of the observation is that the excited state, with increasing pressure, becomes slightly bonded with the formation of a shallow minimum (see Figure 5.6). Considering that the zero-point energy of the deuterated species is smaller than for the hydrogenated alcohol, the change of the bonding character of the excited state affects EtOD substantially, but the hydrogenated species to a much smaller degree.

The photo-activation of the high-pressure chemical reactions of benzene (and other aromatics), ethylene and of butadiene, nicely evidences that the population of appropriately excited states can induce a lowering of the pressure thresholds and open or select specific reaction paths. This is extremely important because it can allow advantage to be taken of the peculiarities and selectivity of solid state high-pressure reactions, in pressure ranges amenable for practical applications. From a more fundamental point of view, identification of the excited state involved in the high-pressure processes opens the way to more detailed studies of reaction mechanisms. The involvement of excited states in the high-pressure benzene amorphization has already been envisaged from the pressure behaviour of the vibrational frequencies in the ground and excited state [741]. Progress in this direction requires a better characterization of the structure and properties of the excited state. To this purpose, experiments have been performed in the diamond anvil cell on the two-photon absorption and two-photon induced fluorescence of benzene as a function of pressure [631]. The two-photon excitation spectrum shown in Figure 6.19 closely reproduces the known Franck–Condon profile of the S_1 state.

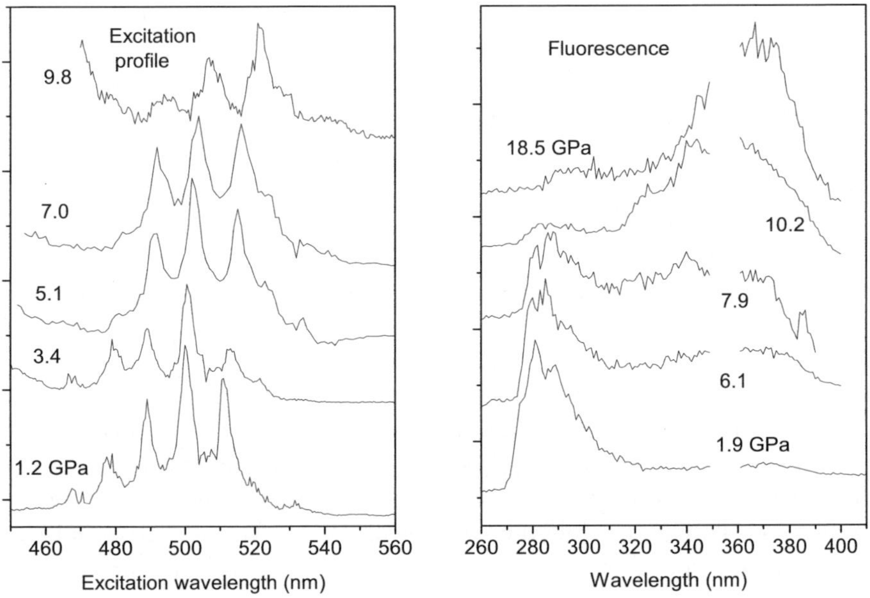

Fig. 6.19 Evolution of the benzene two-photon excitation spectrum as a function of pressure (left). Two-photon induced fluorescence in crystalline benzene as a function of pressure (right).

Increasing the pressure, a red shift of the progression is observed and the shift of the 0-0 transition is found to be -60 cm^{-1}/GPa (see Figure 5.5), and corresponds to the relative vertical shift of the ground and excited state.

This value is considerably smaller than a previous estimate from one-photon absorption spectra [643]. A second important finding is that the intensity distribution of the progression changes upon pressurization, and increases at high pressures in the peaks closer to the origin of the progression. The implication is that the positions of the ground and excited state overlap in the configuration space, and the structures of the two states become similar. This result is actually the first clear experimental demonstration of the lateral shift of the ground and excited states discussed by Drickamer [12].

A further insight on the properties of the excited state have been obtained from the study of the two-photon induced fluorescence. At low pressure, the emission is dominated by the monomer fluorescence centred at 280 nm, with a structure that is not resolved in the experiment. Increasing the pressure, an emission centred at 370 nm develops which is due to the excimer fluorescence. Above 8 GPa, the emission is dominated by the excimer fluorescence which accounts for more than 90% of the emission. Formation of excimers is a common feature in crystals of aromatic hydrocarbons and in particular, in anthracene [8, 999, 1059, 1060]. However, it has generally been found that excimer formation occurs upon irradiation and at defects, since in the pure crystal structure the orientation of the molecules is not optimal for the $\pi \rightarrow \pi*$ interaction and for the formation of dimers. On the contrary, the formation of excimers in benzene is a structural feature of the crystal. This is demonstrated by the observation that when a benzene crystal is pressurized to 13 GPa, without any irradiation, and then decompressed to 5 GPa, fluorescence from excimers is actually observed. Therefore, experiments strongly support suggestions of *ab initio* molecular dynamics studies [642] that the formation of dimers (excimers) is the intermediate of the ring opening and amorphization. It is of interest to note that the pressure red shifts of the monomer and excimer fluorescence have also been measured, obtaining the values of -125 cm^{-1}/GPa and -135 cm^{-1}/GPa, respectively. The excess red shift of the monomer fluorescence compared with the exciton absorption line, results from the combined effect of pressure on the horizontal shift and on the form of the energy surfaces (force constants). It is also seen that the stabilization effect of increasing pressure is larger for the excimer.

One could expect a similar behaviour in the photo-activation of the high-pressure reaction of a heteroaromatic such as furane, but structural peculiarities produce some differences. Experiments have been performed [753] irradiating furane with the 458 nm laser line, which induces a two-photon transition to the 1^1B_2 or 2^1A_1 excited state. The reaction threshold reduces from 10 to 3 GPa, very much as observed in benzene. The peculiar thing is that at 3 GPa, furane is in a glassy disordered phase, different from the ordered crystal phase present at 10 GPa and this has consequences on the reaction. In fact, analysing the reaction product with infrared spectroscopy, it is found that it is the same as that obtained in the pure pressure experiment, but with an increase of the intensity of the band due to saturated -C-H stretching and to the C=O stretching mode. In addition, the formation of carbon dioxide trapped in the polymeric disordered product is observed. A reasonable explanation can be provided on the basis of the disordered nature of the phase in which the ring opening occurs, with some preference for breaking of the C-O bond. Migration of the hydrogen atoms, from the carbon atoms that form CO_2 and carbonyl bond, determines the formation of a greater number of saturated C-H bonds. In fact, the estimated abundance of sp^3 carbon atoms increases with respect to the purely pressure-induced reaction from 67 to 96%. The specific role of the sample structure in the laser-assisted reaction, in the present pressure regime, is fully appreciated considering that increasing the pressure and producing the transition from the glassy to the high-pressure ordered phase, the amount of carbon dioxide produced decreases and finally disappears at 10 GPa. In this way, a general behaviour similar to benzene is fully recovered.

A characterization of the electronic excited states through the study of the two-photon induced fluorescence and two-photon excitation profiles has also been carried out in pyridine crystal phases I and II [645]. It has been found that the high-pressure two-photon induced fluorescence has a qualitative structure similar to benzene, with a high wavelength emission due to the monomer and a low wavelength emission due to excimers, however the excimer emission band is much weaker than in benzene. In fact, the major difference with benzene is that in both phases I and II, the crystal structure and packing of pyridine do not allow for the formation of structural excimers, as in benzene, which requires a stacked geometry. Therefore, excimers form only at defects. This is consistent with the minor high-pressure reactivity of pyridine in comparison with benzene.

The relevance of the structure for high-pressure reactivity is evident in the indole crystal, which has been studied up to 25.5 GPa using infrared spectroscopy and two-photon absorption and fluorescence [258]. Indole has been found to be very stable to pressurization and at the highest pressure of the experiment, only traces of C-H saturated bonds could be detected in the infrared spectrum. This has been ascribed to strong hydrogen bonds between the N-H bond and the π electron distribution of nearest neighbouring molecules, preventing the optimal arrangement for a chemical reaction. In comparison with benzene and pyridine, no excimer emission could be detected at high pressure and absorption to the excited state is unable to induce a photo-activated chemical reaction.

Photoirradiation of butadiene fully exposes the selectivity that can be realized at high pressures. *Trans*-butadiene is itself an unstable molecule at ambient conditions and reacts by three different cycloaddition reactions $[2\pi + 2\pi]$, $[4\pi + 4\pi]$ and $[4\pi + 2\pi]$, to give a mixture of the three dimers 1,2-divinylcyclobutane, 1,5-cyclooctatetraene and 4-vinylcyclohexene, respectively. Liquid butadiene loaded in the diamond anvil cell at room temperature reacts at 0.7 GPa [629]. The reaction is very slow and is completed in several days. Slightly increasing the pressure to 1 GPa, the reaction considerably accelerates. At a slightly lower temperature (280 K) butadiene does not react before crystallizing. The reaction product is the single dimer 4-vinylcyclohexene with no traces of the other dimers or of the polymer. The results show how an extreme selectivity can be realized at high pressures, even in a disordered system like a liquid. When, in the same pressure conditions, liquid butadiene is irradiated with the 488 or 458 nm lines of an Ar^+ laser, the formation of the dimer is completely inhibited and the formation of a *trans* polybutadiene polymer is observed by infrared monitoring. The result is quite remarkable in two respects. Firstly, the full inhibition of the dimer is rather surprising. Secondly, the polymer is a really conformationally pure *trans* polybutadiene with a complete absence, in the infrared spectrum, of the bands of the *cis* polymer that are generally observed in the commercial polymer, as reported in Figure 6.20.

As to the effect of laser irradiation, butadiene is transparent at 488 and 458 nm and only a two-photon transition to the $S_1(2^1A_g)$ excited state is possible. The transition is symmetry allowed with a large cross section. In the ground state, the central and lateral bonds of butadiene have lengths of 1.487 and 1.349 Å, respectively. In the excited state, due to the decrease of the π bonding character, the central bond shortens to 1.472 Å and the outer bonds lengthen to 1.493 Å, whereas the terminal -CH_2 groups are free to reorient to minimize hindrance. For all these reasons, the structure of the excited state is favourable for a head-to-tail condensation. As already described, a molecular dynamics simulation [757] has shown that at high pressure, a lengthening of the outer bonds and a shortening of the central bond occur before the reaction starts, thus resembling the photo-excitation effect. Photo-activation of the butadiene reaction has also been attempted

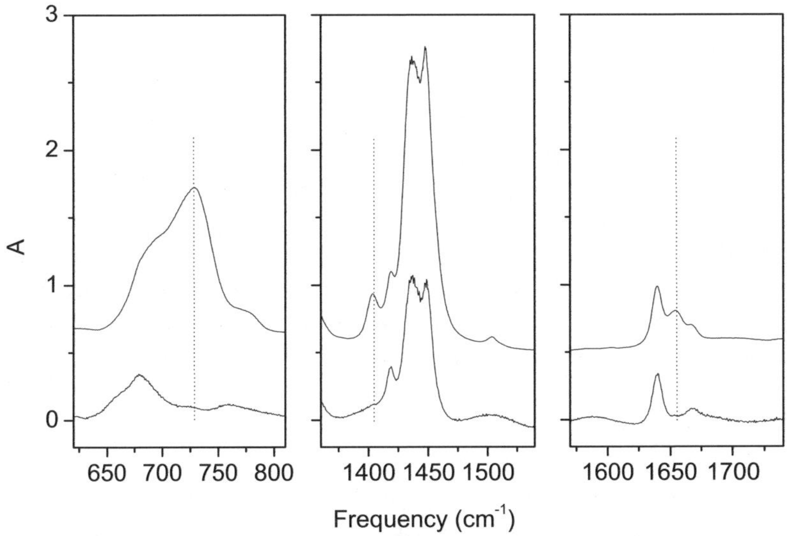

Fig. 6.20 A comparison of selected regions of the infrared spectra of polybutadiene obtained by the high-pressure photo-induced reaction of liquid butadiene at 0.7 GPa (lower spectrum) and of a commercial butadiene (upper spectrum) containing 59% and 36% of the *trans-* and *cis*-polymer, respectively. The bands marked with a vertical line are signatures of the *cis* polymer which is not present in the high-pressure photopolymerized material. (Reprinted from R. Bini, et al. *Chem. Phys.* **398** (2012) 262, Copyright (2012), with permission from Elsevier).

in the crystal [730]. It has been found that the formation of polybutadiene is induced, but the formation of the dimer is not inhibited and the two reaction pathways remain active. When the irradiation is shut down, the formation of the polymer stops and only the dimerization reaction proceeds.

Once the peculiarities of the photo-activation at high pressure in butadiene in comparison with benzene and other heteroaromatics have been illustrated, the question arises as to whether the statement from the benzene experiment that irradiation at an appropriate wavelength is equivalent to an increase of pressure, has a general validity and can apply to butadiene as well. A direct answer from the experiment is not possible because the reactivity of butadiene is more complex, depending on the aggregation state and because more reactive channels are possible (dimerization and polymerization), so that pressure and photo-excitation act as independent activation and regulatory tools of the reaction. However, limiting attention to the already discussed behaviour of solid butadiene [970], a good answer to the question is provided. In fact, in the solid form at low pressure, only the formation of the dimer is observed, but upon increasing the pressure, the formation of the *trans* polymer is observed and the fractional amount of the polymer increases with pressure. Therefore, a kind of equivalence between pumping in the excited state and an increase of pressure, seems valid in the present case and also in general, apart from complications that can arise from steric constraints.

The clearest demonstration of the degree of conformational and steric selectivity that can be induced by photo-activation at high pressure, is given by the polymerization reaction of liquid ethylene. At room temperature and at 0.7 GPa liquid ethylene is stable. If the liquid is irradiated at wavelengths lower than 460 nm (and particularly at 351 and 364 nm), with a power in the range 25–100 mW, a reaction is observed with a significant volume reduction [630]. The reaction product is recoverable and has been analysed by angle resolved X-ray diffraction, which demonstrated that a highly crystalline polyethylene polymer was obtained. In fact, as can be seen from Figure 6.21, the diffraction lines of this polymer are very narrow and the underlying diffuse scattering, characterizing defected (mainly gauche) polyethylene, is completely absent. The result is remarkable considering that generally, polyethylene is obtained as a mixture of an ordered crystalline polymer and amorphous domains, and the crystallinity is increased by thermal annealing. The crystallinity of the polymer, obtained by photo-activation at high pressure, is confirmed by the analysis of the infrared and Raman spectra where the factor group splitting components of the vibrational modes, are well resolved and appear with the correct intensity ratio. The reaction rate increases with pressure in experiments performed up to 1.8 GPa, but it is remarkable that the reaction can be carried out satisfactorily at a pressure as low as 0.2 GPa. The mechanism of the reaction involves the two-photon excitation of the $^1B_{1u}$ excited state, in which the C-C bonds lengthens to 1.38–1.47 Å and the methylene groups are in a twisted configuration with increased sp^3 hybridization, favourable to the formation of linear saturated chains. Considering the very short lifetime (few fsec) of the excited state, it is likely that the reaction occurs after relaxation to the ground state and before structural relaxation, the latter occurring by vibrational relaxation in the picosecond regime.

It has already been noted that the high-pressure reaction of acetylene is also photosensitive and that even the mild irradiation, necessary to monitor the pressure using the ruby fluorescence method or to measure the Raman spectrum, affects the characteristics of the reaction product. Specific ex-

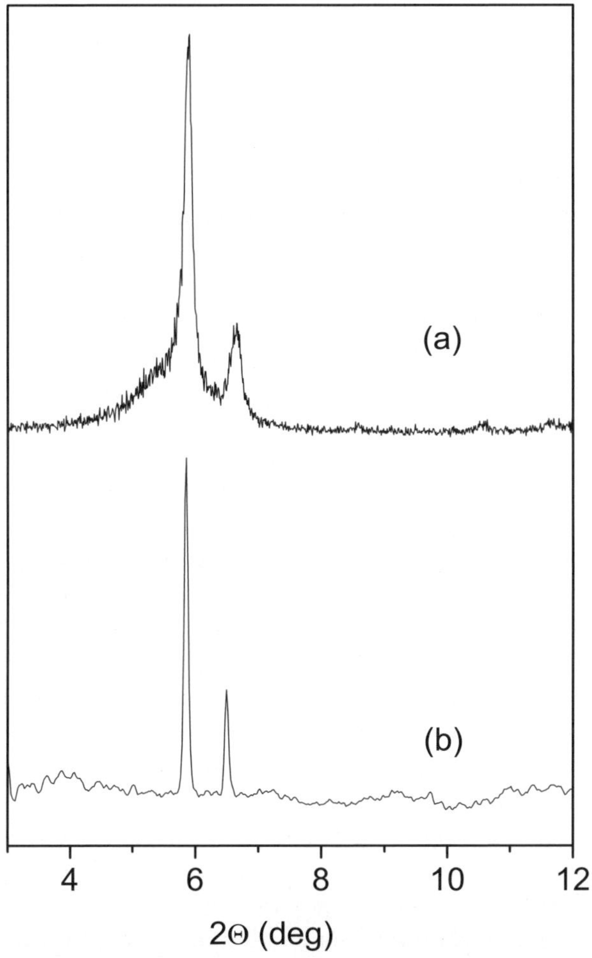

Fig. 6.21 X-ray diffraction pattern of polyethylene obtained by the high-pressure photoactivated reaction (lower) compared with the pattern of commercial highly crystalline polymer (upper). (Reprinted from R. Bini, et al. *Chem. Phys.* **398** (2012) 262, Copyright (2012), with permission from Elsevier).

periments have been carried out at high pressure to elucidate the effect of laser irradiation [376]. It has been found that when the pressure-induced reaction has reached saturation, by irradiation of the sample with the output of an Ar^+ laser with 8 mW power, the reaction starts again producing a decrease of the intensity of the infrared bands of the polymer and an increase of the vibrational bands of the saturated carbon groups. This is a clear indication that laser irradiation, in this case, mostly produces a cross linking of the polymeric chains. When the irradiation is turned off, the reaction readily stops.

Laser irradiation of liquid propene reduces the reaction threshold pressure for polymerization [746]. Irradiation with the 458 nm line of an Ar ion laser (with 20 mW power) reduces the threshold from 3.1 to 1.8 GPa. A larger reduction of the threshold (to 0.75 GPa) is obtained by irradiating with the Ar ion laser multiline (337.5–356.5 nm) with 500 mW power. The reaction product is the same as obtained in the pure pressure-induced reaction. Also in this case, as with ethylene, irradiation produces a two-photon population of the lowest excited singlet state. On the whole, it is seen that the photochemical behaviour of propene parallels that of ethylene and the methyl substitution has no dramatic effects, but the steric hindrance prevents chain lengthening. On the contrary, the methyl substitution of butadiene to isoprene has some peculiarities [953]. The pressure threshold is also reduced in isoprene from 1.1 to 0.5 GPa. Irradiation at 0.5 GPa with the 488 nm line has no effect on the reaction. With the 458 nm line irradiation, the effects are minimal with power up to 250 mW. However, increasing the power accelerates the reaction, and with 610 mW power the reaction is completed in a few hours. This strong dependence on the irradiation power highlights the two-photon nature of the absorption, which has a quadratic dependence on the incident power. Again, irradiation produces both a lowering of the threshold and an acceleration of the reaction. The product is composed of a volatile component and a liquid recoverable sample. The latter is the same polymeric product obtained by pressure activation. The volatile component is a 4-ethenyl-2,4-dimethyl-cyclohexene dimer. It is therefore seen that irradiation opens a new reaction pathway, since the dimer obtained under laser irradiation is different from the dimers obtained by activation under pressurization only. The selectivity obtained by laser irradiation is also seen, considering the reaction at 1.1 GPa. In fact, in these conditions only a polymer is obtained with no traces of dimers, which instead form in the absence of irradiation.

The photo-activated chemical reactions of alcohols previously discussed are induced by the dissociative character of the first excited state along the O-H coordinate. The same dissociative character is shared by the simplest R-OH system, water, whose first excited state falls in the UV at 155–177 nm. A rich chemistry associated with the photodissociation of water at high energy is well known [1061–1064]. Of particular relevance for our purposes is the high-pressure experiment by Mao et al. [1065], where the photodissociation of water at high pressure is induced by X-ray, producing a solid alloy of molecular hydrogen and oxygen. Considering the general lowering of the pressure thresholds for photo-activated reactions described in this section, and the efficiency of two-photon absorption to populate the excited state to a sufficient degree, the challenge of exploiting the water dissociation under mild conditions of pressurization and irradiation seems affordable [154]. Trial experiments have been carried out on mixtures of water with nitrogen or carbon monoxide [465]. Samples were prepared in the diamond anvil cell and the experimental conditions were selected in such a way that among the individual components, only carbon monoxide showed a well-characterized reactivity upon irradiation at 350 nm. This wavelength produces a two-photon absorption to the lowest excited state $\tilde{A}(^1B_1)$ of water, which is also dissociative, as are all the other excited states. The dissociation from this state produces a hydrogen atom and a hydroxyl radical in its electronic ground state. Once loaded in the cell, the liquid mixture was compressed and a phase separation, consisting of small single crystals immersed in a fluid phase, was visually observed. Crystallites were characterized by Raman spectroscopy as clathrate hydrates of nitrogen or carbon monoxide because a splitting into a doublet of the vibrons, accounting for the occupancy by N_2 or CO molecules of both the large and small cages of the clathrate, was observed. The fluid phase was instead identified as N_2 or CO in excess. The sample was uniformly laser irradiated with a power in the 60–600 mW range. Nitrogen hydrate does not react upon irradiation at 0.5 GPa, therefore the pressure was lowered to 0.1 GPa where the clathrate was observed to decompose. In these conditions, irradiation produced a reaction monitored by the appearance of new bands in the Raman spectrum, assigned to molecular hydrogen and to vibrations of single and double-bonded nitrogen atoms to oxygen. Besides the formation of molecular hydrogen, the remarkable result is that in the mild conditions of the experiments (0.1 GPa and irradiation at 350 nm with 600 mW power), a partial destabilization of the N-N triple bond is obtained, recalling that in pure nitrogen, the triple bond opening requires a pressure in the order of 100 GPa and high temperatures. On the contrary, a reaction is also observed in the case of CO when the clathrate hydrate is irradiated. After a few hours of irradiation at 0.4–0.5 GPa, all the water is apparently (IR spectra) consumed and a heterogeneous material composed of a dark solid and gaseous bubbles is obtained. The Raman spectra show the presence of hydrogen in the bubbles, whereas FTIR spectra provide evidence of CO_2 formation. The dark solid material looks like the product obtained in the reaction of pure CO. It could certainly be of interest to carry out these same experiments in pure clathrate hydrates for a finer control of the phase in which the reaction is induced, and in view of the general interest in the physics and chemistry of clathrate hydrates [1066]. It can also be noted that these high-pressure experiments are carried out on two-component systems and this opens novel additional perspectives of the chemistry at very high pressures in solid phases.

Chemical reactions induced by the two-photon absorption of water at moderate pressure have also been studied in mixtures of water with simple hydrocarbons (ethylene, ethane, propene, acetylene) [1067], that in most of the cases are, as already reported for CO, in the form of a clathrate hydrate when the reaction is photochemically induced. In the ethylene/water mixture by irradiation at 350 nm with 200 mW power at 0.3–0.5 GPa, the formation of crystalline polyethylene is observed with only traces of carbon dioxide. Therefore, in this mixture the involvement of the water photodissociation is negligible, the high-pressure photo-induced polymerization being far more efficient. On the contrary, a reaction is observed in the ethane/water mixture. In the conditions of the experiment (0.6 GPa at room temperature), the formation of the clathrate hydrate is first observed visually. A reaction occurring upon irradiation is monitored by the appearance in the infrared spectrum of the bands of carbon dioxide and by the intensity decrease of the bending mode of water. Accurate analysis of the shape of the antisymmetric mode shows that carbon dioxide formed during the reaction is captured as a clathrate hydrate, where both structures of type I and type II are present (see Figure 6.22).

When mixtures of propene/water are prepared at room temperature, the infrared spectrum is a superposition of the two individual spectra, but when the sample is pressurized the bands of water drastically weaken, a peculiarity that is associated with the formation of the clathrate hydrate. The system reacts upon irradiation to give carbon dioxide and a solid recoverable product that turns out to be the same mixture of propene oligomers obtained in the high-pressure reaction of pure propene. In the present case, it can be observed that the propene reaction is greatly facilitated in the presence of water.

Also, in the case of the acetylene/water mixtures, the formation of a clathrate hydrate is observed. Mixtures with an excess of acetylene at pressures of 0.3–0.5 GPa, upon irradiation with the multiline laser emission centred at 350 nm, react to give carbon dioxide and a solid red coloured product that can be recovered at ambient conditions and characterized

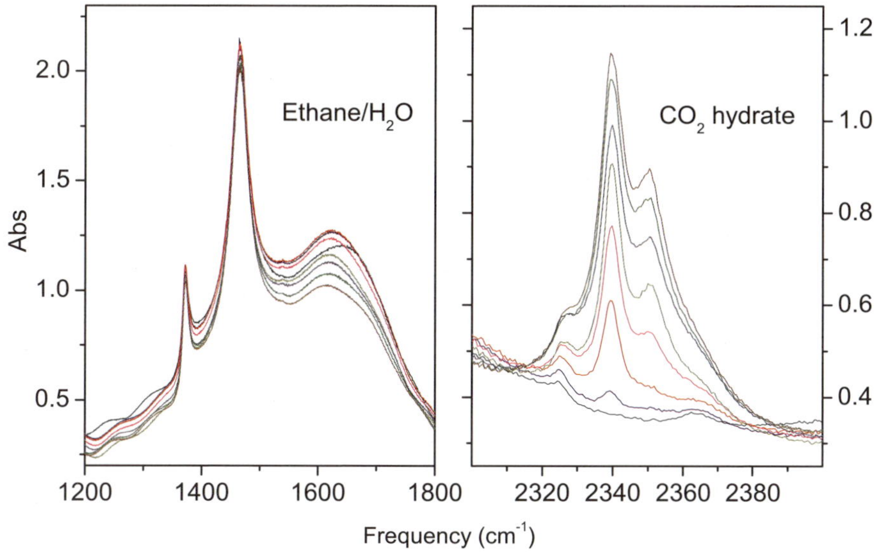

Fig. 6.22 Photo-induced reaction in ethane hydrate at 0.6 GPa monitored by FTIR spectroscopy. On the left panel, the decrease of ethane and water bands with the irradiation time is shown. On the right panel, the corresponding increase of the absorption band (asymmetric stretching) of CO_2 formed in the reaction is shown. The two stronger peaks at 2339 cm^{-1} and 2349 cm^{-1} correspond to the CO_2 occupation of large and small cages, respectively, of the type-I structure of carbon dioxide clathrate hydrate. Nevertheless, because the 3:1 ideal ratio expected for this structure is also not satisfied, the type-II structure, which presents only the peak at 2349 cm^{-1}, could contribute to the absorption profile.

through the several bands appearing in the infrared spectrum. In addition, a small amount of methane is detected. In water-rich mixtures, the reaction is accelerated but the product is the same with only minor difference in the infrared spectrum. The recovered solid product, on the basis of the infrared spectrum, should contain a mixture of molecules such as methyl- and ethyl formate, and 1,2-ethanediol diacetate. From the abundance of carbon dioxide formed in this reaction one would expect the formation of hydrogen as well. Attempts to detect hydrogen in the Raman spectrum failed. Possibly the detection of hydrogen is prevented by the strong fluorescence of the solid product.

Bibliography

[1] Callen, H. B. (1960). *Thermodynamics* (John Wiley, New York).

[2] Hirshfelder, J. O., Curtiss, C. F. and Byron, B. R. (1967). *Molecular Theory of Gases and Liquids* (John Wiley, New York).

[3] Margenau, H. and Kestner, N. (1969). *Theory of Intermolecular Forces* (Pergamon Press, Oxford).

[4] Kihara, T. (1977). *Intermolecular Forces* (John Wiley, New York).

[5] Califano, S., Schettino, V. and Neto, N. (1981). *Lattice Dynamics of Molecular Crystals* (Springer Verlag, Heidelberg).

[6] Kitaigorodsky, A. (1973). *Molecular Crystals and Molecules* (Academic Press, London).

[7] Wright, J. (1995). *Molecular Crystals* (Cambridge University Press, Cambridge).

[8] Jones, W. (1997). *Organic Molecular Solids* (CRC Press Inc., Boca Raton).

[9] Desiraju, G. R., Vittal, J. J. and Ramanan, A. (2011). *Crystal Engineering. A Textbook* (World Scientific Publishing, Singapore).

[10] Bernstein, J. (2002). *Polymorphism in Molecular Crystals* (Oxford University Press, Oxford).

[11] Parr, R. and Yang, W. (1989). *Density-Functional Theory of Atoms and Molecules* (Oxford Science Publications, New York).

[12] Drickamer, H. G. and Frank, C. (1973). *Electronic Transitions and the High Pressure Chemistry and Physics of Solids* (Chapman and Hall, London).

[13] Drickamer, H. G., Frank, C. and Slichter, C. (1972). Optical versus thermal transitions in solids at high pressure, *Proc. Natl. Acad. Sci. USA* **69**, pp. 933–937.

[14] Lin, S. (1973). Effect of high pressures on molecular electronic spectra and electronic relaxation, *J. Chem. Phys.* **59**, pp. 4458–4467.

[15] Tompkins, R. (1978). Effects of high pressures on molecular electronic spectra, *J. Chem. Phys.* **69**, pp. 579–583.

[16] Tompkins, R. (1980). Effects of high pressures on molecular electronic spectra II. Morse potential formulation, *J. Chem. Phys.* **72**, pp. 3449–3453.

[17] Kesarwani, R. and Varshni, Y. (1984). Effects of high pressure on molecular electronic spectra, *J. Chem. Phys.* **81**, pp. 5508–5513.

[18] McMahan, A. K. (1986). Pressure-induced changes in the electronic structure of solids, *Physica B+C* **139-140**, pp. 31–41.

[19] David, W. I. F., Ibberson, R. M., Dennis, T. J. S., Hare, J. P. and Prassides, K. (1992). Structural phase transition in the fullerene C_{60}, *Europhys. Lett.* **18**, pp. 219–225.

[20] Sun, L., Yi, W., Wang, L., Shu, J., Sinogeikin, S., Meng, Y., Shen, G., Bai, L., Li, Y., Liu, J., Mao, H. K. and Mao, W. L. (2009). X-ray diffraction studies and equation of state of methane at 202 GPa, *Chem. Phys. Lett.* **473**, pp. 72–74.

[21] Scandolo, S. and Jeanloz, R. (2003). The centers of planets, *American Scientist* **91**, pp. 516–525.

[22] Jeanloz, R. (1989). Physical chemistry at ultrahigh pressures and temperatures, *Ann. Rev. Phys. Chem.* **40**, pp. 237–259.

[23] Ciabini, L., Santoro, M., Bini, R. and Schettino, V. (2001). High pressure crystal phases of benzene probed by infrared spectroscopy, *J. Chem. Phys.* **115**, pp. 3742–3749.

[24] Citroni, M., Datchi, F., Bini, R., Vaira, M. D., Pruzan, P., Canny, B. and Schettino, V. (2008). Crystal structure of nitromethane up to the reaction threshold pressure, *J. Phys. Chem. B* **122**, pp. 1095–1103.

[25] Boldyreva, E., Shakhtschneider, T., Vasilchenko, M. A., Ahsbahs, H. and Uchtmann, H. (2000). Anisotropic crystal structure distortion of the monoclinic polymorph of acetaminophen at high hydrostatic pressures, *Acta Cryst. B* **56**, pp. 299–309.

[26] Prewitt, C. and Downs, R. (1998). High-pressure crystal chemistry, in Hemley, R. J. (Ed.), *Ultra-high-pressure Mineralogy: Physics and Chemistry of Earth's Deep Interior*, Vol. 37 (Mineralogical Society of America, Review of Mineralogy).

[27] Grochala, W., Hoffmann, R., Feng, J. and Ashcroft, N. (2007). The chemical imagination at work in very tight places, *Ang. Chem. Int. Ed.* **46**, pp. 3620–3642.

[28] Drickamer, H. G. (1993). High pressure chemistry in the solid state: pressure induced molecular rearrangements in rigid media, in Winter, R. and Jonas, J. (Eds.) *High Pressure Chemistry, Biochemistry and Materials Science* (Kluwer Academic Publishers, Netherlands) pp. 67–77.

[29] Mitra, S. (2004). *High Pressure Geochemistry and Mineral Physics* (Elsevier, Amsterdam).

[30] Bridgman, P. (1931). *The Physics of High Pressure* (G. Bell and Sons Ltd, London).

[31] McMillan, P. (2002). New materials from high-pressure experiments, *Nature Mater.* **1**, pp. 19–25.

[32] Hazen, R. (1993). *The New Alchemist* (Random House, New York).

[33] Hazen, R. (1999). *The Diamond Makers* (Cambridge University Press, Cambridge).

[34] Brazhkin, V. V., Lyapin, A. G. and Hemley, R. J. (2002). Harder than diamond: dreams and reality, *Phil. Mag. A* **82**, pp. 231–253.

[35] Banly, C. (1997). High pressure and bioscience, a long story of interdisciplinary research, in Heremans, K. (Ed.), *High Pressure Research in the Biosciences and Biotechnology* (Leuven, Leuven University Press).

[36] Heremans, K. (1997). *High Pressure Research in the Biosciences and Biotechnology* (Leuven University Press, Leuven).

[37] Winter, R. and Jonas, J. (Eds.) (1993). *High Pressure Chemistry, Biochemistry and Materials Science* (Kluver Academic Publishers, Dordrecht).

[38] Winter, R. and Jonas, J. (Eds.) (1999). *High Pressure Molecular Science*, Nato Science Series C: Vol. 401 (Kluver Academic Publishers, Dordrecht).

[39] Winter, R. (2010). Exploring the energy and conformational landscape of biomolecules under extreme conditions, in Boldyreva, E. and Dera, P. (Eds.), *High-Pressure Crystallography* (Springer, Dordrecht) pp. 573–590.

[40] Daniel, I., Oger, P. and Winter, R. (2006). Origin of life and biochemistry under high pressure conditions, *Chem. Soc. Rev* **35**, pp. 858–875.

[41] Doona, C. J. and Feeherry, F. E. (Eds.) (2007). *High-pressure processing of food* (Wiley John & Sons Inc., Oxford).

[42] Schaschke, C. J. (2011). *Developments in High-pressure Food Processing*, (Nova Science Publishers Inc., Hauppauge NY).

[43] Brittain, H. (1999). *Polymorphism in Pharmaceuticals* (Marcel Dekker, New York).

[44] Hilfiker, R. (Ed.) (2006). *Polymorphism in Pharmaceutical Industry* (Wiley-VCH, Weinheim).

[45] Hemley, R. J. (2000). Effects of high pressure on molecules, *Ann. Rev. Phys. Chem.* **51**, pp. 763–800.

[46] Akahama, Y., Kawamura, H., Hausermann, D., Hanfland, M. and Shimomura, O. (1995). New high-pressure structural transition of oxygen at 96 GPa associated with metallization in a molecular solid, *Phys. Rev. Lett.* **74**, pp. 4690–4693.

[47] Weck, G., Desgreniers, S., Loubeyre, P. and Mezouar, M. (2009). Single-crystal structural characterization of the metallic phase of oxygen, *Phys. Rev. Lett.* **102**, p. 255503.

[48] Sakai, N., Takemura, K. I. and Tsuji, K. (1982). Electrical properties of high-pressure metallic modification of iodine, *J. Phys. Soc. Jpn* **51**, pp. 1811–1816.

[49] Takemura, K., Minomura, S., Shimomura, O., Fujii, Y. and Axe, J. D. (1982). Structural aspects of solid iodine associated with metallization and molecular dissociation under high pressure, *Phys. Rev. B* **26**, pp. 998–1004.

[50] San Miguel, A., Libotte, H., Gaspard, J. P., Gauthier, M., Itie, J. P. and Polian, A. (2000). Bromine metallization studied by X-ray absorption spectroscopy, *Eur. Phys. J. B* **17**, pp. 227–233.

[51] Goettel, K. A., Eggert, J. H., Silvera, I. F. and Moss, W. C. (1989). Optical evidence for the metallization of xenon at 132 GPa, *Phys. Rev. Lett.* **62**, pp. 665–668.

[52] Wigner, E. and Huntington, H. B. (1935). On the possibility of a metallic modification of hydrogen, *J. Chem. Phys.* **3**, pp. 764–770.
[53] Lee, K. K. M., Neumann, G. S. and Jeanloz, R. (2004). *Ab initio* pressure alloying of iron and potassium: Implications for the Earth's core, *Geophys. Res. Lett.* **31**, p. L11603.
[54] Dubrovinskaia, N., Dubrovinsky, L., Kantor, I., Crichton, W. A., Dmitriev, V., Prakapenka, V., Shen, G., Vitos, L., Ahuja, R., Johansson, B. and Abrikosov, I. A. (2005). Beating the miscibility barrier between iron group elements and magnesium by high-pressure alloying, *Phys. Rev. Lett.* **95**, p. 245502.
[55] Jephcoat, A. (1998). Rare-gas solids in the Earth's deep interiors, *Nature* **393**, pp. 355–358.
[56] Kelm, H. (Ed.) (1979). *High Pressure Chemistry.* NATO Advanced Study Institute Series (D. Reidel Publishing Company, Dordrecht).
[57] van Eldik, R. and Klarner, F. G. (Eds.) (2002). *High Pressure Chemistry.* (Wiley-VCH, Weinheim).
[58] Manaa, M. R. (Ed.) (2005). *Chemistry at Extreme Conditions* (Elsevier, Amsterdam).
[59] van Eldik, R. and Hubbard, C. D. (1997). *Chemistry Under Extreme or Non Classical Conditions* (Wiley and Spektrum, New York).
[60] Loikowski, W. (2002). *High Pressure Effects in Chemistry, Biology and Materials Science* (Scitec Publications, Totton).
[61] Boldyreva, E. V. and Boldyrev, V. V. (Eds.) (1999). *Reactivity in Molecular Solids* (John Wiley, Chichester).
[62] Coppens, P. (2009). The new photocrystallography, *Ang. Chem. Int. Ed.* **48**, pp. 4280–4281.
[63] Kleinermans, K. and Wolfram, J. (1987). Laser chemistry: what is the current status, *Ang. Chem. Int. Ed.* **26**, pp. 38–58.
[64] Le Chatelier, H. (1884). Sur un énoncé général des lois des équilibres chimiques, *Comptes Rendus* **99**, pp. 786–789.
[65] Le Chatelier, H. (1908). *Leçons sur le Carbon* (Dunod et Pinat Ed, Paris).
[66] van't Hoff, J. (1884). *Etudes de Dynamique Chimique* (Frederik Muller, Amsterdam).
[67] Hamann, S. D. (1980). The role of electrostriction in high pressure chemistry, in Osugi, J., (Ed.), *Modern Aspects of Physical Chemistry at High Pressure Rev. Phys. Chem. Jpn.* **50**, pp. 147–168.
[68] Thomson, J. (1861). On crystallization and liquefaction, as influenced by stresses tending to change of form in the crystal, *Proc. Roy. Soc. London* **11**, p. 473.
[69] Califano, S. (2010). *Storia della Chimica - Dall'alchimia alla Chimica del XIX Secolo* (Bollati Boringhieri, Torino).
[70] Kendall, D. (2000). A short history of high pressure technology from Bridgman to Division 3, *J. Press. Vess. Techn.* **122**, pp. 229–233.
[71] Plato, (2000). *Timaeus* (Hackett Publishing Company, Indianapolis).
[72] Hero, (1851). *The Pneumatics of Hero of Alexandria.* (Taylor Walton and Maberly, London).

[73] Lucretius, (1969). *De Rerum Natura* (Sansoni, Firenze).

[74] Forbes, R. J. (1964). *Studies in Ancient Technology* (E. J. Brill, Leiden).

[75] Wertime, T. (1964). Man's first encounter with metallurgy, *Science* **146**, pp. 1257–1267.

[76] Wertime, T. (1973). The beginning of metallurgy: a new look, *Science* **182**, pp. 875–887.

[77] Charles, J. (1980). The coming of copper and copper-base alloys and iron. A metallurgical sequence, in Wertime, T. and Muhly, J. D. (Eds.), *The coming of the Age of Iron* (Yale University Press, London).

[78] Wheeler, T. S. and Maddin, R. (1980). Metallurgy and Ancient Man, in Wertime, T. and Muhly, J. D. (Eds.), *The Coming of the Age of Iron* (Yale University Press, London).

[79] Tankha, V. (2006). *Ancient Greek Philosophy: Thales to Gorgias.* (Dorling Kindersley Pvt Ltd, Delhi).

[80] Eliade, M. (1977). *Forgerons et Alchimistes* (Flammarion, Paris).

[81] Popov, A. (1933). Consecration ritual among the yakuts, *J. Am. Folkl.* **46**, pp. 257–271.

[82] Braidwood, R. J., Çambel, H. and Watson, P. J. (1969). Prehistoric investigations in southeastern Turkey, *Science* **164**, pp. 1275–1276.

[83] Muhly, J. D. (1988). The beginning of metallurgy in the old world, in Madden, R. (Ed.), *The Beginning and the Use of Metals and Alloys* (MIT Press, London) pp. 2–20.

[84] Vitruvius (1960). *Ten Books of Architecture Book X* (Dover, New York).

[85] Koutsoyiannis, D. and Angelakis, A. N (2003). Hydrology and hydraulic science and technology in ancient greek times, *Encyclopedia of Water Science* (Marcel Dekker, Inc., New York) pp. 415–417.

[86] Agricola, G. (1950). *De Re Metallica Libri XII*, trans. Hoover, H. C. and Hoover, L. H. (Dover Publications Inc., New York).

[87] Macini, P. and Mesini, E. (2004). Hydraulic pumps of agricola's de re metallica, *J. Hydraul. Eng.* **130**, pp. 1051–1054.

[88] Theophrastus (1965). *De Lapidibus*, Eichholz, D. E. (Ed.) (Oxford University Press, Oxford).

[89] Mottana, A. (2001). Il pensiero di Teofrasto sui metalli secondo i frammenti delle sue opere e le testimonianze greche, latine, siriache ed arabe, *Rend. Fis. Acc. Lincei* **12**, pp. 133–241.

[90] Secondo, G. P. (1986). *Storia Naturale, V Mineralogia e Storia dell'Arte, Libri 33-37*, trans. Capitani, U. and Garofalo, I. (Giulio Einaudi Editore, Torino).

[91] Agricola, G. (1955). *De Natura Fossilium Libri X*, trans. Bandy, M. C. and Bandy, J. A. (Geological Society of America).

[92] Boldyrev, V. and Tkáčová, K. (2000). Mechanochemistry of solids: past, present, and prospects, *J. Mater. Synth. Proc.* **8**, pp. 121–132.

[93] Baláž, P. (2003). Mechanical activation in hydrometallurgy, *Int. J. Min. Process* **72**, pp. 341–354.

[94] Takacs, L. (2000). Quicksilver from cinnabar: the first documented mechanochemical reaction, *Jom - J. Min. Met. Mat. S.* **52**, pp. 12–13.

[95] *Saggi di naturali esperienze fatte nell'Accademia del Cimento* (1667) (Florence).

[96] Canton, J. (1762). Experiments to prove that water is not incompressibile, *Trans. Roy. Soc.*, pp. 640–643.

[97] Tait, P. (1889). *Report on the Scientific Results of the Voyage of H.M.S. Challenger* (Physics and Chemistry, Vol. II (H.M.S.O., London).

[98] Neece, G. A. and Squire, D. R. (1968). Tait and related empirical equations of state, *J. Phys. Chem.* **72**, pp. 128–136.

[99] Hayward, A. (1967). Compressibility equations for liquids: a comparative study, *Brit. J. Appl. Phys.* **18**, pp. 965–977.

[100] Dymonds, J. H. and Malhotra, R. (1988). The Tait equation: 100 years on, *Int. J. Thermophys.* **9**, pp. 941–951.

[101] Milne-Edwards, A. and Perrier, E. (1888). *Expéditions Scientifiques du Travailleur et du Takisman Pendant les Années 1880–1883* (G. Masson, Paris).

[102] Fenn, W. (1970). Life under high pressures, *Proc. Am. Philos. Soc.* **114**, pp. 191–197.

[103] Regnard, P. (1891). *Recherches Expérimentales sur les Conditions Physiques de la Vie dans les Eaux* (G. Masson, Paris).

[104] Bridgman, P. (1914). The coagulation of albumen by pressure, *J. Biol. Chem.* **19**, pp. 511–512.

[105] Hite, B. (1899). A method for carrying out chemical reactions under high pressures, *Am. Chem. J.* **22**, pp. 80–85.

[106] Conant, J. B. and Tongberg, C. O. (1930). Polymerization reactions under high pressures. I. Some experiments with isoprene and butyraldehyde, *J. Am. Chem. Soc.* **22**, pp. 1659–1660.

[107] Conant, J. B. and Bridgman, P. (1929). Irreversible transformations of organic compounds under high pressure, *Proc. Natl. Acad. Sci.* **15**, pp. 680–683.

[108] Cohen, E. (1919). *Piezochemie Kondensierter Systeme* (Akademische Verlagsgesellschaft, Leipzig).

[109] Cohen, E. (1926). *Physico-Chemical Metamophosis and Problems in Piezo-chemistry* (McGraw Hill, New York).

[110] Bridgman, P. (1946). Recent work in the field of high pressure, *Rev. Mod. Phys.* **18**, pp. 1–95.

[111] Merrill, L. (1970). *High Pressure Bibliography 1960–1968* (National Standard Reference Data System).

[112] Bradley, R. (1963). A brief historical survey of high pressure research in this century, in Bradley R. (Ed.), *High Pressure Physics and Chemistry*, Vol. I (Academic Press, London).

[113] Born, M. and Wang, K. (1954). *Dynamical Theory of Crystal Lattices* (Oxford University Press, London).

[114] Nye, J. (1955). *Physical Properties of Crystals* (Oxford University Press, Oxford).

[115] Eliezer, S. and Ricci, R. (Eds.) (1991). *High Pressure Equations of State: Theory and Applications* (North Holland, New York).

[116] Wallace, D. (1972). *Thermodynamics of Crystals* (Wiley, New York).

[117] Zharkov, V. N. and Kalinin, V. A. (1971). *Equation of state for solids at high pressures and temperatures* (Consultant Bureau, New York).

[118] Anderson, O. (1955). *Equations of State of Solids for Geophysics and Ceramic Science* (Oxford University Press, Oxford).

[119] Poirier, J. (2000). *Introduction to the Physics of the Earth's Interior* (Cambridge University Press, Cambridge).

[120] Knopoff, L. (1963). Solids: equations of state of solids at moderately high pressure, in Bradley. R. (Ed.), *High Pressure Physics and Chemistry*, Vol. I (Glasgow, Academic Press), pp. 228–245.

[121] Angel, R. (2000). Equations of state, in Hazen, R. and Downs, R (Eds.), *High-temperature and High-pressure Crystal Chemistry* (Mineralogical Society of America, Washington), pp. 35–59.

[122] Angel, R. (2004). Some practical aspects of studying equations of state and structural phase transitions, in Katrusiak, A. and McMillan, P. F. (Eds.), *High-pressure Crystallography* (Kluwer Academic Publishing, Dordrecht), pp. 21–36.

[123] Holzapfel, H. (2004). Equations of state and thermophysical properties of solids under pressure, in Katrusiak, A. and McMillan, P. F. (Eds.), *High-pressure Crystallography* (Kluwer Academic Publishing, Dordrecht), pp. 217–236.

[124] Holzapfel, H. (1996). Physics of solids under strong compression, *Rep. Progr. Phys* **59**, pp. 29–90.

[125] Garai, J. (2007). Semi-empirical pressure-volume-temperature equation of state: $MgSiO_3$ as an example, *J. Appl. Phys* **102**, p. 123506.

[126] Stacey, F. (2001). Finite strain, thermodynamics and the Earth's core, *Phys. Earth Planet. Int.* **128**, pp. 179–193.

[127] Singh, P., Sushil, K., Arunesh, K. and Sharma, B. (2004). Analysis of finite strain equations of state for solids under pressure, *Physica B* **352**, pp. 134–146.

[128] Roy, P. and Roy, S. (2003). Applicability of three-parameters equation of state of solids: compatibility with first principles approaches and application to solids, *J. Phys: Condens Matter* **15**, pp. 1643–1663.

[129] Jeanloz, R. (1988). Universal equation of state, *Phys. Rev. B* **38**, pp. 805–807.

[130] Holzapfel, H. (2001). Equation of state for solids under strong compression, *Z. Kristallogr* **216**, pp. 473–488.

[131] Biot, M. A. (1954). Theory of stress-strain relations in anisotropic viscoelasticity and relaxation phenomena, *J. Appl. Phys.* **25**, pp. 1385–1391.

[132] Levy, M., Bass, H. and Stern, R. (Eds.) (2001). *Handbook of Elastic Properties of Solids, Liquids and Gases Vol I, Dynamic Methods for Measuring the Elastic Properties of Solids* (Academic Press, New York).

[133] Polian, A. (2003). Brillouin scattering at high pressure: an overview, *J. Raman Spectr.* **34**, pp. 633–637.

[134] Hayes, H. and Loudon, R. (1978). *Scattering of Light by Crystals* (Wiley, New York).

[135] Cummins, H. Z. and Schoen, P. E. (1972). Light scattering from thermal fluctuations, in Arecchi, F. T. and Schulz-Dubois, E. (Eds.) *Laser Handbook* (North-Holland, Amsterdam) pp. 1029–1075.

[136] Ramachandran, G. and Wooster, W. (1951). Determination of elastic constants of crystals from diffuse reflection of X-rays. I. Theory of method, *Acta Cryst.* **4**, pp. 335–344.

[137] Ramachandran, G. and Wooster, W. (1951). Determination of elastic constants of crystals from diffuse reflection of X-rays. II. Application to some cubic crystals, *Acta Cryst.* **4**, pp. 431–440.

[138] Prince, E. and Wooster, W. (1953). Determination of elastic constants of crystals from diffuse reflection of X-rays. III. Diamond, *Acta Cryst.* **6**, pp. 450–454.

[139] Day, G. and Price, S. (2001). Properties of crystalline organic solids, in Levy, M. (Ed.), *Handbook of Elastic Properties of Solids, Liquids and Gases, Vol. III, Elastic Properties of Solids: Biological and Organic Materials, Earth and Marine Science* (Academic Press, New York).

[140] Pavlides, P., Pugh, D. and Roberts, K. J. (1991). Elastic-tensor atom-atom potential calculations for molecular crystals: C_6H_6 and $CO(NH_2)_2$, *Acta Cryst.* **A47**, pp. 846–850.

[141] Catti, M. (1985). Calculation of elastic constants by the method of crystal static deformation, *Acta Cryst.* **A41**, pp. 494–500.

[142] Walmsley, S. (1968). Lattice vibrations and elastic constants of molecular crystals in the pair potential approximation, *J. Chem. Phys.* **48**, pp. 1438–1444.

[143] Pawley, G. (1967). A model for the lattice dynamics of naphtalene and anthracene, *Phys. Stat. Sol.* **30**, pp. 347–360.

[144] Frenkel, D. and Smit, B. (1996). *Understanding Molecular Simulations* (Academic Press, San Diego).

[145] Parrinello, M. and Rahman, A. (1982). Strain fluctuations and elastic constants, *J. Chem. Phys.* **76**, pp. 2662–2666.

[146] Bouma, R. H. B., van der Heijden, A. E. D. M., Sewell, T. D. and Thompson, D. L. (2011). Simulation of deformation processes in energetic materials, in Awrejcewicz, J. (Ed.), *Numerical Simulations of Physical and Engineering Processes* (InTech Publisher, Rijeka Croatia).

[147] Haussühl, S. (2001). Elastic and thermoelastic properties of selected organic crystals: acenaphtene, trans-azobenzene, benzophenone,tolane, trans-stilbene, dibenzyl, dephenil sulphone, 2,2'-biphenol, urea, melamine, hexogen, succinimide, penthaerythritol, urotropine, malonic acid, dimethyl malonic acid, maleic acid, hippuric acid, aluminium acetylacetonate, iron acetylacetonate, and tetraphenyl silicon, *Z. Kristallogr* **216**, pp. 339–353.

[148] Day, G. M., Price, S. L. and Leslie, M. (2001). Elastic constants calculations for molecular organic crystals, *Cryst. Growth Des.* **1**, pp. 13–27.

[149] Roberts, R. J., Rowe, R. C. and York, P. (1991). The relationship between Young's modulus of elasticity of organic solids and their molecular structure, *Powder Technol.* **65**, pp. 139–146.

[150] Haussühl, S. (1993). Interpretation of elastic properties of ionic crystals. Validity of quasi-additivity rule? *Z. Kristallogr.* **205**, pp. 205–215.

[151] Cogoni, M., D'Aguanno, B., Kuleshova, L. N. and Hofmann, D. W. M. (2011). A powerful computational crystallography method to study ice polymorphism, *J. Chem. Phys.* **134**, p. 204506.

[152] Gagnon, R. and Jones, S. (2001). Elastic properties of ice, in Levy, M. (Ed.), *Handbook of Elastic Properties of Solids, Liquids and Gases Vol III, Elastic Properties of Solids: Biological and Organic Materials, Earth and Marine Science*, Vol. III (Academic Press, New York).

[153] Shimizu, H., Kumazaki, T., Kume, T. and Sasaki, S. (2002). Elasticity of single-crystal methane hydrate at high pressure, *Phys. Rev. B* **65**, p. 212102.

[154] Ceppatelli, M., Bini, R. and Schettino, V. (2011). High-pressure reactivity of clathrate hydrates by two-photon dissociation of water, *Phys. Chem. Chem. Phys.* **13**, pp. 1264–1275.

[155] Sandstedt, C. A., Michaalski, D. and Eckhardt, C. J. (2000). Quantitative measurement of guest-host interaction in supramolecular systems: a comparative Brillouin scattering study of the dianin compound clathrand and two of its isostructural clathrates, *J. Chem. Phys.* **112**, pp. 7606–7614.

[156] Selbo, J. G., Haycraft, J. J. and Eckhardt, C. J. (2003). Elastic and thermodynamic properties of dianin's inclusion compounds and their guest-host interactions, *J. Phys. Chem. B* **107**, pp. 11163–11169.

[157] Haussühl, S. (1995). Beta-succinic acid, a crystal of extreme elastic anisotropy, *Z. Kristallogr.* **210**, pp. 903–904.

[158] Ward, I. (1982). *Mechanical Properties of Solid Polymers* (Wiley, New York).

[159] Bowden, P. and Young, R. (1974). Deformation mechanism in crystalline polymers, *J. Mater. Sci.* **9**, pp. 2034–2051.

[160] Tashiro, K. (1993). Molecular theory of mechanical properties of crystalline polymers, *Progr. Polym. Sci.* **18**, pp. 374–435.

[161] Tashiro, K. and Kobayashi, M. (1996). Molecular theoretical study of the intimate relationship between structure and mechanical properties of polymer crystals, *Polymer* **37**, pp. 1775–1786.

[162] Tashiro, K., Kobayashi, M. and Tadokoro, H. (1992). Vibrational spectra and theoretical three-dimensional elastic constants of isotactic polypropylene crystal: an important role of anharmonic vibrations, *Polymer J.* **24**, pp. 899–916.

[163] Kruger, J. K., Pietralla, M. and Unruh, H. G. (1983). The elastic properties of hexatriacontane single crystals at their various phase transitions, *Coll. Polym. Sci* **261**, pp. 409–411.

[164] Kruger, J. K., Bastian, H., Asbach, G. I. and Pietralla, M. (1980). The elastic properties of $C_{36}H_{74}$ single crystals from Brillouin spectroscopy, *Polym. Bull.* **3**, pp. 633–640.

[165] Weisshaupt, W., Krbecek, H., Pietralla, M., Hochheimer, H. D. and Mayr, P. (1995). Pressure dependence of the elastic constants of poly(methyl methacrylate), *Polymer* **36**, pp. 3267–3271.

[166] Stevens, L. L., Orler, E. B., Dattelbaum, D. M., Ahart, M. and Hemley, R. J. (2007). Brillouin-scattering determination of the acoustic properties and their pressure dependence for three polymeric elastomers, *J. Chem. Phys.* **127**, p. 104906.

[167] Hinsen, K. (2008). Structural flexibility of proteins: impact of the crystal environment, *Bioinformatics* **24**, pp. 521–528.

[168] Doucet, J. and Benoit, J. (1987). Molecular dynamics studied by analysis of X-ray diffuse scattering from lysozyme crystals, *Nature* **325**, pp. 643–646.

[169] Morozov, V. and Morozova, T. (1981). Viscoelastic properties of protein crystals: triclinic crystals of hen egg white lysozyme in different conditions, *Biopolymers* **20**, pp. 451–467.

[170] Morozov, V. and Morozova, T. (1986). Thermal motion of whole protein molecules in protein solids, *J. Theor. Biol.* **121**, pp. 73–88.

[171] Morozov, V. and Morozova, T. (1993). Elasticity of globular proteins. The relation between mechanics, thermodynamics and mobility, *J. Biomol. Struct. Dyn.* **11**, pp. 459–481.

[172] Caylor, C. L., Speziale, S., Kriminski, S., Duffy, T., Zha, C. S. and Thorne, R. E. (2001). Measuring the elastic properties of protein crystals by Brillouin scattering, *J. Cryst. Growth* **232**, pp. 498–501.

[173] Ike, Y., Hashimoto, E., Aoki, Y., Kanazawa, H. and Kojima, S. (2009). Micro-Brillouin scattering study of low temperature elastic properties of protein crystals, *J. Mol. Struct.* **924–926**, pp. 157–160.

[174] Zamiri, A. and De, S. (2010). Modeling the mechanical response of tetragonal lysozyme crystals, *Langmuir* **26**, pp. 4251–4257.

[175] Paci, E. and Marchi. M. (1996). Intrinsic compressibility and volume compression in solvated proteins by molecular dynamics at high pressure, *Proc. Natl. Acad. Sci. USA* **93**, pp. 11609–11614.

[176] Fabbiani, F. P. A. and Pulham, C. R. (2006). High-pressure studies of pharmaceutical compounds and energetic materials, *Chem. Soc. Rev.* **35**, pp. 932–942.

[177] Coombes, D. S., Catlow, C. R., Gale, J. D., Hardy, M. J. and Saunders, M. R. (2002). Theoretical and experimental investigations on the morphology of pharmaceutical crystals, *J. Pharm. Sci.* **91**, pp. 1652–1658.

[178] Grzesiak, A. L., Lang, M., Kim, K and Matzger, A. J. (2003). Comparison of the four anhydrous polymorphs af carbamazepine and the crystal structure of form I, *J. Pharm. Sci.* **92**, pp. 2260–2271.

[179] Mohapatra, H. and Eckhardt, C. (2008). Elastic constants and related mechanical properties of monoclinic polymorph of the carbamazepine molecular crystal, *J. Phys. Chem. B* **112**, pp. 2293–2298.

[180] Roberts, R. J., Payne, R. S. and Rowe, R. C. (2000). Mechanical property predictions for polymorphs of sulphathiazole and carbamazepine, *Eur. J. Pharm. Sci.* **9**, pp. 277–283.

[181] Beyer, T., Day, G. M. and Price, S. L. (2001). The prediction, morphology, and mechanical properties of the polymorphs of paracetamol, *J. Am. Chem. Soc.* **123**, pp. 5086–5094.

[182] Walker, A. M., Gale, J. D., Slater, B. and Wright, K. (2005). Atomic scale modelling of the cores of dislocations in complex materials part 2: applications, *Phys. Chem. Chem. Phys.* **7**, pp. 3235–3242.

[183] Karki, S., Friscic, T., Fabian, L., Laity, P. R., Day, G. M. and Jones, W. (2009). Improving mechanical properties of crystalline solids by cocrystal formation: new compressible forms of paracetamol, *Adv. Mater.* **21**, pp. 3905–3909.

[184] Joiris, E., Di Martino, P., Berneron, C., Guyot-Hermann, A. M. and Guyot, J. C. (1998). Compression behaviour of orthorhombic paracetamol, *Pharm. Res.* **15**, pp. 1122–1130.

[185] Payne, R. S., Roberts, R. J., Rowe, R. C., McPartlin, M. and Bashal, A. (1996). The mechanical properties of two forms of primidone predicted from their crystal structure, *Int. J. Pharm.* **145**, pp. 165–173.

[186] Stevens, L. and Eckhartdt, C. (2005). The elastic constants and related properties of β-HMX determined by Brillouin scattering, *J. Chem. Phys.* **122**, p. 174701.

[187] Amer, M. and Maguire, J. (2009). On the compressibility of C_{60} individual molecules, *Chem. Phys. Lett.* **476**, pp. 232–235.

[188] Kobelev, N. P., Nikolaev, R. K., Soifer, Ya. M. and Khasanov, S. S. (1998). Elastic moduli of single crystal C_{60}. *Phys. Stat. Solidi* **40**, pp. 154–156.

[189] Burgos, E. H. and Bonadeo, H. (1994). Intermolecular forces and phase transitions in solid C_{60}, *Phys. Rev. B* **76**, pp. 15544–15549.

[190] Heseltine, J. C. W., Elliott, D. W. and Wilson, O. B. Jr. (1964). Elastic constants of single-crystal benzene, *J. Chem. Phys.* **40**, pp. 2584–2587.

[191] Ledbetter, H. (2006). Sound velocities, elastic constants: temperature dependence, *Mat. Sci. Eng. A* **442**, pp. 31–34.

[192] Kobelev, N. P., Nikolaev, R. K., Sidorov, N. S. and Soifer Ya. M. (2002). The temperature dependence of the elastic moduli for solid C_{60}, *Phys. Sol. State* **44**, pp. 429–431.

[193] Ecolivet, C., Toudic, B. and Sanquer, M. (1984). Brillouin scattering in p-polyphenyls. III. The improper ferroelastic phase transition of p-terphenyl, *J. Chem. Phys.* **81**, pp. 599–606.

[194] Duan, D., Liu, Y., Ma, Y., Liu, Z., Cui, T., Liu, B. and Zou G. (2007). *Ab initio* study of solid bromine under high pressure, *Phys. Rev. B* **76**, p. 104113.

[195] San, X., Wang, L., Ma, Y., Liu, Z., Cui, T., Liu, B. and Zou, G. (2008). Theoretical calculations of the phase transitions and optical properties of solid iodine under high pressure, *J. Phys. Condens. Matter* **20**, p. 175225.

[196] Shimizu, H., Kamabuchi, K., Kume, T. and Sasaki, S. (1999). High-pressure elastic properties of the orientationally disordered and hydrogen-bonded phase of solid HCl, *Phys. Rev. B* **59**, pp. 11727–11732.

[197] Hazen, R. M., Downs, R. T. and Prewitt, C. T. (2000). Principles of comparative crystal chemistry, in Hazen, R. M. and Downs R. T. (Eds.), *High-temperature and High-pressure Crystal Chemistry* (Mineralogical Society of America, Blacksburg, USA).

[198] Hazen, R. M. (1977). Temperature, pressure, and composition: structurally analogous variables, *Phys. Chem. Minerals* **1**, pp. 83–94.

[199] Hazen, R. M. and Finger, L. (1982). *Comparative Crystal Chemistry: Temperature, Pressure, Composition and Variation of Crystal Structure* (Wiley, New York).

[200] Boldyreva, E. V. (2004). Molecules in strained environment, in Katrusiak, A. and McMillan, P. F. (Eds.), *High-pressure Crystallography* (Kluwer Academic Publishing, Dordrecht), pp. 495–512.

[201] Voigt, W. (1889). Ueber die beziehung zwischen den beiden elastizitat konstanten isotroper koerper, *Wiedemanns Annalen der Physik und Chemie* **38**, pp. 573–587.

[202] Reuss, A. (1929). Berechnung der fliessgrenze von mischkristallel auf grund der plasticitätbedingung für einkristalle, *Zeit. Ang. Mat. Mech.* **9**, pp. 49–58.

[203] Hill, R. (1952). The elastic behavior of a crystalline aggregate, *Proc. Phys. Soc.* **A65**, pp. 349–354.

[204] Ledbetter, H. (1990). Monocrystal-polycrystal elastic-constant models, in Wolfenden, A. (Ed.), *Dynamic Elastic Modulus Measurements in Materials* (American Society for Testing and Materials, Philadelphia).

[205] Simmons, G. and Wang, H. (1971). *Single Crystal Elastic Constants and Calculated Aggregated Properties* (MIT Press, Cambridge).

[206] Press, W. and Eckert, J. (1976). Structure of solid ethylene-d_4, *J. Chem. Phys.* **65**, pp. 4362–4364.

[207] Mugnai, M., Pagliai, M., Cardini, G. and Schettino, V. (2008). Mechanism of ethylene polymerization at very high pressure, *J. Chem. Theory Comput.* **4**, pp. 2673–2685.

[208] Ciabini, L., Gorelli, F. A., Santoro, M., Bini, R., Schettino, V. and Mezouar, M. (2005). High-pressure and high-temperature equation of state and phase diagram of solid benzene, *Phys. Rev. B* **72**, p. 094108.

[209] Boldyreva, E. V., Shakhtschneider, T. P., Ahsbahs, H., Sowa, H. and Uchtmann, H. (2002). Effect of high pressure on the polymorphs of paracetamol, *J. Therm. Anal. Calorim.* **68**, pp. 437–447.

[210] Shakhtschneider, T. P., Boldyreva, E. V., Vasilchenko, M. A., Ahsbahs, H. and Uchtmann, H. (1999). Anisotropy of crystal structure distortion in organic molecular crystals of drugs induced by hydrostatic compression, *J. Struct. Chem.* **40**, pp. 892–898.

[211] Boldyreva, E. V., Dmitriev, V. and Hancock, B. (2006). Effect of pressure up to 5.5 GPa on dry powder samples of chlorpropamide form a, *Int. J. Pharm.* **327**, pp. 51–57.

[212] Boldyreva, E. V., Sowa, H., Ahsbahs, H., Goryainov, S., Chernyshev, V., Dmitriev, V., Seryotkin, Y., Kolesnik, E., Shakhtshneider, T. P., Ivashevskaya, S. and Drebushchak, T. (2008). Pressure induced phase transitions in organic molecular crystals: a combination of X-ray single crystal and powder diffraction, Raman and IR spectroscopy, *J. Phys: Conference Series* **121**, p. 022023.

[213] Boldyreva, E. V. (1999). Interplay between intra- and intermolecular interactions in solid-state reactions: general overview, in Boldyreva, E. V. and Boldyrev, V. V. (Eds.), *Reactivity of Molecular Solids* (John Wiley, Chichester), pp. 1–50.

[214] Boldyreva, E. V. (2003). High-pressure-induced structural changes in molecular crystals preserving the space group symmetry: anisotropic distortion/isosymmetric polymorphism, *Cryst. Eng.* **6**, pp. 235–254.

[215] Boldyreva, E. V. (2003). High-pressure studies of the anisotropy of structural distortion of molecular crystals, *J. Mol. Struct.* **647**, pp. 159–179.

[216] Boldyreva, E. V. (2004). High pressure and supramolecular systems, *Russ. Chem. Bull.* **53**, pp. 1369–1378.

[217] Boldyreva, E. V. (2008). High-pressure diffraction studies of molecular organic solids. A personal view, *Acta Cryst. A* **64**, pp. 218–231.

[218] Katrusiak, A. (2004). General description of hydrogen-bonded solids at varied pressures and temperatures, in Katrusiak, A. and McMillan, P. F. (Eds.), *High-pressure Crystallography* (Kluwer Academic Publishing, Dordrecht) pp. 513–520.

[219] Katrusiak, A. (1995). High-pressure X-ray diffraction study of pentaerythritol, *Acta Cryst. B* **51**, pp. 873–879.

[220] Katrusiak, A. (1990). High-pressure X-ray diffraction study on the structure and phase transition of 1,3-cyclohexanedione, *Acta Cryst. B* **46**, pp. 246–256.

[221] Katrusiak, A. (1991). High-pressure X-ray diffraction study of 2-methyl-1,3-cyclopentanedione, *High Press. Res.* **6**, pp. 155–167.

[222] Katrusiak, A. (1990). Structure of 1,3-cyclopentanedione, *Acta Cryst. C* **46**, pp. 1289–1293.

[223] Katrusiak, A. (1991). High-pressure X-ray diffraction study of dimedone, *High Press. Res.* **6**, pp. 265–275.

[224] Boldyreva, E. V., Ahsbahs, H. and Weber, H. (2003). A comparative study of the pressure induced lattice strain of $\alpha-$ and $\gamma-$ polymorphs of glycine, *Z. Krist.* **218**, pp. 231–236.

[225] Iitaka, Y. (1960). The crystal structure of $\beta-$glycine, *Acta Cryst.* **13**, pp. 35–45.

[226] Boldyreva, E. V., Ivashevskaya, S., Sowa, H., Ahsbhs, H. and Weber, H. (2004). Effect of high pressure on crystalline glycine: a new high pressure polymorph, *Doklady Phys. Chem.* **396**, pp. 111–114.

[227] Boldyreva, E. V., Kolesnik, E., Drebushchak, T., Ahsbahs, H. and Beukes, J. (2005). A comparative study of the lattice strain induced in the crystal of l-serine by cooling down to 100 K or by increasing pressure up to 4.4 GPa, *Z. Krist.* **220**, pp. 58–65.

[228] Boldyreva, E. V., Shakhshneider, T. P., Ahsbahs, H. and Uchtmann, H. (2002). Pressure effect on the crystal structure of $Na_2C_2O_4$: deformation anisotropy during hydrostatic compression and phase transition at 3.8 GPa, *J. Struct. Chem.* **43**, pp. 101–107.

[229] Blank, V., Buga, S., Serebryanaya, N., Dubitsky, G., Sylyanov, S., Popov, M., Denisov, V., Ivlev, A. and Mavrin, B. (1996). Phase transformation in

solid C_{60} at high-pressure-high-temperature treatment and the structure of 3d polymerized fullerite, *Phys. Lett. A* **220**, pp. 149–157.

[230] Mezouar, M., Marques, L., Hodeau, J., Pischedda, V. and Nunez-Requeiro, M. (2003). Equation of state of an anisotropic three-dimensional C_{60} polymer: The most stable form of fullerene, *Phys. Rev. B* **68**, p. 193414.

[231] Christy, A. (1995). Isosymmetric structural phase transitions: phenomenology and examples, *Acta Cryst. B* **51**, pp. 753–757.

[232] Chanh, N., Clastre, J., Gaultier, J., Haget, Y. and Meresse, A. (1988). The tensor of compositional deformation. A new crystallographic way to analyse syncrystallization, *J. Appl. Cryst.* **21**, pp. 10–14.

[233] Zotov, N. (1990). Review of relationships between different strain tensors, *Acta Cryst A* **46**, pp. 627–628.

[234] Zotov, N. and Petrov, K. (1991). Analysis of the dependence of lattice deformation in Cu^{II}-Co^{II} hydroxide nitrate solid solutions on their composition, *J. Appl. Cryst.* **24**, pp. 227–231.

[235] Jessen, S. and Kuppers, H. (1991). The precision of thermal-expansion tensors of triclinic and monoclinic crystals, *J. Appl. Cryst* **24**, pp. 239–242.

[236] Schlenker, J. L., Gibbs, G. V. and Boisen, M. B. Jr. (1978). Strain tensor components expressed in terms of lattice parameters, *Acta Cryst.* **A34**, pp. 52–54.

[237] Singh, A. K. (1993). The lattice strain in a specimen (cubic system) compressed non-hydrostatically in an opposed anvil device, *J. Appl. Phys.* **73**, pp. 4278–4286.

[238] Singh, A. K. and Balasingh, C. (1994). The lattice strain in a specimen (hexagonal system) compressed non-hydrostatically in an opposed anvil high pressure setup, *J. Appl. Phys.* **75**, pp. 4956–4962.

[239] Singh, A. K., Balasingh, C., Mao, H. K., Hemley, R. J. and Shu, J. (1998). Analysis of lattice strain measured under non-hydrostatic pressure, *J. Appl. Phys.* **63**, pp. 7567–7575.

[240] Uchida, T., Funamori, N. and Yagi, T. (1996). Lattice strain in crystals under uniaxial stress field, *J. Appl. Phys.* **80**, pp. 739–746.

[241] Singh, A. K. (2000). Lattice strain under non-hydrostatic compression, in Manghnani, M. H., Nellis, W. J. and Nicol, M. (Eds.), *Science and Technology of High Pressure* (Universities Press, Hyderabad, India) pp. 62–67.

[242] Catti, M. (1989). Crystal elasticity and inner strain: a computational model, *Acta Cryst.* **A45**, pp. 20–25.

[243] Kinsland, G. and Bassett, W. (1976). Modification of the diamond cell for measuring strain and strength of materials at pressures up to 300 kilobars, *Rev. Sci. Instrum.* **47**, pp. 130–132.

[244] Hemley, R. J., Mao, H. K., Shen, G., Badro, J., Gillet, P., Hanfland, M. and Hausermann, D. (1997). X-ray imaging of stress and strain of diamond, iron and tungsten at megabar pressure, *Science* **276**, pp. 1242–1245.

[245] Singh, A. K., Mao, H. K., Shu, J. and Hemley, R. J. (1998). Estimation of single-crystal elastic moduli from polycrystalline X-ray diffraction at high pressure: application to FeO and iron, *Phys. Rev. Lett.* **80**, pp. 2157–2160.

[246] Wenk, H. R., Lonardelli, I., Merkel, S., Miyagi, L., Pehl, J., Speziale, S. and Tommaseo, C. E. (2006). Deformation textures produced in diamond anvil experiments, analyzed in radial diffraction geometry, *J. Phys. : Condens. Matter* **18**, pp. S933–S947.

[247] Mao, W. L., Struzhkin, V. V., Baron, A. Q. R., Tsutsui, S., Tommaseo, C. E., Wenk, H. R., Hu, M. Y., Chow, P., Sturhahn, W., Shu, J., Hemley, R. J., Heinz, D. L. and Mao, H. K. (2008). Experimental determination of the elasticity of iron at high pressure, *J. Geophys. Res.* **113**, p. B09213.

[248] Chesnut, G. N., Schiferl, D., Streetman, B. D. and Anderson, W. W. (2006). Diamond-anvil cell for radial X-ray diffraction, *J. Phys.: Condens. Matter* **18**, pp. S1083–S1090.

[249] Bassett, W. (2006). Deviatoric stress: a nuisance or a gold mine? *J. Phys.: Condens. Matter* **18**, pp. S921–S931.

[250] Yoneda, A. and Kubo, A. (2006). Simultaneous determination of mean pressure and deviatoric stress based on numerical tensor analysis: a case study for polycrystalline X-ray diffraction of gold enclosed in methanol-ethanol mixture, *J. Phys.: Condens. Matter* **18**, pp. S979–S994.

[251] Ohashi, Y. and Burnham, C. (1973). Clinopyroxene lattice deformations: the roles of chemical substitution and temperature, *Am. Mineral.* **58**, pp. 843–849.

[252] Marques, L., Mezouar, M., Hodeau, J.-L., Nunez-Regueiro, M., Serebryanaya, N. R., Ivdenko, V. A., Blank, V. D. and Dubitsky, G. A. (1999). "Debye-Scherrer ellipses"from 3D fullerene polymers: an anisotropic pressure memory signature, *Science* **283**, pp. 1720–1723.

[253] Mao, H. K., Badro, J., Shu, J., Hemley, R. J. and Singh, A. K. (2006). Strength, anisotropy, and preferred orientation of solid argon at high pressure, *J. Phys.: Condens. Matter* **18**, pp. S963–S968.

[254] Saxena, S. (2004). Pressure-volume equation of state for solids, *J. Phys. Chem. Solids* **65**, pp. 1561–1563.

[255] Li, J., Liang, S., Guo, H. and Liu, B. (2005). Four-parameters equation of state, *Appl. Phys. Lett.* **87**, p. 194111.

[256] Murnaghan, F. (1944). The compressibility of media under extreme pressure, *Proc. Natl. Acad. Sci. USA* **30**, pp. 244–247.

[257] Hofmeister, A. (1991). Pressure derivatives of the bulk modulus, *J. Geophys Res. Sol. Earth* **96**, pp. 21893–21907.

[258] Citroni, M., Bini, R., Costantini, B. and Schettino, V. (2009). Crystalline indole at high pressure: chemical stability, electronic and vibrational properties, *J. Phys. Chem. B* **113**, pp. 13526–13536.

[259] Birch, F. (1947). Finite elastic strain in cubic crystals, *Phys. Rev.* **71**, pp. 809–824.

[260] Cohen, R., Gulseren, O. and Hemley, R. (2000). Accuracy of equation-of-state formulation, *Am. Mineral.* **85**, pp. 338–344.

[261] Poirier, J. and Tarantola, A. (1998). A logaritmic equation of state, *Phys. Earth Plan. In.* **109**, p. 1–8.

[262] Vinet, P., Smith, J., Ferrante, J. and Rose, J. (1987). Temperature effects on the universal equation of state, *Phys. Rev. B* **35**, pp. 1945–1953.

[263] Vinet, P., Rose, J., Ferrante, J. and Smith, J. (1989). Universal features of the equation of state of solids, *J. Phys. Cond. Matter* **1**, pp. 1941–1963.

[264] Hama, J. and Suito, K. (1996). The search for a universal equation of state correct up to very high pressure, *J. Phys.: Condens. Matter* **8**, pp. 67–81.

[265] Duffy, T. and Wang, Y. (1998). Pressure-volume-temperature equation of state, in Hemley, R. J. (Ed.), *Ultrahigh-pressure mineralogy Reviews in Mineralogy*, **37**, pp. 425–458.

[266] Isaacs, N. S. (1981). *Liquid Phase High Pressure Chemistry* (Wiley, Chichester).

[267] Jenner, G. (2002). High pressure organic synthesis: overview of recent applications, Hemley, R. J., Chiarotti, G. L., Bernasconi, M. and Ulivi, L. (Eds.), *High Pressure Phenomena*, Proceedings of the International School of Physics Enrico Fermi, Course CXLVII, (IOS Press, Amsterdam) pp. 373–393.

[268] Sherman, W. F. and Stadmuller, A. A. (1987). *Experimental Techniques in High Pressure Research* (Wiley, Chichester).

[269] Hall, H. T. (1964). High pressure-temperature apparatus, in Gschneidner Jr, K. A., Hepworth, M. T. and Parlee, N. A. D. (Eds.) *Metallurgy at High Pressures and High Temperatures* (Gordon and Breach Science Publishers, New York).

[270] Coes Jr., L. L. (1953). A new dense crystalline silica, *Science* **118**, pp. 131–132.

[271] Coes Jr., L. L. (1962). Synthesis of minerals at high pressures in Wentorf, Jr., R. H. (Ed.) *Modern Very High Pressure Techniques* (Butterworths, London).

[272] Birch, F., Robertson, E. C. and Clark Jr., S. P. (1957). Apparatus for pressures of 27,000 bars and temperatures of 1400 °C, *Ind. and Eng. Chem.* **49**, pp. 1965–1966.

[273] Hall, H. T. (1958). Some high-pressure, high-temperature apparatus design considerations: equipment for use at 100000 atmospheres and 3000 °C, *Rev. Sci. Instr.* **29**, pp. 267–275.

[274] Boyd, F. R. and England, J. L. (1960). Apparatus for phase-equilibrium measurements at pressures up to 50 kilobars and temperatures up to 1750 °C, *J. Geophys. Res.* **65**, pp. 741–748.

[275] Jayaraman, A., Hutson, A. R., McFee, J. H., Coriell, A. S. and Maines, R. G. (1967). Hydrostatic and uniaxial pressure generation using teflon cell container in conventional piston-cylinder device, *Rev. Sci. Instrum.* **38**, pp. 44–49.

[276] Boyd, F. R. and England, J. L. (1958). Experimentation at high pressure, *Yearb. Carnegie Instn.*, **57**, pp. 170–175.

[277] Giardini, A. A., Tydings, J. E. and Levin, S. B. (1960). Very high pressure-high temperature research apparatus and the synthesis of diamond, *Amer. Min.* **45**, pp. 217–220.

[278] Ferraro, J. R. and Basile, L. J. (1974). Spectroscopy at high pressures status report and update of instrumental techniques, *Appl. Spectrosc.* **28**, pp. 505–517.

[279] Eckel, R., Buback, M. and Strobl, G. R. (1981). Untersuchung der druckinduzierten kristallisation von polyäthylen mit hilfe einer neuen Ramanhochdruckzelle, *Colloid. Polym. Sci.* **259**, pp. 326–334

[280] Poulter, T. C. (1932). Apparatus for optical studies at high pressure, *Phys. Rev.* **40**, pp. 860–871.

[281] Eremets, M. I. (1996). *High Pressure Experimental Methods* (Oxford University Press, New York).

[282] Besson, J. M., Pinceaux, J. P. and Piottrzkowski, R. (1974). *High Temp.-High Press.* **6**, p. 101.

[283] Chervin, J. C. Syfosse, G. and Besson, J. M. (1994). Mechanical strength of sapphire windows under pressure, *Rev. Sci. Instrum.* **65**, pp. 2719–2725.

[284] Fitch, R. A., Slykhouse, T. F. and Drickamer, H. G. (1957). Apparatus for optical studies to very high pressures, *J. Opt. Soc. Am.* **47**, pp. 1015–1016.

[285] Drickamer, H. G. and Balchan, A. S. (1962). Optical and electrical studies at high pressures, in Wentorf, R. N. (Ed.), *Modern Very High Pressure Techiques* (Butterworths Scientific Publications Ltd., London).

[286] Kennedy, G. C. and LaMori, P. N. (1961). Some fixed points on the high pressure scale, in Bundy, F. B., Hibbard Jr., W. R. and Strong, H. M. (Eds.), *Progress in Very High Pressure Research* (John Wiley, New York) pp. 304–313.

[287] Holzapfel, W. B. and Isaacs, N. S. (1997). *High-pressure Techniques in Chemistry and Physics, A Practical Approach* (Oxford University Press, New York).

[288] Le Noble, W. J. and Schlott, R. (1977). All quartz optical cell of constant diameter for use in high pressure studies, *Rev. Sci. Instrum.* **47**, pp. 770–771.

[289] Bundy, F. P. (1988). Ultra-high pressure apparatus, *Phys. Rep.* **167**, pp. 133–176.

[290] Walker, I. R. (1999). Nonmagnetic piston-cylinder pressure cell for use at 35 kbar and above, *Rev. Sci. Instrum.* **70**, pp. 3402–3412.

[291] Semerchan, A. A., Kuzin, N. N. and Davydova, T. N. (1981). *Dokl. Akad. Nauk SSSR* **258**, p. 1359.

[292] Fukunaga, O., Yamaoka, S., Akaishi, M., Kanda, H., Osawa, T., Shimomura, O., Nagashima, T. and Yoshikawa, M. (1987). Large-volume flat belt apparatus, *High Pressure Research in Mineral Physics: A Volume in Honor of Syun-iti Akimoto*, Geophysical Monograph Series Vol. 39, pp. 17–28.

[293] Bridgman, P. W. (1935). Effects of high shearing stress combined with high hydrostatic pressure, *Phys. Rev.* **48** pp. 825–847.

[294] Bridgman, P. W. (1941). Explorations toward the limit of utilizable pressures, *J. Appl. Phys.* **12**, pp. 461–469.

[295] Bridgman, P. W. (1950). Bakerian Lecture. Physics above 20,000 kg/cm^2, *Proc. Roy. Soc. A* **203**, pp. 1–17.

[296] Wartman, E. (1859). Proceedings of Learned Societies *Phil. Mag.* **17** p. 441.

[297] Lees, J. (1966). In Bradley, R. S. (Ed.), *Advances in High Pressure Research* Vol. 1. (Academic Press, London).

[298] Fang, L., He, D., Chen, C., Ding, L. and Luo, X. (2007). Effect of precompression on pressure-transmitting efficiency of pyrophyllite gaskets, *High Press. Res.* **27** pp. 367–374.

[299] Griggs, D. T. and Kennedy, G. C. (1956). A simple apparatus for high pressures and temperatures, *Am. J. Sci.* **254**, pp. 722–735.

[300] Yoneda, A. (1987). Highest pressure attainable in Bridgman anvil apparatus, *High Temp. - High Press.* **19**, pp. 531–536.

[301] Fasol, G. and Schilling, J. C. (1978). New hydrostatic pressure cell to 90 kilobars for precise electrical and magnetic measurements at low temperatures, *Rev. Sci. Instrum.* **49**, pp. 1722–1724.

[302] Reghu, M., Vaidyanathan, R. S., Prasad, V. and Subramanyam, S. V. (1990). Miniature high-pressure cell for transport property measurements down to 2 K, *Rev. Sci. Instrum.* **61**, pp. 1336–1338.

[303] Balchan, A. S. and Drickamer, H. G. (1961). High-pressure electrical resistance cell, and calibration points above 100 kilobars, *Rev. Sci. Instrum.* **32**, pp. 308–313.

[304] Sherman, W. F. (1966). A high-pressure absorption cell for obtaining well-resolved infra-red spectra of samples under 50 kb pressures at 90 K, *J. Sci. Instrum.* **43**, pp. 462–465.

[305] Bundy, F. P. (1975). Ultrahigh pressure apparatus using cemented tungsten carbide pistons with sintered diamond tips, *Rev. Sci. Instrum.* **46**, pp. 1318–1324.

[306] Yagi, T., Utsumi, W., Yamakata, W. and Shimomura, O. (1992). High-pressure *in situ* X-ray-diffraction study of the phase transformation from graphite to hexagonal diamond at room temperature, *Phys. Rev. B* **46**, pp. 6031–6039.

[307] Okuchi, T., Sasaki, S., Osakabe, T., Ohno, Y., Odake, S. and Hagi, H. (2010). Large-volume static compression using nano-polycrystalline diamond for opposed anvils in compact cells, *J. Phys.: Conf. Ser.* **215**, p. 012188.

[308] Hall, H. T. (1960). Ultra-high-pressure, high-temperature apparatus: the "Belt", *Rev. Sci. Instrum.* **31**, pp. 125–131.

[309] Khvostantsev, L. G. (1984). Toroidal high-pressure devices with combined pistons, *High Temp. - High Press.* **16**, pp. 171–176.

[310] Khvostantsev, L. G. (1984). A verkh-niz (up-down) toroid device for generation of high pressure, *High Temp. - High Press.* **16**, pp. 165–169.

[311] Semerchan, A. A., Kuzin, N. N., Davydova, T. N. and Bibaev, K. Kh. (1983). High pressure equipment operating at pressures of about 80 Kbar and temperatures of about 1500 °C, *Sov. J. Superhard Mater.* **5**, pp. 8–10.

[312] Besson, J. M., Nelmes, R. J., Hamel, G., Loveday, J. S., Weill, G. and Hull, S. (1992). Neutron powder diffraction above 10 GPa, *Physica B* **180–181** pp. 907–910.

[313] Klotz, S., Besson, J. M., Hamel, G., Nelmes, R. J., Loveday, J. S., Marshall, W. G. and Wilson, R. M. (1994). Neutron powder diffraction at pressures beyond 25 GPa, *Appl. Phys. Lett.* **66**, pp. 1735–1737.

[314] Le Godec, Y., Dove, M. T., Francis, D. J., Kohn, S. C., Marshall, W. G., Pawley, A. R., Price, G. D., Redfern, S. A. T., Rhodes, N., Ross, N. L., Schofield, P. F., Schooneveld, E., Syfosse, G., Tucker, M. G. and Welch, M. D. (2001). Neutron diffraction at simultaneous high temperatures and pressures, with measurement of temperature by neutron radiography, *Mineral. Mag.* **65** pp. 737–748.

[315] Le Godec, Y., Dove, M. T., Redfern, S. A. T., Tucker, M. G., Marshall, W. G., Syfosse, G. and Klotz, S. (2003). Recent developments using the Paris-Edinburgh cell for neutron diffraction at high pressure and high temperature and some applications, *High Press. Res.* **23** pp. 281–287.

[316] Grima, P., Polian, A., Gauthier, M., Itie, J. P., Mezouar, M., Weill, G., Besson, J. M., Hauserman, D. and Hanfland, M. (1995). Phase relationships in mercury telluride under high temperature and pressure, *J. Phys. Chem. Solids* **56**, pp. 525–530.

[317] Klotz, S., Besson, J. M., Hamel, G., Nelmes, R. J., Loveday, J. S., Marshall, W. G. and Wilson, R. M. (1993). Crystal structure studies to 10 GPa with the Paris-Edinburgh cell: high pressure aspects, in Schmidt, S. C., Shaner, J. W., Samara, G. A. and Ross, M. (Eds.), *High-pressure Science and Technology* (AIP Press, New York) pp. 1577–1580.

[318] Marshall, W. G. and Francis, D. J. (2002). Attainment of near-hydrostatic compression conditions using the Paris-Edinburgh cell, *J. Appl. Crystallogr.* **35** pp. 122–125.

[319] Ito, E. (2007). Theory and practice - multianvil cells and high-pressure experimental methods, in Price, G. D. (Ed.), *Mineral Physics* (Elsevier, Amsterdam) pp. 198–230.

[320] Liebermann, R. C. (2011). Multi-anvil, high pressure apparatus: a half-century of development and progress, *High Press. Res.* **31** pp. 493–532.

[321] Wakutsuki, M. and Ichinose, K. (1982). *High Pressure Research in Geophysics* Akimoto, S. and Manghnani, M. M. (Eds.) Vol.12 of *Advances in Earth and Planetary Sciences*, Dodrecht, p.5.

[322] Onodera, A. (1987). Octahedral-anvil high-pressure devices. *High Temp.-High Press.* **19** pp. 579–609.

[323] Yagi, T. (2002). Experimental overview of large volume techniques, in Hemley, R. J., Chiarotti, G. L., Bernasconi, M. and Ulivi, L. (Eds.), *High Pressure Phenomena*, Proceedings of the International School of Physics Enrico Fermi, Course CXLVII (IOS Press, Amsterdam) pp. 41–54.

[324] Kumazawa, M. and Endo, S. (1984). Recent progress in the generation of static high pressure by means of the multiple anvil, in Sunagawa, I. (Ed.), *Materials Science of the Earth's Interior* (Terra Scientific Publishing, Tokyo) pp. 587–603.

[325] Graham, E. K. (1986). Recent developments in conventional high-pressure methods, *J. Geophys. Res.* **91** pp. 4630–4642.

[326] Graham, E. K. (1987). The multianvil press, *Methods in Experimental Physics* **24** pp. 237–270.

[327] Akimoto, S. (1987). High-pressure research in geophysics: past, present and future, in Manghnani, M. H. and Syono, Y. (Ed.), *High-Pressure Research in Mineral Physics* (Terra Scientific Publishing/American Geophysical Union, Tokyo/Washington) pp. 1–13.

[328] Liebermann, R. C. (1979). Impact of technology on high-pressure geophysics, in Newell, H. (Ed.), *Impact of Technology on Geophysics* (National Academy of Sciences, Washington) pp. 96–109.

[329] Irifune, T. and Tsuchiya, T. (2007). Mineralogy of the Earth - phase transitions and mineralogy of the lower mantle, in Price, G. D. (Ed.), *Treatise on Geophysics* (Elsevier, Amsterdam) pp. 33–62.

[330] Hoppertz, H. (2004). Multianvil high-pressure / high-temperature synthesis in solid state chemistry, *Z. Kristallogr.* **219** pp. 330–338.

[331] Osugi, J., Shimizu, K., Inoue, K. and Yasunami, K. (1964). A compact cubic anvil high pressure apparatus, *Rev. Phys. Chem. Jpn.* **34** pp. 1–6.

[332] Kawai, N. (1966). Apparatus with tapering multi-pistons forming a sphere, *Proc. Jpn. Acad.* **42** pp. 385–388.

[333] Kawai, N. and Endo, S. (1970). The generation of ultrahigh hydrostatic pressures by a split sphere apparatus, *Rev. Sci. Instrum.* **41**, pp. 1178–1181.

[334] Yagi, T. (2001). Kawai-type apparatus, *Rev. High Press. Sci. Tech.* **11** pp. 171–172.

[335] Ito, E., Katsura, T., Aizawa, Y., Kawabe, K., Yokoshi, S., Kubo, A., Nozawa, A. and Funakoshi, K. (2005). High-pressure generation in the Kawai-type apparatus equipped with sintered diamond anvils: application to the wurtzite-rocksalt transformation in GaN, in Chen, J., Duffy, T. S., Shen, G. and Dobrzhinetskaya, L.F. (Eds.), *Advances in High-pressure Technology for Geophysical Applications* (Elsevier, Amsterdam) pp. 451–460.

[336] Kawai, N., Togaya, M. and Onodera, A. (1973). A new device for pressure vessels, *Proc. Jpn. Acad.* **49**, pp. 623–626.

[337] Shatskiy, A., Katsura, T., Litasov, K. D., Shcherbakova, A. V., Borzdov, Y. M., Yamazaki, D., Yoneda, A., Ohtani, E. and Ito, E. (2011). High pressure generation using scaled-up Kawai-cell, *Phys. Earth Planet. In.* **189**, pp. 92–108.

[338] Prikhna, A. I. and Borimsky, A. I. (1974). A cubic high pressure chamber, *Sinteticheskie Almazy* **3**, pp. 6–8.

[339] Prikhna, A. I. and Borimsky, A. I. (1974). High temperature-high pressure apparatus for growing diamond monocrystals, *J. Cryst. Growth* **26**, pp. 129–132.

[340] Walker, D., Carpenter, M. A. and Hitch, C. M., (1990). Some simplifications to multianvil devices for high pressure experiments, *Am. Mineral.* **75**, pp. 1020–1028.

[341] Ohtani, E., Irifune, T., Hibberson, W. O. and Ringwood, A. E. (1987). Modified split-sphere guide block for practical operation of a multiple anvil apparatus, *High Temp.-High Press.* **19**, pp. 523–529.

[342] Walker, D. (1991). Lubrication, gasketing, and precision in multianvil experiments, *Am. Mineral.* **76**, pp. 1092–1100.

[343] Bassett, W. A. (2009). Diamond anvil cell, 50th birthday, *High Press. Res.* **29**, pp. 163–186.

[344] Lawson, A. W. and Tang, T. Y. (1950). A diamond bomb for obtaining powder pictures at high pressures, *Rev. Sci. Instrum.* **21**, p. 815.

[345] Jamieson, J. C. (1957). Introductory studies of high-pressure polymorphism to 24.000 bars by X-ray diffraction with some comments on calcite II, *The Journal of Geology* **65**, pp. 334–343.

[346] Jamieson, J. C., Lawson, A. W. and Nachtrieb, N. D. (1959). New device for obtaining X-ray diffraction patterns from substances exposed to high pressure. *Rev. Sci. Instrum.* **30**, pp. 1016–1019.

[347] Weir, C. E., Lippincott, E. R., Van Valkenburg, A. and Bunting, E. N. (1959). Infrared studies in the 1- to 15-micron region to 30000 atmospheres, *J. Res. Natl. Bur. Stand. Sec A* **63**, pp. 55–62.

[348] Jayaraman, A. (1983). Diamond anvil cell and high-pressure physical investigations, *Rev. Mod. Phys.* **55**, pp. 65–108.

[349] Jayaraman, A. (1986). Ultrahigh pressures, *Rev. Sci. Instrum.* **57**, pp. 1013–1031.

[350] Block, S. and Piermarini, G. J. (1976). The diamond cell stimulates high-pressure research, *Physics Today* **29**, p. 44.

[351] Block, S. and Piermarini, G. J. (1984). *Vibrational Spectroscopy at High External Pressure* (Academic Press, Orlando).

[352] Williams, Q. and Jeanloz, R. (1991). Ultra-high-pressure experimental technique, in Gale, R. G. and Lovering, D. G. (Eds.), *Molten Salt Techniques*, Vol. 4 (Plenum Pres, New York) pp. 193–227.

[353] Jephcoat, A. P., Mao, H. K. and Bell, P. M. (1987). Operation of the megabar diamond anvil cell, in Ulmer, G. C. and Barnes, H. E. (Eds.), *Hydrothermal Experimental Techniques* (Wiley-Interscience, New York) pp. 469–506.

[354] Spain, I. L. and Dunstan, D. J. (1989). The technology of diamond anvil high-pressure cells: II. operation and use, *J. Phys. E: Sci. Instrum.* **22**, pp. 923–933.

[355] Dunstan, D. J. and Spain, I. L. (1989). Technology of diamond anvil high-pressure cells: I. principles, design and construction, *J. Phys. E* **22**, pp. 913–923.

[356] Van Valkenburg, A. (1965). *Conference Internationale sur les Hautes Pressions*, LeCreusot, Saone-et-Loire, France

[357] Mao, H. K. and Bell, P. M. (1978). Design and varieties of the megabar cell, *Carnegie Inst. Yearbk* **77**, pp. 904–908.

[358] Yu, S. C., Liu, C. Y., Spain, I. L. and Skelton, E. F. (1979). In Timmerhaus, K. D. and Barkereds, M. S. (Eds.), *High Pressure Science and Technology* (Plenum Publishing, New York) pp. 274–286.

[359] Piermarini, G. J. and Block, S. (1975). Ultrahigh pressure diamond-anvil cell and several semiconductor phase transition pressures in relation to the fixed point pressure scale, *Rev. Sci. Instrum.* **46**, pp. 973–979.

[360] Piermarini, G. J., Forman, R. A. and Block, S. (1978). Viscosity measurements in the diamond anvil pressure cell, *Rev. Sci. Instrum.*, **49**, pp. 1061–1066.

[361] Bassett, W. A., Takahashi, T. and Stock, P. W. (1967). X-ray diffraction and optical observations on crystalline solids up to 300 kbar, *Rev. Sci. Instrum.* **38**, pp. 37–42.

[362] Mao, H. K. (1978). High-pressure physics: sustained static generation of 1.36 to 1.72 Megabars, *Science* **200**, pp. 1145–1147.

[363] Ciezak, J. A. and Jenkins, T. A. (2011). Optical cell for *in situ* vibrational spectroscopic measurements at high pressures and shear, *Rev. Sci. Instrum.* **82**, p. 073905.

[364] Huber, G., Syassen, K. and Holzapfel, W. B. (1977). Pressure dependence of 4f levels in europium pentaphosphate up to 400 kbar, *Phys. Rev. B* **15**, pp. 5123–5128.

[365] Keller, R. and Holzapfel, W. B. (1977). Diamond anvil device for X-ray diffraction on single crystals under pressures up to 100 kilobar, *Rev. Sci. Instrum.* **48**, pp. 517–523.

[366] Hirsch, K. R. and Holzapfel, W. B. (1981). Diamond anvil high-pressure cell for Raman spectroscopy. *Rev. Sci. Instrum.* **52**, pp. 52–55.

[367] Dunstan, J. and Scherrer, M. (1988). Miniature cryogenic diamond-anvil high-pressure cell, *Rev. Sci. Instrum.* **59**, pp. 627–630.

[368] Tozer, S. W. (1993). Miniature diamond-anvil cell for electrical transport measurements in high magnetic fields, *Rev. Sci. Instrum.* **64**, pp. 2607–2610.

[369] Eremets, M. I. and Timofeev, Yu. A. (1992). Miniature diamond anvil cell: incorporating a new design for anvil alignment, *Rev. Sci. Instrum.* **63**, pp. 3123–3126.

[370] Machavariani, G. Yu., Pasternak, M. P., Heame, G. R. and Rozenberg, G. Kh. (1998). A multipurpose miniature piston-cylinder diamond-anvil cell for pressures beyond 100 GPa, *Rev. Sci. Instrum.* **69**, pp. 1423–1425.

[371] Gavriliuk, A. G., Mironovic, A. A. and Struzhkin, V. V. (2009). Miniature diamond anvil cell for broad range of high pressure measurements, *Rev. Sci. Instrum.* **80**, p. 043906.

[372] Merrill, L. and Bassett, W. A. (1974). Miniature diamond anvil pressure cell for single crystal X-ray diffraction studies, *Rev. Sci. Instrum.* **45**, pp. 290–294.

[373] Evans, W. J., Yoo, C. S., Lee, G. W., Cynn, H., Lipp, M. J. and Visbeck, K. (2007). Dynamic diamond anvil cell (dDAC): a novel device for studying the dynamic-pressure properties of materials, *Rev. Sci. Instrum.* **78**, p. 073904.

[374] Bassett, W. A., Shen, A. H., Bucknum, M. and Ming Chou, I. (1993). A new diamond anvil cell for hydrothermal studies to 2.5 GPa and from 190 to 1200 °C, *Rev. Sci. Instrum.* **64**, pp. 2340–2345.

[375] Burchard, M., Zaitsev, A. M. and Maresch, W. V. (2003). Extending the pressure and temperature limits of hydrothermal diamond anvil cells, *Rev. Sci. Instrum.* **74**, pp. 1263–1266.

[376] Ceppatelli, M., Santoro, M., Bini, R. and Schettino, V. (2000). Fourier transform infrared study of the pressure and laser induced polymerization of solid acetylene, *J. Chem. Phys.* **113**, pp. 5991–6000.

[377] Schettino, V. and Bini, R. (2003). Molecules under extreme conditions: chemical reactions at high pressure, *Phys. Chem. Chem. Phys.* **5** pp. 1951–1965.

[378] Schettino, V., Bini, R., Ceppatelli, M., Ciabini, L. and Citroni, M. (2005). Chemical reactions at very high pressure, in Rice S. A. (Ed.), *Advances in Chemical Physics*, Vol. 131 (Wiley, New York) pp. 105–242.

[379] Schettino, V. and Bini, R. (2007). Constraining molecules at the closest approach: chemistry at high pressure, *Chem. Soc. Rev.* **36**, pp. 869–880.

[380] Le Toullec, R., Pinceaux, J. P. and Loubeyre, P. (1988). The membrane diamond anvil cell: a new device for generating continuous pressure and temperature variations, *High Press. Res.* **1**, pp. 77–90.

[381] Le Toullec, R., Loubeyre, P., Pinceaux, J. P., Mao, H. K. and Hu, J. (1993). Single crystal X-ray diffraction with a synchrotron source in a mdac at low temperature, *High Press. Res.* **8**, pp. 691–696.

[382] Chervin, J. C., Canny, B., Besson, J. M. and Pruzan, Ph. (1995). A diamond anvil cell for IR microspectroscopy, *Rev. Sci. Instrum.* **66**, pp. 2595–2598.

[383] Sterer, E. and Silvera, I. F. (2006). The c-DAC: A novel cubic diamond anvil cell with large sample volume/area and multidirectional optics, *Rev. Sci. Instrum.* **77**, p. 115105.

[384] Piermarini, G. J. and Weir, C. E. (1962). A diamond cell for X-ray diffraction studies at high pressures, *J. Res. Natl. Bur. Stand. Sec A*, **66**, pp. 325–332.

[385] Weir, C. E., Block, S. and Piermarini, G. J. (1965). Single-crystal X-ray diffraction at high pressures, *J. Res. Natl. Bur. Stand. Sec C*, **69**, pp. 275–282.

[386] Mao, H. K. and Bell, P. M. (1976). High-pressure physics: the 1-Megabar mark on the ruby R1 static pressure scale, *Science* **191**, pp. 851–852.

[387] Vohra, Y. K., Duclos, S. J., Brister, K. E. and Ruoff, A. L. (1988). Static pressure of 255 GPa (2.55 Mbar) by X-ray diffraction: comparison with extrapolation of the ruby pressure scale, *Phys. Rev. Lett.* **61**, pp. 574–577.

[388] Mao, H. K., Wu, Y., Chen, L. C., Shu, J. F. and Jephcoat, A. P. (1990). Static compression of iron to 300 GPa and $Fe_{0.8}Ni_{0.2}$ alloy to 260 GPa: implications for composition of the core, *J. Geophys. Res.* **95**, pp. 21737–21742.

[389] Ruoff, A. L., Xia, H., Luo, H. and Vohra, Y. (1990). Miniaturization techniques for obtaining static pressures comparable to the pressure at the center of the Earth: X-ray diffraction at 416 GPa, *Rev. Sci. Instrum.* **61**, pp. 3830–3833.

[390] Forman, R. A., Piermarini, G. J., Barnett, J. D. and Block, S. (1972). Pressure measurement made by the utilization of ruby sharp-line luminescence, *Science* **176**, pp. 284–285.

[391] Barnett, J. D., Block, S. and Piermarini, G. J. (1973). An optical fluorescence system for quantitative pressure measurement in the diamond-anvil cell, *Rev. Sci. Instrum.*, **44**, pp. 1–9.

[392] Piermarini, G. J., Block, S., Barnett, J. D. and Forman, R. A. (1975). Calibration of the pressure dependence of the R1 ruby fluorescence line to 195 kbar, *J. Appl. Phys.* **46**, pp. 2774–2780.

[393] Piermarini, G. J., Block, S. and Barnett, J. D. (1973). Hydrostatic limits in liquids and solids to 100 kbar, *J. Appl. Phys.* **44**, pp. 5377–5382.

[394] Mao, H. K. and Bell, P. M. (1979). Design of the diamond window, high pressure apparatus for cryogenic experiments, *Carnegie Inst. Yearbk* **78**, pp. 659–660.

[395] Besson, J. M. and Pinceaux, J. P. (1979). Melting of helium at room temperature and high pressure, *Science*, **206**, pp. 1073–1075.

[396] Hemley, R. J. and Mao, H. K. (2002). Overview of static high pressure science, in Hemley, R. J., Chiarotti, G. L., Bernasconi, M. and Ulivi, L. (Eds.), *High Pressure Phenomena*, Proceedings of the International School of Physics Enrico Fermi, Course CXLVII (IOS Press, Amsterdam) pp. 3–40.

[397] Pravica, M. G. and Silvera, I. F. (1998). NMR study of ortho-para conversion at high pressure in hydrogen, *Phys. Rev. Lett.* **81**, pp. 4180–4183.

[398] Shimizu, K., Suhara, K., Ikumo, M., Eremets, M. I. and Amaya, K. (1998). Superconductivity in oxygen, *Nature* **393**, pp. 767–769.

[399] Struzhkin, V. V., Gregoryanz, E., Mao, H.,K., Hemley, R. J. and Timofeev, Y. A. (2002). New methods for investigating superconductivity, in Hemley, R. J., Chiarotti, G. L., Bernasconi, M. and Ulivi, L. (Eds.), *High Pressure Phenomena*, Proceedings of the International School of Physics Enrico Fermi, Course CXLVII (IOS Press, Amsterdam) pp. 275–296.

[400] Ruoff, A. L., Xia, H. and Xia, Q. (1992). The effect of a tapered aperture on X-ray diffraction from a sample with a pressure gradient: studies on three samples with a maximum pressure of 560 GPa, *Rev. Sci. Instrum.* **63**, pp. 4342–4348.

[401] Moshary, F., Chen, N. H. and Silvera, I. F. (1993). Remarkable high pressure phase line of orientational order in solid hydrogen deuteride, *Phys. Rev. Lett.* **71**, pp. 3814–3817.

[402] Goncharov, A. F., Hemley, R. J., Mao, H. K. and Shu, J. (1998). New high-pressure excitations in parahydrogen, *Phys. Rev. Lett.* **80**, pp. 101–104.

[403] Webb, A. W., Gubser, D. U. and Towle, L. C. (1976). Cryostat for generating pressures to 100 kilobar and temperatures to 0.03 K, *Rev. Sci. Instrum.* **47**, pp. 59–62.

[404] Shaw, R. W. and Nicol, M. (1981). Simple low-temperature press for diamond-anvil high pressure cells, *Rev. Sci. Instrum.* **52**, pp. 1103–1104.

[405] Eremets, M. I., Gregoryanz, E., Mao, H. K., Hemley, R. J., Mulders, N. and Zimmerman, N. M. (2000). Electrical conductivity of xenon at megabar pressures, *Phys. Rev. Lett.* **85**, pp. 2797–2800.

[406] Haselwimmer, R. K. W., Tyer, A. W. and Pugh, E. (1998). Millikelvin diamond anvil cell for the study of quantum critical phenomena, *Review of High Pressure Science and Technology* **7**, pp. 481–483.

[407] Boehler, R. (1993). Temperatures in the Earth's core from melting-point measurements of iron at high static pressures, *Nature* **363**, pp. 534–536.

[408] Ming, L. C. and Bassett, W. A. (1974). Laser heating in the diamond anvil press up to 2000 °C sustained and 3000 °C pulsed at pressures up to 260 kilobars, *Rev. Sci. Instrum.* **45**, pp. 1115–1118.

[409] Blank, V., Popov, M., Pivovarov, G., Lvova, N., Gogolinsky, K. and Reshetov, V. (1998). Ultrahard and superhard phases of fullerite C_{60}: comparison with diamond on hardness and wear, *Diam. Relat. Mater.* **7**, pp. 427–431.

[410] Pan, Z., Sun, H., Zhang, Y. and Chen, C. (2009). Harder than diamond: superior indentation strength of wurtzite BN and lonsdaleite, *Phys. Rev. Lett.* **102** p. 055503.

[411] Ruoff, A. L. and Wanagel, J. (1977). High pressures on small areas, *Science* **198** pp. 1037–1038.

[412] Telling, R. H., Pickard, C. J., Payne, M. C. and Field, J. E. (2000). Theoretical strength and cleavage of diamond, *Phys. Rev. Lett.* **84**, pp. 5160–5163.

[413] Luo, X., Liu, Z., Xu, B., Yu, D., Tian, Y., Wang, H.-T. and He, J. (2010). Compressive strength of diamond from first-principles calculation, *J. Phys. Chem. C* **114**, pp. 17851–17853.

[414] Nelson, D. A. and Ruoff, A. L. (1979). The compressive strength of perfect diamond, *J. Appl. Phys.* **50**, pp. 2763–2765.

[415] Zhang, Y., Sun, H. and Chen, C. F. (2006). Structural deformation, strength, and instability of cubic BN compared to diamond: a first-principles study, *Phys. Rev. B* **73**, pp. 144115–144121.

[416] Walker, J. (1979). Optical absorption and luminescence in diamond, *Rep. Prog. Phys.* **42**, pp. 1605–1660.

[417] Landstrass, M. I. and Ravi, K. V. (1989). Hydrogen passivation of electrically active defects in diamond, *Appl. Phys. Lett.* **55**, pp. 1391–1393.

[418] Anthony, T. R., Banholzer, W. F., Fleischer, J. F., Wei, L., Kuo, P. K., Thomas, R. L. and Pryor, R. W. (1990). Thermal diffusivity of isotopically enriched ^{12}C diamond, *Phys. Rev. B* **42**, pp. 1104–1111.

[419] Born, M. and Wolf, E. (1993). *Principles of Optics* (Pergamon Press Ltd., London).

[420] Boheler, R. and De Hantesetters, K. (2004). New anvil designs in diamond-cells, *High Press. Res.* **24**, pp. 391–396.

[421] Kenichi, T. and Satoshi, N. (2003). Performance of a synthetic diamond backing-plate for the diamond-anvil cell at ultrahigh pressures, *Rev. Sci. Instrum.* **74**, pp. 3017–3020.

[422] Seal, M. (1984). Diamond anvils, *High Temp.- High Press.* **16**, pp. 573–579.

[423] Jelezko, F. and Wrachtrup, J. (2006). Single defect centres in diamond: a review, *Phys. Status Solidi A* **203**, pp. 3207–3225.

[424] Davies, G. (Ed.), (1994). *Properties and Growth of Diamond* (EMIS Data Review Series, London).

[425] Adams, D. M. and Sharma, S. K. (1977). Selection of diamonds for infrared and Raman spectroscopy, *J. Phys. E: Sci. Instrum.* **10**, pp. 680–682.

[426] Dadashev, A., Pasternak, M. P., Rozenberg, G. K. and Taylor, R. D. (2001). Applications of perforated diamond anvils for very high-pressure research, *Rev. Sci. Instrum.* **72**, pp. 2633–2637.

[427] Yan, C. S., Vohra, Y. K., Mao, H.-K. and Hemley, R. J. (2002). Very high growth rate chemical vapor deposition of single-crystal diamond, *Proc. Natl. Acad. Sci. USA* **99**, pp. 12523–12525.

[428] Vohra, Y. K. and Vagarali, S. S. (1992). Isotopically pure diamond anvil for ultrahigh pressure research, *Appl. Phys. Lett.* **61**, pp. 2860–2862.

[429] Vohra, Y. K., McCauley, T. S., Gu, G. and Vagarali, S. S. (1993). Isotopically pure ^{12}C diamond anvil at megabar pressure, in Schmidt, S. C., Shaner, J. W., Samara, G. A. and Ross, M. (Eds.), *High Pressure Science and Technology* (AIP Press, New York) pp. 515–518.

[430] Goncharov, A. F., Struzhkin, V. V., Mao, H. K. and Hemley, R. J. (1999). Raman spectroscopy of dense H_2O and the transition to symmetric hydrogen bonds, *Phys. Rev. Lett.* **83**, pp. 1998–2001.

[431] Merkel, S., Goncharov, A. F., Mao, H. K., Gillet, P. and Hemley, R. J. (2000). Raman spectroscopy of iron to 152 Gigapascals: implications for Earth's inner core, *Science* **288**, pp. 1626–1629.

[432] Xu, J., Mao, H.-K. and Hemley, R. J. (2002). The gem anvil cell: high-pressure behaviour of diamond and related materials, *J. Phys. Condens. Matter.* **14**, pp. 11549–11552.

[433] Xu, J., Yeh, S., Yen, J. and Huang, E. (1996). Raman study of D_2O at high pressures in a cubic zirconia anvil cell, *J. Raman Spectroscopy* **27**, pp. 823–827.

[434] Russell, T. P. and Piermarini, G. J. (1997). A high pressure optical cell utilizing single crystal cubic zirconia anvil windows, *Rev. Sci. Instrum.* **68**, pp. 1835–1840.

[435] Hawke, R. S., Syassen and K., Holzapfel W. B. (1972). An apparatus for high pressure Raman spectroscopy, *Rev. Sci. Instrum.* **45**, pp. 1598–1601.

[436] Chervin, J. C., Syfosse, G. and Besson, J. M. (1994). Mechanical strength of sapphire windows under pressure, *Rev. Sci. Instrum.* **65**, pp. 2719–2725.

[437] Takano, K. J. and Wakatsuki, M. (1991). An optical high pressure cell with spherical sapphire anvils, *Rev. Sci. Instrum.* **62**, pp. 1576–1580.

[438] Xu, J., Yen, J., Wang, Y. and Huang, E. (1996). Ultrahigh pressures in gem anvil cells, *High Press. Res.* **15**, pp. 127–134.

[439] Onodera, A., Hasegawa, M., Furuno, K., Kobayashi, M., Nisida, Y., Sumija, H. and Yazu, S. (1991). Pressure dependence of the optical-absorption edge of diamond, *Phys. Rev. B* **44**, pp. 12176–12179.

[440] Onodera, A., Nakatani, A., Kobayashi, M., Nisida, Y. and Mishima, O. (1993). Pressure dependence of the optical-absorption edge of cubic boron nitride, *Phys. Rev. B* **48**, pp. 2777–2780.

[441] Trojan, I. A., Eremets, M. I., Korolik, M. Y., Struzhkin, V. V. and Utjuzh, A. N. (1993). Fundamental gap of diamond under hydrostatic pressure, *Jpn. J. Appl. Phys.* **32**, pp. 282–284.

[442] Xu, J. and Mao, H. K. (2000). Moissanite: a window for high-pressure experiments, *Science* **290**, pp. 783–785.

[443] Xu, J., Mao, H. K., Hemley, R. J. and Hines, E. (2002). The moissanite anvil cell: a new tool for high-pressure research, *J. Phys: Condens. Matter* **14**, pp. 11543–11548.

[444] Liu, Z., Xu, J., Scott, H. P., Williams, Q., Mao, H.-K. and Hemley, R. J. (2004). Moissanite (SiC) as windows and anvils for high-pressure infrared spectroscopy, *Rev. Sci. Instrum.* **75**, pp. 5026–5029.

[445] Dunstan, D. J. (1989). Theory of the gasket in diamond anvil high-pressure cells, *Rev. Sci. Instrum.* **60**, pp. 3789–3795.

[446] Orloff, J., Narajana, C. and Ruoff A. L. (2000). Use of focused ion beams for making tiny sample holes in gaskets for diamond anvil cells, *Rev. Sci. Instrum.* **71**, pp. 216–219.

[447] Hemley, R. J., Mao, H. K., Shen, G., Badro, J., Gillet, Ph., Hanfland, M. and Hausermann, D. (1997). X-ray imaging of stress and strain of diamond, iron, and tungsten at megabar pressures, *Science* **276**, pp. 1242–1245.

[448] Mao, H. K. and Hemley, R. J. (1996). Energy dispersive X-ray diffraction of micro-crystals at ultrahigh pressures, *High Press. Res.* **14**, pp. 257–267.

[449] Leong, D., Feyrit, H., Prins, A. D., Wilkinson, V. A., Homewood, K. P. and Dunstan, D. J. (2003). Laminated gaskets for absorption and electrical measurements in the diamond anvil cell, *Rev. Sci. Instrum.* **63**, pp. 5760–5763.

[450] Weck, G., Loubeyre, P., Eggert, J., Mezouar, M. and Hanfland, M. (2007). Melting line and fluid structure factor of oxygen up to 24 GPa, *Phys. Rev. B* **76**, p. 054121

[451] Datchi, F., Loubeyre, P. and LeToullec, R. (2000). Extended and accurate determination of the melting curves of argon, helium, ice (H_2O), and hydrogen (H_2), *Phys. Rev. B* **61**, pp. 6535–6546.

[452] He, D., Zhao, Y., Sheng, T. D., Schwarz, R. B., Qian, J., Lokshin, K. A., Bobev, S., Daemen, L. L., Mao, H. K., Hu, J. Z., Shu, J. and Xu, J. (1992). Bulk metallic glass gasket for high pressure, *in situ* X-ray diffraction, *Rev. Sci. Instrum.* **74**, pp. 3012–3016.

[453] Boehler, R., Ross, M. and Boercker, D. B. (1997). Melting of LiF and NaCl to 1 Mbar: systematics of ionic solids at extreme conditions, *Phys. Rev. Lett.* **78**, pp. 4589–4592.

[454] Zou, G., Ma, Y., Mao, H. K., Hemley, R. J. and Gramsh, S. A. (2001). A diamond gasket for the laser-heated diamond anvil cell, *Rev. Sci. Instrum.* **72**, pp. 1298–1301.

[455] Pravica, M. and Remmers, B. (2003). A simple and efficient cryogenic loading technique for diamond anvil cells. *Rev. Sci. Instrum.* **74**, pp. 2782–2783.

[456] Baggen, M., Manuputy, R., Scheltema, R. and Lagendijk, A. (1988). Diamond anvil cell and loading system for liquid CO_2, *Rev. Sci. Instrum.* **59**, pp. 2592–2595.
[457] Schouten, J. A., Trappeniers, N. J. and van den Bergh, L. C. (1983). Diamond-anvil system for the investigation of phase equilibria in mixtures at high pressures, *Rev. Sci. Instrum.* **54**, pp. 1209–1212.
[458] Mills, R. L., Liebenberg, D. H., Bronson, J. C. and Schmidt, L. C. (1980). Procedure for loading diamond cells with high-pressure gas, *Rev. Sci. Instrum.* **51**, pp. 891–895.
[459] Yagi, T., Yusa, H. and Yamakata, M. (1996). An apparatus to load gaseous materials to the diamond anvil cell, *Rev. Sci. Instrum.* **67**, pp. 2981–2984.
[460] Kenichi, T., Sahub, P. Ch., Yoshiyasu, K. and Yasuo, T. (2001). Versatile gas-loading system for diamond-anvil cells, *Rev. Sci. Instrum.* **72**, pp. 3873–3876.
[461] Kurnosov, A., Kantor, I., Boffa-Ballaran, T., Lindhardt, S., Dubrovinsky, L., Kuznetsov, A. and Zehnder, B. H. (2008). A novel gas-loading system for mechanically closing of various types of diamond anvil cells, *Rev. Sci. Instrum.* **79**, p. 045110.
[462] Couzinet, B., Dahan, N., Hamel, G. and Chervin, J. C. (2003). Optically monitored high-pressure gas loading apparatus for diamond anvil cells, *High Press. Res.* **23**, pp. 409–415.
[463] Bocian, A., Bull, C. L., Hamidov, H., Loveday, J. S., Nelmes, R. J. and Kamenev, K. V. (2010). Gas loading apparatus for the Paris-Edinburgh press, *Rev. Sci. Instrum.* **81**, p. 093904.
[464] Liedenberg, D. H. (1979). A new hydrostatic medium for diamond anvil cells to 300 kbar pressure, *Phys. Lett.* **73A**, pp. 74–76.
[465] Ceppatelli, M., Bini, R. and Schettino, V. (2009). High-pressure photodissociation of water as a tool for hydrogen synthesis and fundamental chemistry, *Proc. Natl. Acad. Sci. USA* **106**, pp. 11454–11459.
[466] Takemura, K. (2007). Pressure scales and hydrostaticity, *High Press. Res.* **27** pp. 465–472.
[467] Fujishiro, I., Piermarini, G. J., Block, S. and Munro, R. G. (1981). Viscosities and glass transition pressures in the methanol-ethanol-water system, in Backman, C. M., Johannisson, T. and Tegner, L. (Eds.), *High Pressure in Research and Industry*, Proceedings of the 8th AIRAPT Conference Uppsala, p. 608.
[468] Besson, J. M. and Pinceaux, J. P. (1979). Uniform stress conditions in the diamond anvil cell at 200 kilobars, *Rev. Sci. Instrum.* **50**, pp. 541–543.
[469] Shen, Y., Kumar, R. S., Pravica, M. and Nicol, M. F. (2004). Characteristics of silicone fluid as a pressure transmitting medium in diamond anvil cells, *Rev. Sci. Instrum.* **75**, pp. 4450–4454.
[470] Decremps, F., Fischer, M., Polian, A., Itie, J. P. and Sieskind, M. (1999). Ionic layered PbFCl-type compounds under high pressure, *Phys. Rev. B* **59**, pp. 4011–4022.
[471] Mao, H. K. (1989). Static compression of simple molecular system in the megabar range, in Polian, A., Loubeyre, P. and Boccara, N. (Eds.), *Simple*

Molecular System at Very High Density (Plenum Press, New York) pp. 221–236.

[472] Loubeyre, P., Le Toullec, R., Hausermann, D., Hanfland, M., Hemley, R. J., Mao, H. K. and Finger, L. W. (1996). X-ray diffraction and equation of state of hydrogen at megabar pressures, *Nature* **383**, pp. 702–704.

[473] Mao, H. K., Xu, J. and Bell, P. M. (1986). Calibration of the ruby pressure gauge to 800 kbar under quasi-hydrostatic conditions, *J. Geophys. Res.* **91**, pp. 4673–4676.

[474] Zha, C. S., Boehler, R., Young, D. A. and Ross, M. (1986). The argon melting curve to very high pressures, *J. Chem. Phys.* **85**, pp. 1034–1036.

[475] Gillet, Ph., Fiquet, G., Daniel, I. and Reynard, B. (1999). Equations of state of ^{12}C and ^{13}C diamond, *Phys. Rev. B* **60**, pp. 14660–14664.

[476] Asaumi, K. and Ruoff, A. L. (1986). Nature of the state of stress produced by xenon and some alkali iodides when used as pressure media, *Phys. Rev. B* **33**, pp. 5633–5636.

[477] Richet, P., Xu, J. A. and Mao, H. K. (1988). Quasi-hydrostatic compression of ruby to 500 Kbar, *Phys. Chem. Miner.* **16**, pp. 207–211.

[478] Angel, R. J., Bujak, M., Zhao, J., Gatta, G. D. and Jacobsen, S. D. (2008). Effective hydrostatic limits of pressure media for high-pressure crystallographic studies, *J. Appl. Cryst.* **40**, pp. 26–32.

[479] Angel, R. J., Allan, D. R., Miletich, R. and Finger, L. W. (1997). The use of quartz as an internal pressure standard in high-pressure crystallography, *J. Appl. Cryst.* **30**, pp. 461–466.

[480] Varga, T., Wilkinson, A. P. and Angel, R. J. (2003). Fluorinert as a pressure-transmitting medium for high-pressure diffraction studies, *Rev. Sci. Instrum.* **74**, pp. 4564–4566.

[481] Chervin, J. C., Canny, B., Gauthier, M. and Pruzan, Ph. (1993). Micro-Raman at variable low-temperature and very high pressure, *Rev. Sci. Instrum.* **64**, pp. 203–206.

[482] Bini, R., Ballerini, R., Pratesi, G. and Jodl, H. J. (1997). Experimental setup for Fourier transform infrared spectroscopy studies in condensed matter at high pressure and low temperatures, *Rev. Sci. Instrum.* **68**, pp. 3154–3160.

[483] Gorelli, F. A., Ulivi, L., Santoro, M. and Bini, R. (1999). The ϵ phase of solid oxygen: evidence of an O_4 molecule lattice, *Phys. Rev. Lett.* **83**, pp. 4093–4096.

[484] Gorelli, F. A., Giordano, V. M., Salvi, P. R. and Bini, R. (2004). Linear carbon dioxide in the high-pressure high-temperature crystalline phase IV, *Phys Rev. Lett.* **93**, p. 205503.

[485] Webb, A. W., Gubser, D. U. and Towle, L. C. (1975). Cryostat for generating pressures to 100 kilobar and temperatures to 0.03 K, *Rev. Sci. Instrum.* **47**, pp. 59–62.

[486] Pobell, F. (2007). *Matter and Methods at Low Temperatures* (Springer-Verlag, Berlin).

[487] Dubrovinskaia, N. and Dubrovinsky, L. (2003). Whole-cell heater for the diamond anvil cell, *Rev. Sci. Instrum.* **74**, pp. 3433–3437.

[488] Heinz, D. L., Sweeney, J. S. and Miller, P. A. (1991). A laser heating system that stabilizes and controls the temperature: diamond anvil cell applications, *Rev. Sci. Instrum.* **62**, pp. 1568–1575.

[489] Liu, L. and Bassett, W. A. (1975). The melting of iron up to 200 kbar, *J. Geophys. Res.* **80**, pp. 3777–3782.

[490] Boehler, R. (1986). The phase diagram of iron to 430 kbar, *Geophys. Res. Lett.* **13**, pp. 1153–1156.

[491] Mao, H. K., Bell, P. M. and Hadidiacos, C. (1987). Experimental phase relations of iron to 360 kbar, 1400°C, determined in an internally heated diamond anvil apparatus, in Manghnani, M. H. and Syono Y. (Eds.), *High-pressure Research in Mineral Physics* (Terra Scientific Publishing Co., Tokyo, and American Geophysical Union, Washington DC) pp. 135–138.

[492] Boehler, R., Nicol, M., Zha, C. S. and Johnson, M. L. (1986). Resistance heating of Fe and W in diamond-anvil cell, *Physica B+C* **139–140**, pp. 916–918.

[493] Zha, C. S. and Bassett, W. A. (2003). Internal resistive heating in diamond anvil cell for *in situ* X-ray diffraction and Raman scattering, *Rev. Sci. Instrum.* **74**, pp. 1255–1262.

[494] Schiferl, D., Sharma, S. K., Cooney, T. F., Wang, S. Y. and Mohanan, K. (1993). Multichannel Raman spectrometry system for weakly scattering materials at simultaneous high pressures and high temperatures, *Rev. Sci. Instrum.* **64**, pp. 2821–2827.

[495] Schiferl, D. (1987). Temperature compensated high-temperature/high-pressure Merrill–Bassett diamond anvil cell, *Rev. Sci. Instrum.* **58**, pp. 1316–1317.

[496] Zinn, A. S., Schiferl, D. and Nicol, M. F. (1987). Raman spectroscopy and melting of nitrogen between 290 and 900 K and 2.3 and 18 GPa, *J. Chem. Phys.* **87**, pp. 1267–1271.

[497] Miletich, R., Allan, D. R. and Kuhs, W. F. (2000). High-pressure single-crystal techniques, in Hazen, R. M. and Downs, R. T. (Eds.), *High-Temperature and High Pressure Crystal Chemistry, Review in Mineralogy and Geochemistry* **41**, pp. 445–520.

[498] Arashi, H. (1987). Raman spectroscopic studies at high temperatures and high pressure: application to deternmination of P-T diagram of Zr_2O, in Manghnani, M. H. and Syono, Y. (Eds.), *High-pressure Research in Mineral Physics* (American Geophysical Union, Washington DC) pp. 335–339.

[499] Bassett, W. A., Shen, A. H., Bucknum, M. and Chou, I. M. (1993). A new diamond anvil cell for hydrothermal studies to 2.5 GPa and from 190 to 1200 °C, *Rev. Sci. Instrum.* **64**, pp. 2340–2345.

[500] Ming, L. C., Manghnani, M. H. and Balogh, J. (1987). Resistive heating in the diamond anvil cell under vacuum conditions, in Manghnani, M. H. and Syono, Y. (Eds.), *High-pressure Research in Mineral Physics* (American Geophysical Union, Washington DC) pp. 69–74.

[501] Gregoryanz, E., Goncharov, A. F., Matsuishi, K., Mao, H. K. and Hemley,

R. J. (2003). Raman spectroscopy of hot dense hydrogen, *Phys. Rev. Lett.* **90**, p. 175701.

[502] Kikegawa, T. (1987). X-ray diamond anvil press for structural studies at high pressures and high tenperatures, in Manghnani, M. H. and Syono, Y. (Eds.), *High-pressure Research in Mineral Physics* (American Geophysical Union, Washington DC) pp. 61–68.

[503] Ming, L. C., Manghnani, M. H., Balogh, J., Qadri, S. B., Skelton, E. F. and Jamieson, J. C. (1983). Gold as a reliable internal pressure calibrant at high temperatures, *J. Appl. Phys.* **54**, pp. 4390–4397.

[504] Fan, D., Zhou, W., Wei, S., Liu, Y., Ma, M. and Xie, H. (2010). A simple external resistance heating diamond anvil cell and its application for synchrotron radiation X-ray diffraction, *Rev. Sci. Instrum.* **81**, p. 053903.

[505] Bassett, W. A. (2001). The birth and development of laser heating in diamond anvil cells, *Rev. Sci. Instrum.* **72**, pp. 1270–1272.

[506] Bassett, W. A. and Ming, L. C. (1972). Disproportionation of Fe_2SiO_4 to $2FeO+SiO_2$ at pressures up to 250 kbar and temperatures up to 3000 °C, *Phys. Earth Planet. Inter.* **6**, pp. 154–160.

[507] Hemley, R. J., Mao, H. K. and Gramsch, S. A. (2000). Pressure-induced transformations in deep mantle and core minerals, *Mineral. Mag.* **64**, pp. 157–184.

[508] Nguyen, J. H. and Holmes, N. C. (2004). Melting of iron at the physical conditions of the Earth's core, *Nature* **427**, pp. 339–342.

[509] Shen, G., Mao, H. K., Hemley, R. J., Duffy, T. S. and Rivers, M. L. (1998). Melting and crystal structure of iron at high pressures and temperatures, *Geophys. Res. Lett.* **25**, pp. 373–376.

[510] Andrault, D., Fiquet, G., Charpin, T. and Le Bihan, T. (2000). Structure analysis and stability field of β-iron at high P and T, *Am. Mineral.* **85**, pp. 364–371.

[511] Williams, Q., Jeanloz, R., Bass, R., Svendsen, B. and Ahrens, T. J. (1987). The melting curve of iron to 250 Gigapascals: a constraint on the temperature at Earth's center, *Science* **236**, pp. 181–182.

[512] Eremets, M. I., Gavriliuk, A. G., Trojan, I. A., Dzivenko, D. A. and Boehler, R. (2004). Single-bonded cubic form of nitrogen, *Nat. Mater.* **3**, pp. 558–563.

[513] Santoro, M., Gorelli, F. A., Bini, R., Haines, J., Cambon, O., Levelut, C., Montoya, J. A. and Scandolo, S. (2012). Partially collapsed cristobalite structure in the non-molecular phase V in CO_2, *Proc. Natl. Acad. Sci. USA* **109**, pp. 5176–5179.

[514] Chandra Shekar, N. V., Sahu, P. Ch. and Govinda Rajan, K. (2003). Laser-Heated Diamond-Anvil Cell (LHDAC) in *Materials Science Research*, *J. Mater. Sci. Technol.* **19**, pp. 518–525.

[515] Goncharov, A. F. and Hemley, R. J. (2006). Probing hydrogen-rich molecular systems at high pressures and temperatures, *Chem. Soc. Rev.* **35**, pp. 899–907.

[516] Boehler, R., Errandonea, D. and Ross, M. (2002). The laser-heated diamond cell: high P-T phase diagrams, in Hemley, R. J., Chiarotti, G. L.,

Bernasconi, M., Ulivi, L. (Eds.), *High Pressure Phenomena*, Proceedings of the International School of Physics Enrico Fermi, Course CXLVII (IOS Press, Amsterdam) pp. 55–72.

[517] Rekhi, S., Tempere, J. and Silvera, I. F. (2003). Temperature determination for nanosecond pulsed laser heating, *Rev. Sci. Instrum.* **74**, pp. 3820–3825.

[518] Lin, J., Santoro, M., Struzhkin, V. V., Mao, H. K. and Hemley, R. J. (2004). *In situ* high pressure-temperature Raman spectroscopy technique with laser-heated diamond anvil cells, *Rev. Sci. Instrum.* **75**, pp. 3302–3306

[519] Mao, H. K., Shen, G., Hemley, R. J. and Duffy, T. S. (1998). X-ray diffraction with a double hot-plate laser-heated diamond cell, in Manghnani, M. H. and Yagi, T. (Eds.), *Properties of Earth and Planetary Materials at High Pressure and Temperature*, **101**, pp. 27–34.

[520] Heinz, D. L. and Jeanloz, R. (1987). Temperature measurements in the laser-heated diamond anvil cell, in Manghnani, M. H. and Syono, Y. (Eds.), *High-pressure Research in Mineral Physics* (American Geophysical Union, Washington DC) pp. 113–128.

[521] Boehler, R., von Bargen, N. and Chopelas, A. (1990). Melting, thermal expansion, and phase transitions of iron at high pressures, *J. Geophys. Res. B* **95**, pp. 21731–21736.

[522] Williams, Q., Jeanloz, R., Bass, J., Svendsen, B. and Ahrens, T. J. (1991). The high-pressure melting curve of iron: a technical discussion, *J. Geophys. Res. B* **96**, pp. 2171–2184.

[523] Jeanloz, R. and Heinz, D. L. (1984). Experiments at high temperature and pressure: laser heating through the diamond cell, *J. Phys. Coll.* **45** C8, pp. 83–92.

[524] Campbell, A. J. (2008). Measurement of temperature distributions across laser heated samples by multispectral imaging radiometry, *Rev. Sci. Instrum.* **79**, p. 015108.

[525] Kavner, A. and Nugent, C. (2008). Precise measurements of radial temperature gradients in the laser-heated diamond anvil cell, *Rev. Sci. Instrum.* **79**, p. 024902.

[526] Santoro, M., Lin, J., Mao, H. K. and Hemley, R. J. (2004). *In situ* high P-T Raman spectroscopy and laser heating of carbon dioxide, *J. Chem. Phys.* **121**, pp. 2780–2787.

[527] Lin, J. F., Sturhahn, W., Zhao, J., Shen, G., Mao, H. K. and Hemley, R. J. (2004). Absolute temperature measurement in a laser-heated diamond anvil cell, *Geophys. Res. Lett.* **31**, L14611.

[528] Zhao, J., Sturhahn, W., Lin, J. F., Shen, G., Alp, E. E. and Mao, H. K. (2004). Nuclear resonant scattering at high pressure and high temperature, *High Press. Res.* **24**, pp. 447–457.

[529] Decker, D. L., Bassett, W. L., Merrill, L., Hall, H. T. and Barnett, J. D. (1972). High-pressure calibration: a critical review, *J. Phys. Chem. Ref. Data* **1**, pp. 773–836.

[530] Bean, V. E., Akimoto, S., Bell, P. M., Block, S., Holzapfel, W. B., Manghnani, M. H., Nicol, M. F. and Stishov, S. M. (1986). Toward an international practical presure scale: 2nd AIRAPT IPPS task group report, *Physica B+C* **139–140**, pp. 52–54.

[531] Holzapfel, W. B., Hartwig, M. and Sievers, W. (2001). Equations of state for Cu, Ag, and Au for wide ranges in temperature and pressure up to 500 GPa and above, *J. Phys. Chem. Ref. Data* **30**, pp. 515–529.

[532] Dorogokupets, P. L. and Oganov, A. R. (2003). Equations of state of Cu and Ag and the revised ruby pressure scale, *Dokl. Earth Sci.* **391A**, pp. 854–857.

[533] Chijioke, A. D., Nellis, W. J. and Silvera, I. F. (2005). High-pressure equations of state of Al, Cu, Ta, and W, *J. Appl. Phys.* **98**, p. 73526.

[534] Holzapfel, W. B. and Nicol, M. F. (2007). Refined equations of state for Cu, Ag, and Au in the sub-TPa region, *High Press. Res.* **27**, pp. 377–392.

[535] Nellis, W. J. (2007). Adiabat-reduced isotherms at 100 GPa pressures, *High Press. Res.* **27**, pp. 393–407.

[536] Anderson, O. L., Isaak, D. G. and Yamamoto, S. (1989). Anharmonicity and the equation of state for gold, *J. Appl. Phys.* **65**, pp. 1534–1543.

[537] Heinz, D. L. and Jeanloz, R. (1994). The equation of state of the gold calibration standard, *J. Appl. Phys.* **55**, pp. 885–893.

[538] Chen, G., Liebermann, R. C. and Weidner, D. J. (1998). Elasticity of single-crystal MgO to 8 Gigapascals and 1600 Kelvin, *Science* **280**, pp. 1913–1916.

[539] Zha, C. S., Mao, H. K. and Hemley, R. J. (2000). Elasticity of MgO and a primary pressure scale to 55 GPa, *Proc. Nat. Acad. Sci.* **97**, pp. 13494–13499.

[540] Mao, H. K., Bell, P. M., Shaner, J. W. and Steinberg, D. J. (1978). Specific volume measurements of Cu, Mo, Pd, and Ag and calibration of the ruby R1 fluorescence pressure gauge from 0.06 to 1 Mbar, *J. Appl. Phys.* **49**, pp. 3276–3283.

[541] Piermarini, G. J. and Block, S. (1975). Ultrahigh pressure diamond-anvil cell and several semiconductor phase transition pressures in relation to the fixed point pressure scale, *Rev. Sci. Instrum.* **46**, pp. 973–979.

[542] Chervin, J. C., Canny, B. and Mancinelli, M. (2001). Ruby-spheres as pressure gauge for optically transparent high pressure cells, *High Press. Res.* **21**, pp. 305–314.

[543] Syassen, K. (2008). Ruby under pressure, *High Press. Res.* **28**, pp. 75–126.

[544] Moss, S. C. and Newnhan, R. E. (1964). The chromium position in ruby, *Z. Kristallogr.* **120**, pp. 359–363.

[545] McCauley, J. W. and Gibbs, G. V. (1972). Redetermination of the chromium position in ruby, *Z. Kristallogr.* **135**, pp. 453–455.

[546] Jephcoat, A. P., Hemley, R. J. and Mao, H. K. (1988). X-ray diffraction of ruby (Al_2O_3:Cr^{3+}) to 175 GPa, *Physica B* **150**, pp. 115–121.

[547] d'Amour, H., Schiferl, D., Denner, W., Schulz, H. and Holzapfel, W. B. (1978). High-pressure single-crystal structure determinations for ruby up to 90 kbar using an automatic diffractometer, *J. Appl. Phys* **49**, pp. 4411–4416.

[548] Finger, L. W. and Hazen, R. M. (1978). Crystal structure and compression of ruby to 46 kbar, *J. Appl. Phys* **49**, pp. 5823–5826.

[549] Kim-Zajonz, J., Werner, S. and Schulz, H. (1999). High pressure single crystal X-ray diffraction study on ruby up to 31 GPa, *Z. Kristallogr.* **214**, pp. 331–336.

[550] Lin, J., Degtyareva, O., Prewitt, C., Dera, P., Sata, N., Gregoryanz, E., Mao, H. K. and Hemley, R. J. (2004). Crystal structure of a high-pressure/high-temperature phase of alumina by *in situ* X-ray diffraction, *Nat. Mater.* **3**, pp. 389–393.

[551] Macfarlane, R. M. (1963). Analysis of the spectrum of d3 ions in trigonal crystal fields. *J. Chem. Phys.* **39**, pp. 3118–3126.

[552] Nelson, D. F. and Sturge, M. D. (1965). Relation between absorption and emission in the region of the R lines of ruby, *Phys. Rev.* **137**, pp. A1117–A1130.

[553] Sugano, S. and Tanabe, Y. (1970). *Multiplets of Transition Metal Ions in Crystals* (Academic, New York).

[554] Eggert, J. H., Goettel, K. A. and Silvera, I. F. (1989). Ruby at high pressure. I. Optical line shifts to 156 GPa, *Phys. Rev. B* **40**, pp. 5724–5732.

[555] Eggert, J. H., Goettel, K. A. and Silvera, I. F. (1989). Ruby at high pressure. II. Fluorescence lifetime of the R line to 130 GPa, *Phys. Rev. B* **40**, pp. 5733–5738.

[556] Kaplyanskii, A. A., Przhevuskii, A. K. and Rozenbaum, R. B. (1969). Concentration dependent line shift in optical spectra of ruby, *Sov. Phys. Solid State* **10**, p. 1864.

[557] Noack, R. A. and Holzapfel, W. B. (1979). Calibration of the ruby-pressure scale at low temperatures, in Timmerhaus, K. D. and Barber, M. S. (Eds.), *High-Pressure Science Technolgy* (Plenum, New York) pp. 748–753.

[558] Wunder, S. L. and Schoen, P. E. (1981). Pressure measurement at high temperatures in the diamond anvil cell, *J. Appl. Phys.* **52**, pp. 3772–3775.

[559] McCumber, D. E. and Sturge, M. D. (1963). Linewidth and temperature shift of the R lines in ruby, *J. Appl. Phys.* **34**, pp. 1682–1684.

[560] Duclos, S. J., Vohra, Y. and Ruoff, A. L. (1990). Pressure dependence of the $^4T_2$ and $^4T_1$ absorption bands of ruby to 35 GPa, *Phys. Rev. B* **41**, pp. 5372–5381.

[561] Eggert, J. H., Moshry, F., Evans, W. J., Goettel, K. A. and Silvera, I. F. (1991). Ruby at high pressure. III. A pumping scheme for the R lines up to 230 GPa, *Phys. Rev. B* **44**, pp. 7202–7208.

[562] Chen, N. H. and Silvera, I. F. (1996). Excitation of ruby fluorescence at multimegabar pressures, *Rev. Sci. Instrum.* **67**, pp. 4275–4278.

[563] Chai, M. and Brown, J. M. (1996). Effects of static non-hydrostatic stress on the R lines of ruby single crystals, *Geophys. Res. Lett.* **23**, pp. 3539–3542.

[564] Nakano, K., Akahama, Y., Ohishi, Y. and Kawamura, H. (2000). Ruby scale at low temperatures calibrated by the NaCl gauge: wavelength shift of ruby R1 fluorescence line at high pressure and low temperature, *Jpn. J. Appl. Phys.* **39**, pp. 1249–1251.

[565] Hemley, R. J. and Mao, H. K. (1988). Phase transition in solid molecular hydrogen at ultrahigh pressures, *Phys. Rev. Lett.* **61**, pp. 857–860.

[566] Holzapfel, W. B. (2003). Refinement of the ruby luminescence pressure scale, *J. Appl. Phys.* **93**, pp. 1813–1818.

[567] Occelli, F., Loubeyre, P. and Le Toullec, R. (2003). Properties of diamond under hydrostatic pressures up to 140 GPa, *Nature Mater.* **2**, pp. 151–154.

[568] Dewaele, A., Loubeyre, P. and Mezouar, M. (2004). Equations of state of six metals above 94 GPa, *Phys. Rev. B* **70**, p. 94112.

[569] Chijioke, A. D., Nellis, W. J., Soldatov, A. and Silvera, I. F. (2005). The ruby pressure standard to 150 GPa, *J. Appl. Phys.* **98**, p. 114905.

[570] Holzapfel, W. B. (2005). Progress in the realization of a practical pressure scale for the range 1300 GPa, *High Press. Res.* **25**, pp. 87–99.

[571] Sherman, W. F. (1985). The diamond Raman band as a high-pressure calibrant, *J. Phys. C Solid State Phys.* **18**, p. L973.

[572] Kunc, K., Loa, I. and Syassen, K. (2003). Equation of state and phonon frequency calculations of diamond at high pressures, *Phys. Rev. B* **68**, p. 094107.

[573] Kunc, K., Loa, I. and Syassen, K. (2004). Diamond under pressure: *Ab initio* calculations of the equation of state and optical phonon frequency revisited, *High Press. Res.* **24**, pp. 101–110.

[574] Ragan, D. D., Gustavsen, R. and Schiferl, D. (1992). Calibration of the ruby R1 and R2 fluorescence shifts as a function of temperature from 0 to 600 K, *J. Appl. Phys.* **72**, pp. 5539–5544.

[575] Vos, W. L. and Schouten, J. A. (1991). On the temperature correction to the ruby pressure scale, *J. Appl. Phys.* **69**, pp. 6744–6746.

[576] Rekhi, S., Dubrovinski, S. and Saxena, S. (1999). Temperature-induced ruby fluorescence shifts up to a pressure of 15 GPa in an externally heated diamond anvil cell, *High Temp. - High Press.* **31**, pp. 299–305.

[577] Lacam, A. and Chateau, C. (1989). High-pressure measurements at moderate temperatures in a diamond anvil cell with a new optical sensor: $SrB_4O_7:Sm^{2+}$, *J. Appl. Phys.* **66**, pp. 366–372.

[578] Datchi, F., LeToullec, R. and Loubeyre, P. (1997). Improved calibration of the $SrB_4O_7:Sm^{2+}$ optical pressure gauge: advantages at very high pressures and high temperatures, *J. Appl. Phys.* **81**, pp. 3333–3339.

[579] Hess, N. J. and Schiferl, D. (1992). Comparison of the pressure-induced frequency shift of Sm:YAG to the ruby and nitrogen vibron pressure scales from 6 to 820 K and 0 to 25 GPa and suggestions for use as a high-temperature pressure calibrant, *J. Appl. Phys.* **71**, pp. 2082–2086.

[580] Liu, J. and Vohra, Y. K. (1996). Photoluminescence and X-ray diffraction studies on Sm-doped yttrium aluminum garnet to ultrahigh pressures of 338 GPa, *J. Appl. Phys.* **79**, pp. 7978–7982.

[581] Arashi, H. and Ishigame, M. (1982). Diamond anvil pressure cell and pressure sensor for high-temperature use, *Jpn. J. Appl. Phys.* **21**, pp. 1647–1649.

[582] Lorenz, B., Shen, Y. R. and Holzapfel, W. B. (1994). Characterization of the new luminescence pressure sensor $SrFCl:Sm^{2+}$, *High Press. Res.* **12**, pp. 91–99.

[583] Schmidt, S. C., Schiferl, D., Zinn, A. S., Ragan, D. D. and Moore, D. S. (1991). Calibration of the nitrogen vibron pressure scale for use at high temperatures and pressures, *J. Appl. Phys.* **69**, pp. 2793–2799.

[584] Hanfland, M. and Syassen, K. (1985). A Raman study of diamond anvils under stress, *J. Appl. Phys.* **57**, pp. 2752–2756.

[585] Popov, M. (2004). Pressure measurements from Raman spectra of stressed diamond anvils, *J. Appl. Phys.* **95**, pp. 5509–5514.

[586] Sun, L., Ruoff, A. L. and Stupian, G. (2005). Convenient optical pressure gauge for multimegabar pressures calibrated to 300 GPa, *Appl. Phys. Lett.* **86**, p. 014103.

[587] Akahama, Y. and Kawamura, H. (2010). Pressure calibration of diamond anvil Raman gauge to 410 GPa, *J. Phys.: Conf. Ser.* **215**, p. 012195.

[588] Schiferl, D., Nicol, M., Zaug, J. M., Sharma, S. K., Cooney, T. F., Wang, S. Y., Anthony, T. R. and Fleischer, J. F. (1997). The diamond $^{13}C/^{12}C$ isotope Raman pressure sensor system for high-temperature/pressure diamond-anvil cells with reactive samples, *J. Appl. Phys.* **82**, pp. 3256–3265.

[589] Qiu, W., Baker, P. A., Velisavljevic, N., Vohra, Y. K. and Weir, S. T. (2006). Calibration of an isotopically enriched carbon-13 layer pressure sensor to 156 GPa in a diamond anvil cell, *J. Appl. Phys.* **99**, p. 064906.

[590] Kawamoto, T., Matsukage, K. N., Nagai, T., Nishimura, K., Mataki, T., Ochiai, S. and Taniguchi, T. (2004). Raman spectroscopy of cubic boron nitride under high temperature and pressure conditions: a new optical pressure marker, *Rev. Sci. Instrum.* **75**, pp. 2451–2454.

[591] Bini, R. (2004). Laser-assisted high-pressure chemical reactions, *Acc. Chem. Res.* **37**, pp. 95–101.

[592] Klug, D. and Whalley, E. (1983). Nitrite and nitrate ions as infrared pressure gauges for diamond anvils, *Rev. Sci. Instrum.* **54**, pp. 1205–1208.

[593] Ceppatelli, M., Serdyukov, A., Bini, R. and Jodl, H. J. (2009). Pressure induced reactivity of solid CO by FTIR studies, *J. Phys. Chem. B* **113**, pp. 6652–6660.

[594] Grzechnik, A., Simon, P., Gillet, P. and McMillan, P. (1999). An infrared study of $MgCO_3$ at high pressure, *Physica B* **262**, pp. 67–73.

[595] Moss, W. C., Hallquist, J. O., Reichlin, R., Goettel, K. A. and Martin, S. (1986). Finite element analysis of the diamond anvil cell: achieving 4.6 Mbar, *Appl. Phys. Lett.* **48**, pp. 1258–1260.

[596] Xu, J. A., Mao, H. K. and Bell, P. M. (1986). High-pressure ruby and diamond fluorescence: observations at 0.21 to 0.55 Terapascal, *Science* **232**, pp. 1404–1406.

[597] Jamieson, J. C., Fritz, J. N. and Manghnani, M. H. (1982). Pressure measurements at high temperature in X-ray diffraction studies: gold as a primary standard, in Akimoto S. and Manghnani, M. H. (Eds.), *High Pressure Research in Geophysics* (Center for Academic Publications Japan, Tokyo) pp. 27–47.

[598] Decker, D. L. (1965). Equation of state of NaCl and its use as a pressure gauge in high-pressure research, *J. Appl. Phys.* **36**, pp. 157–161.

[599] Zha, C. S., Bassett, W. A. and Shim, S. H. (2004). Rhenium, an *in situ* pressure calibrant for internally heated diamond anvil cells, *Rev. Sci. Instrum.* **75**, pp. 2409–2418.
[600] Holmes, N. C., Moriarty, J. A., Gathers, G. R. and Nellis, W. J. (1989). Anharmonicity and the equation of state for gold, *J. Appl. Phys.* **65**, pp. 1534–1543.
[601] Wang, L., Ding, Y., Yang, W., Liu, W., Cai, Z., Kung, J., Shu, J., Hemley, R. J., Mao, W. L. and Mao, H. K. (2010). Nanoprobe measurements of materials at megabar pressures, *Proc. Natl. Acad. Sci. USA* **107**, pp. 6140–6145.
[602] Boldyreva, E. and Dera, P. (Eds.) (2010). *High-Pressure Crystallography: From Fundamental Phenomena to Technological Applications*, NATO Science for Peace and Security Series B: Physics and Biophysics (Springer, Dordrecht).
[603] Vočadlo, L., Poirer, J. P. and Price, G. D. (2000). Grüneisen parameters and isothermal equations of state, *American Min.* **85**, pp. 390–396
[604] Hofmeister, A. M. and Mao, H. K. (2002). Redefinition of the mode Grüneisen parameter for polyatomic substances and thermodynamic implications, *Proc. Natl. Acad. Sci. USA* **99**, pp. 559–564.
[605] Bini, R., Ulivi, L., Jodl, H. J. and Salvi, P. R. (1995). High pressure crystal phases of solid CH_4 probed by Fourier transform infrared spectroscopy, *J. Chem. Phys.* **103**, pp. 1353–1360.
[606] Kobayashi, M. (2001). Infrared spectroscopy of pressure-induced metallization in semiconductors, *Phys. Stat. Sol. (b)* **223**, pp. 55–64.
[607] Chen, X. J., Struzhkin, V. V., Song, Y., Goncharov, A. F., Ahart, M., Liu, Z., Mao, H. K. and Hemley, R. J. (2008). Pressure-induced metallization of silane, *Proc. Natl. Acad. Sci. USA* **108**, pp. 20–23.
[608] Miller, L. M. and Smith, R. J. (2005). Synchrotrons versus globars, point-detectors versus focal plane arrays: Selecting the best source and detector for specific infrared microspectroscopy and imaging applications, *Vibrat. Spectrosc.* **38**, pp. 237–240.
[609] Zha, C. S., Liu, Z. and Hemley, R. J. (2012). Synchrotron infrared measurements of dense hydrogen to 360 GPa, *Phys. Rev. Lett.* **108**, p. 146402.
[610] Ferraro, J. R. and Basile, L. J. (1980). Interfacing a diamond anvil cell with a commercial interferometer, *Appl. Spectrosc.* **34**, pp. 217–219.
[611] Miller, P. J., Piermarini, G. J. and Block, S. (1984). An FT-IR microspectroscopic method for kinetic measurements at high temperatures and high pressures, *Appl. Spectrosc.* **38**, pp. 680–686.
[612] Johannsen, P. G., Wetringhaus, C. and Holzapfel, W. B. (1987). The effect of pressure on FIR spectra of solid chlorine, *J. Phys. C* **20**, p. L151.
[613] Challenger, W. A. and Thompson, J. D. (1986). Far-infrared spectroscopy in diamond anvil cells, *Appl. Spectrosc.* **40**, pp. 298–303.
[614] Chen, R. J. and Weinstein, B. A. (1996). New diamond-anvil cell design for far infrared magnetospectroscopy featuring *in situ* cryogenic pressure tuning, *Rev. Sci. Instrum.* **67**, pp. 2883–2889.
[615] Hofmeister, A. M., Xu, J., Mao, H. K., Bell, P. M. and Hoering, T. C.

(1989). Thermodynamics of Fe-Mg olivines at mantle pressures: mid- and far-infrared spectroscopy at high pressure, *Am. Mineral.* **74**, pp. 281–306.

[616] Chervin, J. C., Canny, B., Besson, J. M. and Pruzan, P. (1995). A diamond anvil cell for IR microspectroscopy, *Rev. Sci. Instrum.* **66**, pp. 2595–2598.

[617] Panchala, V., Segura, A. and Pellicer-Porresa, J. (2011). Low-cost set-up for Fourier-transform infrared spectroscopy in diamond anvil cell from 4000 to 400 cm^{-1}, *High Press. Res.* **31**, pp. 445–453.

[618] Chen, N. H., Sterer, E. and Silvera, I. F. (1996). Extended infrared studies of high pressure hydrogen, *Phys. Rev. Lett.* **76**, pp. 1663–1666.

[619] Li, F., Cui, Q., He, Z., Cui, T., Zhang, J., Zhou, Q., Zou, G. and Sasaki, S. (2005). High pressure-temperature Brillouin study of liquid water: evidence of the structural transition from low-density water to high-density water, *J. Chem. Phys.* **123**, p. 174511.

[620] Li, F., Li, M., Cui, Q., Cui, T., He, Z., Zhou, Q. and Zou, G. (2009). The velocity, refractive index, and equation of state of liquid ammonia at high temperatures and high pressures, *J. Chem. Phys.* **131**, p. 134502.

[621] Li, M., Li, F., Gao, W., Ma, C., Huang, L., Zhou, Q. and Cui, Q. (2010). Brillouin scattering study of liquid methane under high pressures and high temperatures, *J. Chem. Phys.* **133**, p. 044503.

[622] Qin, J., Li, M., Li, J., Chen, R., Duan, Z., Zhou, Q., Li, F. and Cui, Q. (2010). High temperatures and high pressures Brillouin scattering studies of liquid H_2O+CO_2 mixtures, *J. Chem. Phys.* **133**, p. 154513.

[623] Shimizu, H. and Sasaki, S. (1992). High-pressure Brillouin studies and elastic properties of single-crystal H_2S grown in a diamond cell, *Science* **257**, pp. 514–516.

[624] Gregoryanz, E., Hemley, R. J., Mao, H. K. and Gillet, P. (2000). High-pressure elasticity of α-quartz: instability and ferroelastic transition, *Phys. Rev. Lett.* **84**, pp. 3117–3120.

[625] Shimizu, H., Tashiro, H., Kume, T. and Sasaki, S. (2001). High-pressure elastic properties of solid argon to 70 GPa, *Phys. Rev. Lett.* **86**, pp. 4568–4571.

[626] Tkachev, S. N., Manghnani, M. H. and Williams, Q. (2005). *In situ* Brillouin spectroscopy of a pressure-induced apparent second-order transition in a silicate glass, *Phys. Rev. Lett.* **95**, p. 057402.

[627] Subramanian, N., Goncharov, A. F., Struzhkin, V. V., Somayazulu, M. and Hemley, R. J. (2011). Bonding changes in hot fluid hydrogen at megabar pressures, *Proc. Natl. Acad. Sci. USA* **108**, pp. 6014–6019.

[628] Goncharov, A. F. and Crowhurst, J. C. (2005). Pulsed laser Raman spectroscopy in the laser-heated diamond anvil cell, *Rev. Sci. Instrum.* **76**, p. 063905.

[629] Citroni, M., Ceppatelli, M., Bini, R. and Schettino, V. (2002). Laser-induced selectivity for dimerization versus polymerization of butadiene under pressure, *Science* **295**, pp. 2058–2060.

[630] Chelazzi, D., Ceppatelli, M., Santoro, M., Bini, R. and Schettino, V. (2004). High-pressure synthesis of crystalline polyethylene using optical catalysis, *Nat. Mater.* **3**, pp. 470–475.

[631] Citroni, M., Bini, R., Foggi, P. and Schettino, V. (2008). Role of excited electronic states in the high-pressure amorphization of benzene, *Proc. Natl. Acad. Sci. USA* **105**, pp. 7658–7663.

[632] Liu, B., Jin, M., Liu, H., He, C., Jiang, D. and Ding, D. (2008). Femtosecond time-resolved measurement of LDS698 molecular processes under high pressure, *Appl. Phys. Lett.* **92**, p. 241916.

[633] Liu, W. L., Zheng, Z. R., Zhang, J. P., Wu, W. Z., Li, A. H., Zhang, W., Huo, M. M., Liu, Z. G., Zhu, R. B., Zhao, L. C. and Su, W. H. (2012). White-light continuum probed femtosecond time-resolved absorption spectroscopic measurement of β-carotene under high pressure, *Chem. Phys. Lett.* **532**, pp. 47–51.

[634] Baggen, M., van Exter, M. and Lagendijk, A. (1987). Time-resolved stimulated Raman scattering in a diamond anvil cell, *J. Chem. Phys.* **86**, pp. 2423–2427.

[635] Baggen, M. and Lagendijk, A. (1991). Time resolved study of vibrational relaxation in solid CO_2 at high pressures, *Chem. Phys. Lett.* **177**, pp. 361–365.

[636] Crowell, R. A. and Chronister, E. L. (1992). A time-resolved Raman study of inhomogeneous dephasing in high-pressure molecular crystals, *Chem. Phys. Lett.* **195**, pp. 602–608.

[637] Crowell, R. A. and Chronister, E. L. (1992). Low-temperature vibrational relaxation in crystalline carbon disulfide at high pressure, *J. Phys. Chem.* **96**, pp. 1045–1051.

[638] Rice, S. F. and Costantino, M. S. (1989). Single-pulse, broad-band coherent anti-Stokes Raman spectroscopy of nitromethane within a diamond anvil cell, *J. Phys. Chem.* **93**, pp. 536–538.

[639] Hellwig, H., Daniels, W. B., Hemley, R. J., Mao, H. K. and Gregoryanz, E. (2001). Coherent anti-Stokes Raman scattering spectroscopy of solid nitrogen to 22 GPa, *J. Chem. Phys.* **115**, pp. 10876–10882.

[640] Baer, B. J. and Yoo, C. S. (2005). Laser heated high density fluids probed by coherent anti-Stokes Raman spectroscopy, *Rev. Sci. Instrum.* **76**, p. 013907.

[641] Baer, B. J., Evans, W. J. and Yoo, C. S. (2007). Coherent anti-Stokes Raman spectroscopy of highly compressed solid deuterium at 300 K: evidence for a new phase and implications for the band gap, *Phys. Rev. Lett.* **98**, p. 235503.

[642] Ciabini, L., Santoro, M., Gorelli, F. A., Bini, R., Schettino, V. and Raugei, S. (2007). Triggering dynamics of the high-pressure benzene amorphization, *Nat. Mater.* **6**, pp. 39–43.

[643] Ciabini, L., Santoro, M., Gorelli, F. A., Bini, R. and Schettino, V. (2002). High pressure photoinduced ring opening of benzene, *Phys. Rev. Lett.* **88**, p. 085505.

[644] Citroni, M., Costantini, B., Bini, R. and Schettino, V. (2009). Crystalline indole at high pressure: chemical stability, electronic, and vibrational properties, *J. Phys. Chem. B* **113**, pp. 13526–13535.

[645] Fanetti, S., Citroni, M. and Bini, R. (2011). Pressure-induced fluorescence of pyridine, *J. Phys. Chem. B* **115**, pp. 12051–12058.

[646] Cohn, W. M. (1933). X-ray investigations at high pressures, *Phys. Rev.* **44**, pp. 326–327.

[647] Smith, J. S. and Desgreniers, S. (2012). Exploiting area detectors to reduce measured Compton scattering via energy selection, *Nucl. Instrum. Meth. A* **668**, pp. 9–13.

[648] Eggert, J. H., Weck, G., Loubeyre, P. and Mezouar, M. (2002). Quantitative structure factor and density measurements of high-pressure fluids in diamond anvil cells by X-ray diffraction: Argon and water, *Phys. Rev. B* **65**, p. 174105.

[649] Shen, G., Prakapenka, V. B., Rivers, M. L. and Sutton, S. R. (2003). Structural investigation of amorphous materials at high pressures using the diamond anvil cell, *Rev. Sci. Instrum.* **74**, pp. 3021–3026.

[650] Ma, Y., Mao, H. K., Hemley, R. J., Gramsch, S. A., Shen, G. and Somayazulu, M. (2001). Two-dimensional energy dispersive X-ray diffraction at high pressures and temperatures, *Rev. Sci. Instrum.* **72**, pp. 1302–1305.

[651] Weck, G., Loubeyre, P. and Le Toullec, R. (2002). Observation of structural transformations in metal oxygen, *Phys. Rev. Lett.* **88**, p. 035504.

[652] Jephcoat, A. P., Mao, H. K., Finger, L. W., Cox, D. E., Hemley, R. J. and Zha, C. S. (1987). Pressure-induced structural phase transitions in solid xenon, *Phys. Rev. Lett.* **59**, pp. 2670–2673.

[653] Yoo, C. S., Akella, J., Campbell, A. J., Mao, H. K. and Hemley, R. J. (1995). Phase diagram of iron by *in situ* X-ray diffraction: implications for Earth's core, *Science* **270**, pp. 1473–1475

[654] Vohra, Y. K. and Spencer, P. T. (2001). Novel γ-phase of titanium metal at megabar pressures, *Phys. Rev. Lett.* **86**, pp. 3068–3071.

[655] Hemley, R. J., Jephcoat, A. P., Mao, H. K., Ming, L. C. and Manghnani, M. H. (1988). Pressure-induced amorphization of crystalline silica, *Nature* **334**, pp. 52–54.

[656] Meade, C., Hemley, R. J. and Mao, H. K. (1992). High-pressure X-ray diffraction of SiO_2 glass, *Phys. Rev. Lett.* **69**, pp. 1387–1390.

[657] Dewaele, A., Mezouar, M., Guignot, N. and Loubeyre, P. (2010). High melting points of tantalum in a laser-heated diamond anvil cell, *Phys. Rev. Lett.* **104**, p. 255701.

[658] Fiquet, G., Auzende, A. L., Siebert, J., Corgne, A., Bureau, H., Ozawa, H. and Garbarino, G. (2010). Melting of peridotite to 140 Gigapascals, *Science* **329**, pp. 1516–1518.

[659] Wenk, H. R., Matthies, S., Hemley, R. J., Mao, H. K. and Shu, J. (2000). The plastic deformation of iron at pressures of the Earth's inner core, *Nature* **405**, pp. 1044–1047.

[660] McMahon, M. I., Nelmes, R. J. and Rekhi, S. (2001). Complex crystal structure of cesium-III, *Phys. Rev. Lett.* **87**, p. 255502.

[661] Lundegaard, L. F., Weck, G., McMahon, M. I., Desgreniers, S. and Loubeyre, P. (2006). Observation of an O_8 molecular lattice in the ϵ phase of solid oxygen, *Nature* **443**, pp. 201–204.

[662] Datchi, F., Giordano, V. M., Munsch, P. and Saitta, A. M. (2009). Structure of carbon dioxide Phase IV: breakdown of the intermediate bonding state scenario, *Phys. Rev. Lett.* **103**, p. 185701.

[663] Pascarelli, S., Ruffoni, M. P., Trapananti, A., Mathon, O., Aquilanti, G., Ostanin, S., Staunton, J. B. and Pettifer, R. F. (2007). Effect of pressure on magnetoelastic coupling in 3d metal alloys studied with X-ray absorption spectroscopy, *Phys. Rev. Lett.* **99**, p. 237204.

[664] Itie, J. P., Polian, A., Calas, G., Petiau, J., Fontaine, A. and Tolentino, H. (1989). Pressure-induced coordination changes in crystalline and vitreous GeO_2, *Phys. Rev. Lett.* **63**, pp. 398–401.

[665] Itie, J. P., Baudelet, F., Dartyge, E., Fontaine, A., Tolentino, H. and San Miguel, A. (1992). X-ray absorption spectroscopy and high pressure, *High Pres. Res.* **8**, pp. 697–702.

[666] Vaccari, M., Garbarino, G., Yannopoulos, S. N., Andrikopoulos, K. S. and Pascarelli, S. (2009). High pressure transition in amorphous As_2S_3 studied by EXAFS, *J. Chem. Phys.* **131**, p. 224502.

[667] Baldini, M., Aquilanti, G., Mao, H. K., Yang, W., Shen, G., Pascarelli, S. and Mao, W. L. (2010). High-pressure EXAFS study of vitreous GeO_2 up to 44 GPa, *Phys. Rev. B* **81**, p. 024201.

[668] Sapelkin, A. V. and Bayliss, S. C. (2001). X-ray absorption spectroscopy under high pressures in diamond anvil cells, *High Press. Res.* **21**, pp. 315–329.

[669] Baldini, M., Yang, W., Aquilanti, G., Zhang, L., Ding, Y., Pascarelli, S. and Mao, W. L. (2011). High-pressure EXAFS measurements of crystalline Ge using nanocrystalline diamond anvils, *Phys. Rev. B* **84**, p. 014111.

[670] Krisch, M. and Sette, F. (2007). Inelastic X-ray scattering from phonons, in Cardona, M. and Merlin, R. (Eds.), *Light Scattering in Solid IX, Novel Materials and Techniques*; *Topics Appl. Physics* **108** (Springer-Verlag, Berlin Heidelberg) pp. 317–370.

[671] Rueff, J. P. and Shukla, A. (2010). Inelastic X-ray scattering by electronic excitations under high pressure, *Rev. Mod. Phys.* **82**, pp. 847–896.

[672] Fiquet, G., Badro, J., Guyot, F., Requardt, H. and Krisch, M. (2001). Sound velocities in iron to 110 Gigapascals, *Science* **291**, pp. 468–471.

[673] Gorelli, F. A., Santoro, M., Scopigno, T., Krisch, M., Byrk, T., Ruocco, G. and Ballerini, R. (2009). Inelastic X-ray scattering from high pressure fluids in a diamond anvil cell, *Appl. Phys. Lett.* **94**, p. 074102.

[674] Krisch, M., Loubeyre, P., Ruocco, G., Sette, F., Cunsolo, A., D'Astuto, M., Le Toullec, R., Lorenzen, M., Mermet, A., Monaco, G. and Verbeni, R. (2002). Pressure evolution of the high-frequency sound velocity in liquid water, *Phys. Rev. Lett.* **89**, p. 125502.

[675] Ghose, S., Krisch, M., Oganov, A., Beraud, A., Bossak, A., Gulve, R., Seelaboyina, R., Yang, H. and Saxena, S. K. (2006). Lattice dynamics of MgO at high pressure: theory and experiment, *Phys. Rev. Lett.* **96**, p. 035507.

[676] Mao, W. L., Mao, H. K., Eng, P. J., Trainor, T. P., Newille, M., Kao, C. C., Heinz, D. L., Shu, J., Meng, Y. and Hemley, R. J. (2003). Bonding changes in compressed superhard graphite, *Science* **302**, pp. 425–427.

[677] Pravica, M., Grubor-Urosevic, O., Hu, M., Chow, P., Yulga, B. and Liermann, P. (2007). X-ray Raman spectroscopic study of benzene at high pressure, *J. Phys. Chem. B* **111**, pp. 11635–11637.

[678] Kumar, R. S., Pravica, M. G., Cornelius, A. L., Nicol, M. F., Hu, M. Y. and Chow, P. C. (2007). X-ray Raman scattering studies on C_{60} fullerenes and multi-walled carbon nanotubes under pressure, *Diamond Relat. Mater.* **16**, pp. 1250–1253.

[679] Meng, Y., Eng, P. J., Tse, J. S., Shaw, D. M., Hu, M. Y., Shu, J., Gramsch, S. A., Kao, C. C., Hemley, R. J. and Mao, H. K. (2008). Inelastic X-ray scattering of dense solid oxygen: evidence for intermolecular bonding, *Proc. Natl. Acad. Sci. USA* **105**, pp. 11640–11644.

[680] Lee, S. K., Eng, P. J., Mao, H. K., Mend, Y., Newille, M., Hus, M. Y. and Shu, J. (2005). Probing of bonding changes in B_2O_3 glasses at high pressure with inelastic X-ray scattering, *Nature Mater.* **4**, pp. 851–854.

[681] Seto, M., Yoda, Y., Kikuta, S., Zhang, X. W. and Ando, M. (1995). Observation of nuclear resonant scattering accompanied by phonon excitation using synchrotron radiation, *Phys. Rev. Lett.* **74**, pp. 3828–3831.

[682] Sturhahn, W., Toellner, T. S., Alp, E. E., Zhang, X., Ando, M., Yoda, Y., Kikuta, S., Seto, M., Kimball, C. W. and Dabrowski, B. (1995). Phonon density of states measured by inelastic nuclear resonant scattering, *Phys. Rev. Lett.* **74**, pp. 3832–3835.

[683] Mao, H. K., Xu, J., Struzhkin, V. V., Shu, J., Hemley, R. J., Sturhahn, W., Hu, M. Y., Alp, E. E., Vocadlo, L., Alf, D., Price, G. D., Gillan, M. J., Schwoerer-Bhning, M., Husermann, D., Eng, P., Shen, G., Giefers, H., Lbbers, R. and Wortmann, G. (2001). Phonon density of states of iron up to 153 Gigapascals, *Science* **292**, pp. 914–916.

[684] Ding, Y., Haskel, D., Ovchinnikov, S. G., Tseng, Y. C., Orlov, Y. S., Lang, J. C. and Mao, H. K. (2008). Novel pressure-induced magnetic transition in Magnetite (Fe_3O_4), *Phys. Rev. Lett.* **100**, p. 045508.

[685] Liu, H., Wang, L., Xiao, X., De Carlo, F., Feng, J., Mao, H. K. and Hemley, R. J. (2008). Anomalous high-pressure behavior of amorphous selenium from synchrotron X-ray diffraction and microtomography, *Proc. Natl. Acad. Sci. USA* **105**, pp. 13229–13234.

[686] Belushkin, A. V., Kozlenko, D. P. and Rogachev, A. V. (2011). Synchrotron and neutron-scattering methods for studies of properties of condensed matter: Competition or complementarity? *J. Surf. Investig.-X-Ray Synchro* **5**, pp. 828–855.

[687] Klotz, S., Braden, M. and Besson, J. M. (2000). Inelastic neutron scattering to very high pressures, *Hyp. Interact.* **128**, pp. 245–254.

[688] Klotz, S. and Braden, M. (2000). Phonon dispersion of bcc iron to 10 GPa, *Phys. Rev. Lett.* **85**, pp. 3209–3212.

[689] Strassle, Th., Saitta, A. M., Le Godec, Y., Hamel, G., Klotz, S., Loveday, J. S. and Nelmes, R. J. (2006). Structure of dense liquid water by neutron scattering to 6.5 GPa and 670 K, *Phys. Rev. Lett.* **96**, p. 067801.

[690] Strassle, Th., Saitta, A. M., Klotz, S. and Braden, M. (2004). Phonon dispersion of ice under pressure, *Phys. Rev. Lett.* **93**, p. 225901.

[691] Klotz, S., Hamel, G., Loveday, J. S., Nelmes, R. J., Guthrie, M. and Soper, A. K. (2003). Structure of high-density amorphous ice under pressure. *Phys. Rev. Lett.* **89**, p. 285502.

[692] Strassle, Th., Klotz, S., Hamel, G., Koza, M. M. and Schober, H. (2007). Experimental evidence for a crossover between two distinct mechanisms of amorphization in ice Ih under pressure, *Phys. Rev. Lett.* **99**, p. 175501.

[693] Klotz, S., Strassle, Th., Nelmes, R. J., Loveday, J. S., Hamel, G., Rousse, G., Canny, B., Chervin, J. C. and Saitta, A. M. (2005). Nature of the polyamorphic transition in ice under pressure, *Phys. Rev. Lett.* **94**, p. 025506.

[694] Bove, L. E., Klotz, S., Philippe, J. and Saitta, A. M. (2011). Pressure-induced polyamorphism in salty water, *Phys. Rev. Lett.* **106**, p. 125701.

[695] Klotz, S., Bove, L. E., Strassle, Th., Hansen, T. C. and Saitta, A. M. (2009). The preparation and structure of salty ice VII under pressure, *Nature Mater.* **8**, pp. 405–409.

[696] Goncharenko, I. N., Mignot, J. M., Andre, G., Lavrova, O. A., Mirebeau, I. and Somenkov, V. A. (1995). Neutron diffraction studies of magnetic structure and phase transitions at very high pressures, *High Pres. Res.* **14**, pp. 41–53.

[697] Goncharenko, I. N. (2004). Neutron diffraction experiments in diamond and sapphire anvil cells, *High Pres. Res.* **24**, pp. 193–204.

[698] Goncharenko, I. N., Makarova, O. L. and Ulivi, L. (2004). Direct determination of the magnetic structure of the delta phase of oxygen, *Phys. Rev. Lett.* **93**, p. 055502.

[699] Goncharenko, I. N. (2005). Evidence for a magnetic collapse in the epsilon phase of solid oxygen, *Phys. Rev. Lett.* **94**, p. 205701.

[700] Gorelli, F. A., Ulivi, L., Santoro, M. and Bini, R. (2000). Antiferromagnetic order in the δ phase of solid oxygen, *Phys. Rev. B* **62**, pp. R3604–R3607.

[701] Klotz, S., Strassle, Th., Cornelius, A. L., Philippe, J. and Hansen, Th. (2010). Magnetic ordering in solid oxygen up to room temperature, *Phys. Rev. Lett.* **104**, p. 115501.

[702] Silvera, I. F. and Wijingaarden, R. J. (1985). Diamond anvil cell and cryostat for low-temperature optical studies, *Rev. Sci. Instrum.* **56**, pp. 121–124.

[703] Lee, S. H., Luszczynski, K. and Norberg, R. E. (1987). NMR in a diamond anvil cell, *Rev. Sci. Instrum.* **58**, pp. 415–417.

[704] Bertani, R., Mali, M. and Brinkmann, D. (1992). A diamond anvil cell for high-pressure NMR investigations, *Rev. Sci. Instrum.* **63**, pp. 3303–3306.

[705] Yarger, J. L., Nieman, R. A., Wolf, G. H. and Marzke, R. F. (1995). High-pressure ^{1}H and ^{13}C nuclear magnetic resonance in a diamond anvil cell, *J. Magn. Reson. A* **114**, pp. 255–257.

[706] Okuchi, T. (2004). A new type of non-magnetic diamond anvil cell for nuclear magnetic resonance spectroscopy, *Phys. Earth Planet. In.* **143–144**, pp. 611–616.

[707] Pravica, M. G. and Silvera, I. F. (1998). Nuclear magnetic resonance in

a diamond anvil cell at very high pressures, *Rev. Sci. Instrum.* **69**, pp. 479–484.

[708] Pravica, M. G. and Silvera, I. F. (1998). NMR study of ortho-para conversion at high pressure in hydrogen, *Phys. Rev. Lett.* **81**, pp. 4180–4183.

[709] Lee, S. H., Conradi, M. S. and Norberg, R. E. (1992). Improved NMR resonator for diamond anvil cells, *Rev. Sci. Instrum.* **63**, pp. 3674–3676.

[710] Okuchi, T., Hemley, R. J. and Mao, H. K. (2005). Radio frequency probe with improved sensitivity for diamond anvil cell nuclear magnetic resonance, *Rev. Sci. Instrum.* **76**, p. 026111.

[711] Okuchi, T., Cody, G. D., Mao, H. K. and Hemley, R. J. (2005). Hydrogen bonding and dynamics of methanol by high-pressure diamond-anvil cell NMR, *J. Chem. Phys.* **122**, p. 244509.

[712] Okuchi, T., Takigawa, M., Shu, J., Mao, H. K., Hemley, R. J. and Hagi, T. (2007). Fast molecular transport in hydrogen hydrates by high-pressure diamond anvil cell NMR, *Phys. Rev. B* **75**, p. 144104.

[713] Okuchi, T. (2012). Collision and diffusion dynamics of dense molecular hydrogen by diamond anvil cell nuclear magnetic resonance, *J. Phys. Chem. C* **116**, pp. 2179–2182.

[714] Kitagawa, K., Gotou, H., Yagi, T., Yamada, A., Matsumoto, T., Uwatoko, Y. and Takigawa, M. (2010). Space efficient opposed-anvil high-pressure cell and its application to optical and NMR measurements up to 9 GPa, *J. Phys. Soc. Jpn.* **79**, p. 024001.

[715] Asano, T. and Le Noble, W. J. (1978). Activation and reaction volumes in solution, *Chem. Rev.* **78**, pp. 407–489.

[716] Van Eldik, R., Asano, T. and Le Noble, W. J. (1989). Activation and reaction volumes in solution. 2, *Chem. Rev.* **89**, pp. 549–688.

[717] Jenner, G. (1988). Pericyclic reactions, in Le Noble, W. J. (Ed.), *Organic High Pressure Chemistry* (Elsevier Science & Technology, Amsterdam) pp. 143–203.

[718] Chandler, D. (1978). Effects of liquid structures on chemical reactions and conformational changes of non-rigid molecules in condensed phases, *Faraday Discuss. Chem. Soc.* **66**, pp. 184–190.

[719] Hamann, S. D. (1963). Chemical equilibria in condensed systems, in Badley, R. S. (Ed.), *High Pressure Physics and Chemistry* (Academic Press, London and New York) p. 139.

[720] Slichter, C. P. and Drickamer, H. G. (1972). Pressure-induced electronic changes in compounds of iron, *J. Chem. Phys.* **56**, pp. 2142–2160.

[721] Mills, R. L., Olinger, B. and Cromer, D. T. (1986). Structures and phase diagrams of N_2 and CO to 13 GPa by X-ray diffraction, *J. Chem. Phys.* **84**, pp. 2837–2845.

[722] Jenner, G. (1998). *High Pressure Molecular Science* in Winter, R. and Jonas, J. (Eds.), Nato Science Series E: Applied Science - Vol. 358 (Kluver Academic Publishers, Dordrecht), p. 291 and references therein.

[723] Klaerner, F. G., Krawczyk, B., Ruster, V. and Deiters, U. K. (1994). Evidence for pericyclic and stepwise processes in the cyclodimerization of chloroprene and 1,3-butadiene from pressure dependence and stereochem-

istry. Experimental and theoretical volumes of activation and reaction, *J. Am. Chem. Soc.* **116**, pp. 7646–7657.

[724] Dumas, F., Mezrhab, B., D'Angelo, J., Riche, C. and Chiaroni, A. (1996). Investigating the π-facial discrimination phenomenon in the conjugate addition of amines to chiral crotonates: a convenient basis for the rational design of chiral auxiliaries, *J. Org. Chem.* **61**, pp. 2293–2304.

[725] Jenner, G. (1991). Volume profiles of pericyclic reactions involving rigid transition states. A pressure study, *New J. Chem.* **15**, pp. 897–899.

[726] Le Noble, W. J. and Asano, T. (1975). Kinetics of reactions in solutions under pressure. XXX. Special effect of pressure on highly hindered reactions as a possible manifestation of the Hammond postulate, *J. Am. Chem. Soc.* **97**, pp. 1778–1782.

[727] Drude, P. and Nernst, W. (1894). Uber elektrostriktion durch freie Ionen, *Z. Phys. Chem.* **15**, pp. 79–85.

[728] Owen, B. B. and Brinkley, S. R. (1943). The effect of pressure upon the dielectric constants of liquids, *Phys. Rev. B* **64**, pp. 32–36.

[729] Drljaca, A., Hubbard, C. D., Van Eldik, R., Asano, T., Basilevsky, M. V. and Le Noble, W. J. (1998). Activation and reaction volumes in solution. 3, *Chem. Rev.* **98**, pp. 2167–2290.

[730] Citroni, M., Ceppatelli, M., Bini, R. and Schettino, V. (2003). The high-pressure chemistry of butadiene crystal, *J. Chem. Phys.* **118**, pp. 1815–1820.

[731] Buchachenko, A. L., Motyakin, M. V. and Aliev, I. I. (1999). Radical solid-state reactions at high pressure, in Boldyreva, E. V. and Boldyrev, V. V. (Eds.), *Reactivity of Molecular Solids* (John Wiley & Sons Ltd., Chichester), pp. 221–239.

[732] Baughman, R. H. (1978). Solid-state reaction kinetics in single-phase polymerizations, *J. Chem. Phys.* **68**, pp. 3110–3121.

[733] Hulbert, S. F. (1969). Models for solid-state reactions in powdered compacts: a review, *J. Br. Ceram. Soc.* **6**, pp. 11–20.

[734] Jander, W. (1927). Reaktionen im festen zustande bei höheren temperaturen. Reaktionsgeschwindigkeiten endotherm verlaufender umsetzungen, *Z. Anorg. Chem.* **163**, pp. 1–30.

[735] Carter, R. E. (1961). Kinetic model for solid-state reactions, *J. Chem. Phys.* **34**, pp. 2010–2015.

[736] Avrami, M. (1939). Kinetics of phase change. I General theory, *J. Chem. Phys.* **7**, pp. 1103–1112. Avrami, M. (1940). Kinetics of phase change. II Transformation-time relations for random distribution of nuclei, *J. Chem. Phys.* **8**, pp. 212–224. Avrami, M. (1940). Granulation, phase change, and microstructure kinetics of phase change. III, *J. Chem. Phys.* **9**, pp. 177–184.

[737] Chelazzi, D., Ceppatelli, M., Santoro, M., Bini, R. and Schettino, V. (2005). Pressure-induced polymerization in solid ethylene, *J. Phys. Chem. B* **109**, pp. 21658-21663.

[738] Aoki, K., Baer, B. J., Cynn, H. C. and Nicol, M. (1990). High-pressure Raman study of one-dimensional crystals of the very polar molecule hydrogen cyanide, *Phys. Rev. B* **42**, pp. 4298–4303.

[739] Gavezzotti, A. and Simonetta, M. (1982). Crystal chemistry in organic solids, *Chem. Rev.* **82**, pp. 1–13.

[740] Cohen, M. D. (1975). The photochemistry of organic solids, *Angew. Chem. Int. Edn Engl.* **14**, pp. 386–393.

[741] Ciabini, L., Santoro, M., Bini, R. and Schettino, V. (2002). High pressure reactivity of solid benzene probed by infrared spectroscopy, *J. Chem. Phys.* **116**, pp. 2928–2935.

[742] Dwarakanath, K. and Prasad, P. N. (1980). Raman phonon spectroscopy of solid-state reactions: thermal rearrangement of methyl p-dimethylaminobenzenesulfonate in solid state, *J. Am. Chem. Soc.* **102**, pp. 4254–4256.

[743] Scheffer, J. R. (1980). Crystal lattice control of unimolecular photorearrangements. *Acc. Chem. Res.* **13**, pp. 283–290.

[744] Yamawaki, H., Sakashita, M., Aoki, K. and Takemura, K. (1996). Reversible phase transition between the metastable phases of tetracyanoethylene under high pressure, *Phys. Rev. B* **53**, pp. 11403–11407.

[745] Chizmeshya, A. V. G., Coffman, P., Robinson, S., Petuskey, W. T., Wolf, G. H. and McMillan, P. F. (1997). Bonding trends in $(NSF)_3$ at high pressures, *Eur. J. Solid State Inorg. Chem.* **34**, pp. 715–731.

[746] Citroni, M., Ceppatelli, M., Bini, R. and Schettino, V. (2005). High-pressure reactivity of propene, *J. Chem. Phys.* **123**, p. 194510.

[747] Fanetti, S., Citroni, M. and Bini, R. (2011). Structure and reactivity of pyridine crystal under pressure, *J. Chem. Phys.* **134**, p. 204504.

[748] Luty, T. and Eckhardt, C. J. (1995). General theoretical concepts for solid state reactions: quantitative formulation of the reaction cavity, steric compression, and reaction-induced stress using an elastic multipole representation of chemical pressure, *J. Am. Chem. Soc.* **117**, pp. 2441–2452.

[749] Brazier, D. W. and Freeman, G. R. (1969). The effects of pressure on the density, dielectric constants, and viscosity of several hydrocarbons and other organic liquids, *Can. J. Chem.* **47**, pp. 893–899.

[750] Reichlin, R., Schiferl, D., Martin, S., Vanderborgh, C. and Mills, R. L. (1985). Optical studies of nitrogen to 130 GPa, *Phys. Rev. Lett.* **55**, pp. 1464–1467.

[751] Aust, R. B., Bentley, W. H. and Drickamer, H. G. (1964). Behavior of fused-ring aromatic hydrocarbons at very high pressure, *J. Chem. Phys.* **41**, pp. 1856–1864.

[752] Sonnenschein, R., Syassen, K. and Otto, A. (1981). Effect of pressure on the first singlet exciton in crystalline anthracene, *J. Chem. Phys.* **74**, pp. 4315–4319.

[753] Santoro, M., Ceppatelli, M., Bini, R. and Schettino, V. (2003). High-pressure photochemistry of furane crystal, *J. Chem. Phys.* **118**, pp. 8321–8325.

[754] Fanetti, S., Ceppatelli, M., Citroni, M. and Bini, R. (2011). Changing the dissociative character of the lowest excited state of ethanol by pressure *J. Phys. Chem. B* **115**, pp. 1523615240.

[755] Ceppatelli, M., Fanetti, S., Citroni, M. and Bini, R. (2010). Photoinduced reactivity of liquid ethanol at high pressure, *J. Phys. Chem. B* **114**, pp. 1543715444.

[756] Mugnai, M., Cardini, G. and Schettino, V. (2004). High-pressure reactivity of propene by first principles molecular dynamics calculations, *J. Chem. Phys.* **120**, pp. 5327–5333.

[757] Mugnai, M., Cardini, G. and Schettino, V. (2004). Charge separation and polymerization of hydrocarbons at an ultrahigh pressure, *Phys. Rev. B* **70**, p. 020101.

[758] Citroni, M., Bini, R., Pagliai, M., Cardini, G. and Schettino, V. (2010). Nitromethane decomposition under high static pressure, *J. Phys. Chem. B* **114**, pp. 9420–9428.

[759] Zharov, A. A. (1984). The polymerisation of solid monomers under conditions of deformation at a high pressure, *Rus. Chem. Rev.* **53**, pp. 140–149.

[760] Brazhkin, V. V. and Lyapin, A. G. (2004). Metastable high-pressure phases of low-z compounds: creation of a new chemistry or a prompt for old principles, *Nat. Mater.* **3**, pp. 497–500.

[761] Brazhkin, V. V. (2006). Metastable phases and 'metastable'phase diagrams, *J. Phys.: Condens. Matter* **18**, pp. 9643–9650.

[762] McMillan, P. F., Shebanova, O., Daisenberger, D., Quesada Cabrera, R., Bailey, E., Hector, A., Lees, V., Machon, D., Sella, A. and Wilson, M. (2007). Metastable phase transitions and structural transformations in solid-state materials at high pressure, *Phase Trans.* **80**, pp. 1003–1032.

[763] Zarko, V. E. (2010). Searching for ways to create new energetic materials based on polynitrogen compounds, *Combust. Explos. Shock Waves* **46**, pp. 121–131.

[764] Talawar, M. B., Sivabalan, R., Asthana, S. N. and Singh, H. (2005). Novel ultrahigh-energy materials, *Combust. Explos. Shock Waves* **41**, pp. 264–277.

[765] Glukhovtsev, M. N., Jiao, H. and von Ragué Schleyer, P. (1996). Besides N_2: what is the most stable molecule composed only of nitrogen atoms, *Inorg. Chem.* **35**, pp. 7124–7133.

[766] Bartlett, R. (2000). Exploring the mysteries of nitrogen, *Chem. Ind.* **4**, pp. 140–143.

[767] Manaa, M. R. (2000). Toward new energy-rich molecular systems: from N_{10} to N_{60}, *Chem. Phys. Lett.* **331**, pp. 262–268.

[768] Smirnov, A., Lempert, D., Pivina, T. and Khakimov, D. (2011). Basic characteristics for estimation polynitrogen compounds efficiency, *Centr. E. J. Energ. Mater.* **201**, pp. 233–247.

[769] Abou-Rachid, Song, Y., Hu, A., Dudly, S., Zybin, V. and Goddard III, W. A. (2008). Predicting solid-state heats of formation of newly synthesized polynitrogen materials by using quantum mechanical calculations, *J. Phys. Chem. A* **112**, pp. 11914–11920.

[770] Rice, B. M., Byrd, E. F. C. and Mattson, W. D. (2007). Computational aspects of nitrogen-rich HEM, *High Ener. Density Mater. Struct. Bond.* **125**, pp. 153–194.

[771] Gagliardi, L., Orlandi, G., Evangelisti, S. and Roos, B. O. (2001). A theoretical study of the nitrogen clusters formed from the ions N_3^-, N_5^+ and N_5^-, *J. Chem. Phys.* **114**, pp. 10733–10737.

[772] Guthrie, J. A., Chaney, R. C. and Cunningham, A. J. (1991). Temperature dependencies of ternary ion-molecule association reactions yielding N_3^+, N_4^+ and $(CO)_2^+$, *J. Chem. Phys.* **95**, pp. 930–936.

[773] Tian, R., Facelli, J. C. and Michl, J. (1988). Vibrational and electronic spectra of matrix isolated nitrogen trimer radical and azide, *J. Phys. Chem.* **92**, pp. 4073–4079.

[774] Thompson, W. E. and Jacox, M. E. (1990). The vibrational spectra of molecular ions isolated in solid neon. III N_4^+, *J. Chem. Phys.* **93**, pp. 3856–3862.

[775] Christe, K. O., Wilson, W. W., Sheehy, J. A. and Boatz, J. A. (1999). N_5^+: a novel homoleptic polynitrogen ion as a high energy density material, *Ang. Chem. Int. Ed.* **38**, pp. 2004–2009.

[776] Xu, W. G., Li, G. L., Wang, L. J., Li, S., Li, Q. S. and Li, S. (1999). *Ab initio* and density functional theory study of the mechanism of synthesis of the N_5^+ cation, *Chem. Phys. Lett.* **314**, pp. 300–306.

[777] Nguyen, M. T. and Ha, T. K. (2000). Theoretical study of the pentanitrogen cation (N_5^+), *Chem. Phys. Lett.* **317**, pp. 135–141.

[778] Christe, K. O. (2007). Recent advances in the chemistry of N_5^+ and N_5^- and high-oxygen compounds, *Propel. Expl. Pyrotech.* **32**, pp. 194–204.

[779] Workentin, M. S., Wagner, B. D., Negri, F., Zgierski, M. Z., Lusztyk, J., Siebrand, W. and Wayner, D. D. M. (1995). N_6^-. Spectroscopic and theoretical studies of an unusual pseudohalogen radical anion, *J. Phys. Chem.* **99**, pp. 94–101.

[780] Workentin, M. S., Wagner, B. D., Lusztyk, J. and Wayner, D. D. M. (1995). Azidyl radical reactivity: N_6^- as a kinetic probe for the addition reaction of azidyl radicals with olefins, *J. Am. Chem. Soc.* **117**, pp. 119–126.

[781] Li, Q. S., Wang, L. J. and Xu, W. G. (2000). Structure and stability of N_9, N_9^- and N_9^+ clusters, *Theor. Chem. Acc.* **104**, pp. 67–77.

[782] Zheng, J. P., Waluk, J., Spanget-Larsen, J., Blake, D. M. and Radziszewski, J. G (2000). Tetrazete (N_4). Can it be prepared and observed? *Chem. Phys. Lett.* **328**, pp. 227–233.

[783] Trinquier, G., Mairieu, J. P. and Daudey, J. P. (1981). *Ab initio* study of regular polyhedral molecules N_4, P_4, As_4, N_8, P_8, As_8, *Chem. Phys. Lett.* **80**, pp. 552–557.

[784] Glukhovtsev, M. N. and Laiter, S. L. (1996). Thermochemistry of tetrazete and tetraazatetrahedrane: a high-level computational study, *J. Phys. Chem.* **100**, pp. 1569–1577.

[785] Leininger, M. L., Van Huis, T. J. and Schaefer, H. F. III (1997). Protonated high energy density material: N_4 tetrahedron and N_8 octahedron, *J. Phys. Chem.* **101**, pp. 4460–4464.

[786] Lauderdale, W. J., Stanton, J. F. and Bartlett, R. J. (1992). Stability and energetics of metastable molecules tetrahazatetrahetrane (N_4), hex-

azabenzene (N_6), octaazacubane (N_8), *J. Phys. Chem.* **96**, pp. 1173–1178.

[787] Nguyen, M. T., Nguyen, T. L., Mebel, A. M. and Flammang, R. (2003). Azido-nitrene is probably the N_4 molecule observed in mass spectrometric experiments, *J. Phys. Chem. A* **107**, pp. 5452–5460.

[788] Nguyen, M. T. (2003). Polynitrogen compounds: 1. Structure and stability of N_4 and N_5 systems, *Coord. Chem. Rev.* **244**, pp. 93–113.

[789] Cacace, F., De Petris, G. and Troiani, A. (2002). Experimental detection of tetranitrogen, *Science* **295**, pp. 480–481.

[790] Leininger, M. L., Sherrill, C. D. and Schaefer, H. F. III (1995). N_8 : a structure analog to pentalene, and other high energy density minima, *J. Phys. Chem.* **99**, pp. 2324–2328.

[791] Tian, A., Ding. F., Zhang, L., Xie, Y. and Schaefer, H. F. III (1997). New isomers of N_8 without double bonds, *J. Phys. Chem.* **101**, pp. 1946–1950.

[792] Gagliardi, L., Evangelisti, S., Widmark, P. O. and Roos, B. O. (1997). A theoretical study of the N_8 cubane to N_8 pentalene isomerization reaction, *Theor. Chem. Acc.* **97**, pp. 136–142.

[793] Engelke, R. and Stine, J. R. (1990). Is N_8 cubane stable? *J. Phys. Chem.* **94**, pp. 5689–5694.

[794] Ha, T.-K., Suleimenov, O. and Nguyen, M. T. (1999). A quantum chemical study of three isomers of N_{20}, *Chem. Phys. Lett.* **315**, pp. 327–334.

[795] Samartzis, P. C. and Wodtke, A. M. (2006). All-nitrogen chemistry: how far are we from N_{60}? *Int. Rev. Phys. Chem.* **25**, pp. 527–552.

[796] Uddin, J., Barone, V. and Scuseria, G. E. (2006). Energy storage capacity of polymeric nitrogen, *Mol. Phys.* **104**, pp. 745–749.

[797] Manzhellii, V. G. and Freiman, Y. A. (Eds.), (1997). *Physics of Cryocrystals* (American Institute of Physics, Woodbury).

[798] Hemley, R. J. and Dera, P. (2001). Molecular crystals, in Hazen, R. (Ed.), *Comparative Crystal Chemistry*, Vol. 42 (Mineral. Soc. Am., Reviews in Mineralogy).

[799] Bini, R., Ulivi, L., Kreutz, J. and Jodl, H. J. (2000). High-pressure phases of solid nitrogen by Raman and infrared spectroscopy, *J. Chem. Phys.* **112**, pp. 8522–8529.

[800] Gregoryanz, E., Goncharov, A. F., Hemley, R. J., Mao, H.-K., Somayazulu, M. and Shen, G. (2002). Raman, infrared and X-ray evidence for new phases of nitrogen at high pressures and temperatures, *Phys. Rev. B* **66**, p. 224108.

[801] Gregoryanz, E., Goncharov, A. F., Sanloup, C., Somayazulu, M., Mao, H. K. and Hemley, R. J. (2007). High P-T transformations of nitrogen to 170 GPa, *J. Chem. Phys.* **126**, p. 184505.

[802] Gregoryanz, E., Sanloup, C., Bini, R., Kreutz, J., Jodl, H. J., Somayazulu, M., Mao, H. K. and Hemley, R. J. (2006). On the $\epsilon - \zeta$ transition of nitrogen, *J. Chem. Phys.* **124**, p. 116102.

[803] Trojan, I. A., Eremets, M. I., Medvedev, S. A., Gavriliuk, A. G. and Prakapenka, V. B. (2008). Transformation from molecular to polymeric nitrogen at high pressures and temperatures: *in situ* X-ray diffraction study, *Appl. Phys. Lett.* **93**, p. 091907.

[804] Nellis, W. J. and Mitchell, A. C. (1980). Shock compression of liquid argon, nitrogen, and oxygen to 90 GPa (900 Kbar), *J. Chem. Phys.* **73**, pp. 6137–6145.

[805] Nellis, W. J., Holmes, N. C., Mitchell, A. C. and van Thiel, M. (1984). Phase transition in fluid nitrogen at high densities and temperatures, *Phys. Rev. Lett.* **53**, pp. 1661–1664.

[806] Radousky, H. B., Nellis, W. J, Ross, M., Hamilton, D. C. and Mitchell, A. C. (1986). Molecular dissociation and shock-induced cooling in fluid nitrogen at high densities and temperatures, *Phys. Rev. Lett.* **57**, pp. 2419–2422.

[807] Ross, M. (1987). The dissociation of dense liquid nitrogen, *J. Chem. Phys.* **86**, pp. 7110–7118.

[808] Hamilton, D. C. and Ree, F. H. (1989). Chemical equilibrium calculations on the molecular-to-nonmolecular transition of shock compressed liquid nitrogen, *J. Chem. Phys.* **90**, pp. 4972–4981.

[809] McMahan, A. K. and LeSar, R. (1985). Pressure dissociation of solid nitrogen under 1 Mbar, *Phys. Rev. Lett.* **54**, pp. 1929–1932.

[810] Mailhiot, C., Yang, L. H. and McMahan, A. K. (1992). Polymeric nitrogen, *Phys. Rev. B* **46**, pp. 14419–14435.

[811] Mitas, L. and Martin, R. M. (1994). Quantum Monte Carlo of nitrogen: atom, dimer and molecular solid, *Phys. Rev. Lett.* **72**, pp. 2438–2441.

[812] Bell, P., Mao, H. K. and Hemley, R. J. (1986). Observations of solid H_2, D_2 and N_2 at pressure around 1.5 Mbar at 25°, *Physica B* **139–140**, pp. 16–20.

[813] Goncharov, A. F., Gregoryanz, E., Mao, H. K., Liu, Z. and Hemley, R. J. (2000). Optical evidence for a non-molecular phase of nitrogen above 150 GPa, *Phys. Rev. Lett.* **85**, pp. 1262–1265.

[814] Eremets, M. I., Hemley, R. J., Mao, H. K. and Gregoryanz, E. (2001). Semiconducting non-molecular nitrogen up to 240 GPa and its low pressure stability, *Nature* **411**, pp. 170–174.

[815] Gregoryanz, E., Goncharov, A. F., Hemley, R. J. and Mao, H.-K. (2001). High-pressure amorphous nitrogen, *Phys. Rev. B* **64**, p. 052103.

[816] Lipp, M. J., Klepeis, J. P., Baer, B. J., Cynn, H., Evans, W. J., Iota, V. and Yoo, C.-S. (2007). Transformation of molecular nitrogen to non-molecular phases at megabar pressures by direct laser heating, *Phys. Rev. B* **76**, p. 014113.

[817] Eremets, M. I., Gavriliuk, A. G. and Trojan, I. A. (2007). Single-crystalline polymeric nitrogen, *Appl. Phys. Lett.* **90**, p. 171904.

[818] Chen, X. Q., Fu, C. L. and Podloucky, R. (2008). Bonding and strength of solid nitrogen in cubic gauche (cg-N) structure, *Phys. Rev. B* **77**, p. 064103.

[819] Yu, H. L., Wang, G. W., Yan, X. H., Xiao, Y., Mao, Y. L., Yang, Y. R. and Cheng, M. X. (2006). First-principle calculations of the single-bonded cubic phase of nitrogen, *Phys. Rev. B* **73**, p. 012101.

[820] Wang, X., Tian, F., Wang, L., Cui, T., Liu, B. and Zou, G. (2010). Structural stability of polymeric nitrogen: a first principle investigation, *J. Chem. Phys* **132**, p. 024502.

[821] Zhao, J. (2007). First-principles study of atomic nitrogen solid with cubic gauche structure, *Phys. Lett. A* **360**, pp. 645–648.

[822] Kotakoski, J. and Albe, K. (2008). First-principles calculations on solid nitrogen. a comparative study of high-pressure phases, *Phys. Rev. B* **77**, p. 144109.

[823] Ma, Y., Oganov, A. R., Li, Z., Xie, Y. and Kotakoski, J. (2009). Novel high pressure phases of polymeric nitrogen, *Phys. Rev. Lett.* **102**, p. 065501.

[824] Barbee, T. W. III (1993). Metastability of atomic phase of nitrogen, *Phys. Rev. B* **48**, pp. 9327–9330.

[825] Caracas, R. (2007). Raman spectra and lattice dynamics of cubic gauche nitrogen, *J. Chem. Phys.* **127**, p. 144510.

[826] Mattson, W. D. and Balu, R. (2011). Shock-induced behavior of cubic gauche polymeric nitrogen, *Phys. Rev. B* **83**, p. 174105.

[827] Zhang, T., Zhang, S., Chen, Q. and Peng, L. M. (2006). Metastability of single-bonded cubic-gauche structure of N under ambient pressure, *Phys. Rev. B* **73**, p. 094105.

[828] Mattson, W. D., Sanchez-Portal, D., Chiesa, S. and Martin, R. M. (2004). Prediction of new phases of nitrogen at high pressure from first-principles simulations, *Phys. Rev. Lett.* **93**, p. 125501.

[829] Zahariev, F., Hu, A., Hooper, J., Zhang, F. and Woo, T. (2005). Layered single-bonded non-molecular phase of nitrogen from first principles simulation, *Phys. Rev. B* **72**, p. 214108.

[830] Zahariev, F., Dudly, S. V., Hooper, J., Zhang, F. and Woo, T. (2006). Systematic method to new phases of polymeric nitrogen under high pressure, *Phys. Rev. Lett.* **97**, p. 155503.

[831] Zahariev, F., Hooper, J., Alavi, S., Zhang, F. and Woo, T. (2007). Low-pressure metastable phase of single-bonded polymeric nitrogen from a helical structure motif and first-principles calculations, *Phys. Rev. B* **75**, p. 140101.

[832] Pickard, C. J. and Needs, R. J. (2009). High-pressure phases of nitrogen, *Phys. Rev. Lett.* **102**, p. 125702.

[833] Yao, Y., Tse, J. S. and Tanaka, K. (2008). Metastable high-pressure phases of single-bonded nitrogen predicted via genetic algorithm, *Phys. Rev. B* **77**, p. 052103.

[834] Wang, X. L., He, Z., Ma, Y. M., Cui, T., Liu, Z. M., Liu, B. B., Li, J. F. and Zou, G. (2007). Prediction of a new layered phase of nitrogen from first-principles simulations, *J. Phys.: Condens. Matter* **19**, p. 425226.

[835] Oganov, A. R. and Glass, C. W. (2006). Crystal structure prediction using *ab initio* evolutionary techniques: principles and applications, *J. Chem. Phys.* **124**, p. 244704.

[836] Erba, A., Maschio, L., Pisani, C. and Casassa, S. (2011). Pressure-induced transitions in solid nitrogen: role of dispersive interactions, *Phys. Rev. B* **84**, p. 012101.

[837] Eremets, M. I., Gavriliuk, A. G., Serebryanaya, N. R., Trojan, I. A., Dzivenko, D. A., Boehler, R., Mao, H. K. and Hemley, R. J. (2004). Structural transformation of molecular nitrogen to a single-bonded atomic

state at high pressure, *J. Chem. Phys.* **121**, pp. 11296–11300.

[838] Katzke, H. and Tolédano, P. (2008). Theoretical description of pressure- and temperature-induced structural phase transition mechanisms of nitrogen, *Phys. Rev. B* **78**, p. 064103.

[839] Popov, M. (2005). Raman and infrared study of high-pressure atomic phase of nitrogen, *Phys. Lett. A* **334**, pp. 317–325.

[840] Eremets, M. I., Trojan, I. A., Gavriliuk, A. G. and Medvedev, S. A. (2008). Synthesis of high-nitrogen energetic material, in Peiris, S. M. and Piermarini, G. J. (Eds.), *Static Compression of Energetic Materials. Shock Wave and High-pressure Phenomena* (Springer-Verlag, Berlin Heidelberg) pp. 75–97.

[841] Peiris, S. M. and Russel, T. P. (2003). Photolysis of compressed sodium azide (NaN_3) as a synthetic pathway to nitrogen materials, *J. Phys. Chem. A* **107**, pp. 944–947.

[842] Hou, D., Zhang, F., Ji, C., Hannon, T., Zhu, H., Wu, J. and Ma, Y. (2011). Series of phase transitions in cesium azide under high pressure studied by *in situ* X-ray diffraction, *Phys. Rev. B* **84**, p. 064127.

[843] Eremets, M. I., Popov, M. Y., Trojan, I. A., Denisov, V. N., Boehler, R. and Hemley, R. J. (2004). Polymerization of nitrogen in sodium azide, *J. Chem. Phys.* **120**, pp. 10618–10623.

[844] Medvedev, S. A., Trojan, I. A., Eremets, M. I., Palasyuk, T., Klapotke, T. M. and Evers, J. (2009). Phase stability of lithium azide up to 60 GPa, *J. Phys. Condens. Matter* **21**, p. 195404.

[845] Long, C. A. and Ewing, G. E. (1973). Spectroscopic investigation of van der waals molecules. I. The infrared and visible spectra of $(O_2)_2$, *J. Chem. Phys.* **58**, pp. 4824–4834.

[846] Aquilanti, V., Ascenzi, D., Bartolomei, M., Cappelletti, D., Cavalli, S., de Castro Vtores, M. and Pirani, F. (1999). Molecular beam scattering of aligned oxygen molecules. The nature of the bond in O_2-O_2, *J. Am. Chem. Soc.* **121**, pp. 10794–10802.

[847] Freiman, Y. A. and Jodl, H. J.(2004). Solid oxygen, *Phys.Rep.* **401**, pp. 1–228.

[848] Akahama, Y. and Kawamura, H. (1996). High-pressure Raman spectroscopy of solid oxygen, *Phys. Rev. B* **54**, pp. R15602–R15605.

[849] Akahama, Y. and Kawamura, H. (2000). High-pressure infrared spectroscopy of solid oxygen, *Phys. Rev. B* **61**, pp. 8801–8805.

[850] Santoro, M., Gorelli, F. A., Ulivi, L. and Bini, R. (2001). Spectroscopic study of the ϵ phase of solid oxygen. *Phys. Rev. B* **63**, p. 104110.

[851] Neaton, J. B. and Ashcroft, N. W. (2002). Low-energy linear structures in dense oxygen: implications for the ϵ phase, *Phys. Rev. Lett.* **88**, p. 205503.

[852] Fujihisa, H., Akahama, Y., Kawamura, H., Ohishi, Y., Shimomura, O., Yamawaki, H., Sakashita, M., Gotoh, Y., Takeya, S. and Honda, K. (2006). O_8 cluster structure of the epsilon phase of solid oxygen, *Phys. Rev. Lett.* **97**, p. 085503.

[853] Steudel, R. and Wong, M. W. (2007). Dark-red O_8 molecules in solid oxygen: rhomboid clusters, not S_8-like rings, *Ang. Chem. Int. Ed.* **46**, pp. 1768–1771.

[854] Santoro, M. and Gorelli, F. A. (2006). High pressure solid state chemistry of carbon dioxide, *Chem. Soc. Rev.* **35**, pp. 918–931.

[855] Iota, V. and Yoo, C. S. (2001). Phase diagram of carbon dioxide: evidence for a new associated phase, *Phys. Rev. Lett.* **86**, pp. 5922–5925.

[856] Yoo, C. S., Iota, V. and Cynn, H. (2001). Non-linear carbon dioxide at high pressures and temperatures, *Phys. Rev. Lett.* **86**, pp. 444–447.

[857] Park, J. H., Yoo, C. S., Iota, V., Cynn, H., Nicol, M. F. and Le Bihan, T. (2003). Crystal structure of bent carbon dioxide phase IV, *Phys. Rev. B* **68**, p. 014107.

[858] Tschauner, O., Mao, H. K. and Hemley, R. J. (2001). New transformations of CO_2 at high pressures and temperatures, *Phys. Rev. Lett.* **87**, p. 075701.

[859] Iota, V., Yoo, C. S. and Cynn, H. (1999). Quartzlike carbon dioxide: an optically non-linear extended solid at high pressures and temperatures, *Science* **283**, pp. 1510–1513.

[860] Yoo, C. S., Cynn, H., Gygi, F., Galli, G., Iota, V., Nicol, M., Carlson, S., Hausermann, D. and Mailhiot, C. (1999). Crystal structure of carbon dioxide at high pressure: superhard polymeric carbon dioxide, *Phys. Rev. Lett.* **83**, pp. 5527–5530.

[861] Serra, S., Cavazzoni, C., Chiarotti, G. L., Scandolo, S. and Tosatti, E. (1999). Pressure-induced solid carbonates from molecular CO_2 by computer simulation, *Science* **284**, pp. 788–790.

[862] Dong, J., Tomfohr, J. K. and Sankey, O. F. (2000). Rigid intertetrahedron angular interaction of non-molecular carbon dioxide solids, *Phys. Rev. B* **6**, pp. 5967–5971.

[863] Holm, B., Ahuja, R., Belonoshko, A. and Johansson, B. (2000). Theoretical investigation of high pressure phases of carbon dioxide, *Phys. Rev. Lett.* **85**, pp. 1258–1261.

[864] Dong, J., Tomfohr, J. K., Sankey, O. F., Leinenweber, K., Somayazulu, M. and McMillan, P. F. (2000). Investigation of hardness in tetrahedrally bonded non-molecular CO_2 solids by density-functional theory, *Phys. Rev. B* **62**, pp. 14685–14689.

[865] Gracia, L., Marqués, M., Beltrán, A., Martin-Pendas, A. and Recio, J. M. (2004). Bonding and compressibility in molecular and polymeric phases of solid CO_2, *J. Phys.: Condens. Matter* **16**, pp. S1263–S1270.

[866] Togo, A., Oba, F. and Tanaka, I. (2008). Transition pathway of CO_2 crystals under high pressures, *Phys. Rev. B* **77**, p. 184101.

[867] Santoro, M., Gorelli, F. A., Bini, R., Ruocco, G., Scandolo, S. and Crichton, W. A. (2006). Amorphous silica-like carbon dioxide, *Nature* **441**, pp. 857–860.

[868] McMillan, P. F. (2006). A glass of carbon dioxide, *Nature* **441**, p. 823.

[869] Montoya, J. A., Rousseau, R., Santoro, M., Gorelli, F. A. and Scandolo, S. (2008). Mixed threefold and fourfold carbon coordination in compressed CO_2, *Phys. Rev. Lett.* **100**, p. 163002.

[870] Iota, V., Yoo, C. S., Klepeis, J. H., Jenei, Z., Evans, W. and Cynn, H. (2007). Six-fold coordinated carbon dioxide VI, *Nature Mater.* **6**, pp. 34–38.

[871] Sun, J., Klug, D. D., Martokc, R., Montoya, J. A., Lee, M. S., Scandolo, S. and Tosatti, E. (2009). High-pressure polymeric phases of carbon dioxide, *Proc. Natl. Acad. Sci. USA* **106**, pp. 6077–6081.

[872] Sengupta, A. and Yoo, C. S. (2010). Coesite-like CO_2: an analog to SiO_2, *Phys. Rev. B* **82**, p. 012105.

[873] Litasov, K. D., Goncharov, A. F. and Hemley, R. J. (2011). Crossover from melting to dissociation of CO_2 under pressure: implications for the lower mantle, *Earth Planet. Sci. Lett.* **309**, pp. 318–323.

[874] Fellay, C., Yan, N., Dyson, P. J. and Laurenczy, G. (2009). Selective formic acid decomposition for high-pressure hydrogen generation: a mechanistic study, *Chem. Eur. J.* **15**, pp. 3752–3760.

[875] Montgomery, W., Zaug, J. M., Howard, W. M., Goncharov, A. F., Crowhurst, J. C. and Jeanloz, R. (2005). Melting curve and high-pressure chemistry of formic acid to 8 GPa and 600 K, *J. Phys. Chem. B* **109**, p. 19443.

[876] Allan, D. R. and Clark, S. J. (1999). Impeded dimer formation in the high-pressure crystal structure of formic acid, *Phys. Rev. Lett.* **82**, pp. 3464–3467.

[877] Nahringbauer, I. (1978). A reinvestigation of the structure of formic acid (at 98 K), *Acta Cryst B* **34**, pp. 315–318.

[878] Albinati, A., Rouse, K. D and Thomas, M. W. (1978). Neutron diffraction analysis of hydrogen-bonded solids. ii. Structural study of formic acid at 4.5 K, *Acta Cryst. B* **34**, pp. 2188–2190.

[879] Goncharov, A. F., Manaa, M. R., Zaug, J. M., Gee, R. H., Fried, L. E. and Montgomery, W. B. (2005). Polymerization of formic acid under high pressure, *Phys. Rev. Lett.* **94**, p. 065505.

[880] Retting, S. J. and Trotter, J. (1987). Refinement of the structure of orthorhombic sulfur, α-S_8, *Acta Cryst. C* **43**, pp. 2260–2262.

[881] Akahama, Y., Kobayashi, M. and Kawamura, H. (1993). Pressure-induced structural phase transition in sulfur at 83 GPa, *Phys. Rev. B* **48**, pp. 6862–6864.

[882] Degtyareva, O., Gregoryanz, E., Somayazulu, M., Dera, P., Mao, H. K. and Hemley, R. J. (2005). Novel chain structures in group VI elements, *Nat. Mater.* **4**, pp. 152–155.

[883] Fujihisa, H., Akahama, Y., Kawamura, H., Yamawali, H., Sakashita, M., Yamada, T., Honda, K. and Le Bihan, T. (2004). Spiral chain structure of high pressure selenium-II and sulfur-II from powder X-ray diffraction, *Phys. Rev. B* **70**, p. 134106.

[884] Crapanzano, L., Crichton, W. A., Monaco, G., Bellissent, R. and Mezouar, M. (2005). Alternating sequence of ring and chain structures in sulphur at high pressure and temperature, *Nat. Mater.* **4**, pp. 550–552.

[885] Degtyareva, O., Hernandez, E. R., Serrano, J., Somayazulu, M., Mao, H. K., Gregoryanz, E. and Hemley, R. J. (2007). Vibrational dynamics and stability of the high-pressure chain and ring phases in S and Se, *J. Chem. Phys.* **126**, p. 084503.

[886] Struzhkin, V. V., Hemley, R. J., Mao, H. K. and Timofeev, Y. A. (1997). Superconductivity at 1017 K in compressed sulphur, *Nature* **390**, pp. 382–384.

[887] Luo, H. and Ruoff, A. L. (1994). High-pressure Raman scattering of sulfur, *AIP Conf. Proc.* **309**, pp. 1527–1530.

[888] Mao, H. K. and Hemley, R. J. (1994). Ultrahigh-pressure transitions in solid hydrogen, *Rev. Mod. Phys.* **66**, pp. 671–692.

[889] Hemley, R. J., Soos, Z. G., Hanfland, M. and Mao, H. K. (1994). Charge-transfer states in dense hydrogen, *Nature* **369**, pp. 384–387.

[890] Pinnick, D. A., Agnew, S. F. and Swanson, B. I. (1992). Fluid dinitrogen tetroxide at very high pressure and high temperature: observation of the nitrite isomer, *J. Phys. Chem.* **96**, pp. 7092–7096.

[891] Bolduan, F. and Jodl, H. J. (1982). Raman spectroscopy on matrix-isolated NO^+, NO_3^- and N_2O_4 in Ne, *Chem. Phys. Lett.* **85**, pp. 283–286.

[892] Bolduan, F., Jodl, H. J. and Loewenschuss, A. (1984). Raman study of solid N_2O_4: temperature induced ionization, *J. Chem. Phys.* **80**, pp. 1739–1743.

[893] Givan, A. and Loewenschuss, A. (1990). Raman studies on molecular and ionic forms in solid layers of nitrogen dioxide; temperature and light induced effects, *J. Chem. Phys.* **93**, pp. 7592–7600.

[894] Sihachakr, D. and Loubeyre, P. (2006). High-pressure transformation of N_2/O_2 mixtures into ionic compounds, *Phys. Rev. B* **74**, p. 064113.

[895] Meng, Y., Von Dreele, R. B., Toby, B. H., Chow, P., Hu, M. Y., Shen, G. and Mao, H. K. (2006). Hard X-ray radiation induced dissociation of N_2 and O_2 molecules and the formation of ionic nitrogen oxide phases, *Phys. Rev. B* **74**, p. 214107.

[896] Agnew, S. F., Swanson, B. I., Jones, L. H., Mills, R. L. and Schiferl, D. (1983). Chemistry of N_2O_4 at high pressure: observation of a reversible transformation between molecular and ionic crystalline forms, *J. Phys. Chem.* **87**, pp. 5065–5068.

[897] Song, Y., Hemley, R. J., Mao, H. K., Liu, Z. and Herschbach, D. R. (2003). New phases of N_2O_4 at high pressures and high temperatures, *Chem. Phys. Lett.* **382**, pp. 686–692.

[898] Song, Y., Somayazulu, M., Mao, H. K., Hemley, R. J. and Herschbach, D. R. (2003). High-pressure structure and equation of state study of nitrosonium nitrate from synchrotron X-ray diffraction, *J. Chem. Phys.* **118**, pp. 8350–8356.

[899] Yoo, C. S., Iota, V., Cynn, H., Nicol, M., Park, J. H., Le Bihan, T. and Mezouar, M. (2003). Disproportionation and other transformation of N_2O at high pressures and temperatures to lower energy, denser phases, *J. Phys. Chem. B* **107**, pp. 5922–5925.

[900] Somayazulu, M., Madduri, A., Goncharov, A. F., Tschauner, O., McMillan, P. F., Mao, H. K. and Hemley, R. J. (2001). Novel broken symmetry phase from N_2O at high pressures and high temperatures, *Phys. Rev. Lett.* **87**, p. 135504.

[901] Yoo, C. S. (2000). High-pressure chemistry of molecular solids: evidences

for novel extended phases of carbon dioxide, in Manghnani, M. H., Nellis, W. J. and Nicol, M. (Eds.), *Science and Technology of High Pressure* (Universities Press, Hyderabad, India) pp. 86–89.

[902] Pickard, C. J. and Needs, R. J. (2008). Highly compressed ammonia forms an ionic crystal, *Nature Mater.* **7**, pp. 775–779.

[903] Fortes, A. D., Brodholt, J. P., Wood, I. G., Vočadlo, L. and Jenkins, H. D. B. (2001). *Ab initio* simulation of ammonia monohydrate ($NH_3{\cdot}H_2O$) and ammonium hydroxide (NH_4OH), *J. Chem. Phys.* **115**, pp. 7006–7014.

[904] Loubeyre, P., Occelli, F. and LeToullec, R. (2002). Optical studies of solid hydrogen to 320 GPa and evidence for black hydrogen, *Nature* **416**, pp. 613-617.

[905] Akahama, Y., Kawamura, H., Hirao, N., Ohishi, Y. and Takemura, K. (2010). Raman scattering and X-ray diffraction experiments for phase III of solid hydrogen, *J. Phys.: Conf. Ser.* **215**, p. 012056.

[906] Eremets, M. I. and Troyan, I. A. (2011). Conductive dense hydrogen *Nature Mater.* **10**, pp. 927–931.

[907] Lebégue, S., Araujo, C. M., Kim, D. Y., Ramzan, M., Mao, H. K. and Ahuja, R. (2012). Semimetallic dense hydrogen above 260 GPa, *Proc. Natl. Acad. Sci. USA* **109**, pp. 9766–9769.

[908] Buzea, C. and Robbie, K. (2005). Assembling the puzzle of superconducting elements: a review, *Supercond. Sci. Technol.* **18**, pp. R1–R8.

[909] Eremets, M. I., Shimizu, K., Kobayashi, T. C. and Amaya, K. (1998). Metallic CsI at pressures of up to 220 GPa, *Science* **281**, pp. 1333–1335.

[910] Xu, Y., Tse, J. S., Oganov, A. R., Cui, T., Wang, H., Ma, Y. and Zou, G. (2009). Superconducting high pressure phase of cesium iodide, *Phys. Rev. B* **79**, p. 144110.

[911] Edwards, P. P. and Hensel, F. (2002). Metallic oxygen, *ChemPhysChem* **3**, pp. 53–56.

[912] Shimizu, K., Eremets, M. I., Suhara, K. and Amaya, K. (1998). Oxygen under high pressure - temperature dependence of electrical resistance, *Rev. High Pressure Sci. Technol.* **7**, pp. 784–786.

[913] Desgreniers, S., Vohra, Y. K. and Ruoff, A. L. (1990). Optical response of very high density solid oxygen to 132 GPa, *J. Phys. Chem.* **94**, pp. 1117–1122.

[914] Serra, S., Chiarotti, G. L., Scandolo, S. and Tosatti, E. (1998). Pressure-induced magnetic collapse and metallization of molecular oxygen; the ζ-O_2 phase, *Phys. Rev. Lett.* **80**, pp. 5160–5163.

[915] Ma, Y., Oganov, A. R. and Glass, C. W. (2007). Structure of the metallic ζ phase of oxygen and isosymmetric nature of the $\epsilon - \zeta$ phase transition: *ab initio* simulations, *Phys. Rev. B* **76**, p. 064101.

[916] Kim, D. Y., Lebègue, S., Moysés Araújo, C., Arnaud, B., Alouani, M. and Ahuja, R. (2008). Structurally induced insulator-metal transition in solid oxygen: a quasi-particle investigation, *Phys. Rev. B* **77**, p. 092104.

[917] Zhu, L., Wang, Z., Wang, Y., Zou, G., Mao, H. K. and Ma, Y. (2012). Spiral chain O_4 form of dense oxygen, *Proc. Natl. Acad. Sci. USA* **109**, pp. 751–753.

[918] Luo, H., Desgreniers, S., Vohra, Y. H. and Ruoff, A. L. (1991). High-pressure optical studies on sulfur to 121 GPa: optical evidence for metallization, *Phys. Rev. Lett.* **67**, pp. 2998–3001.

[919] Struzhkin, V. V., Hemley, R. J., Mao, H. K. and Timofeev, Y. A. (1997). Superconductivity at 10-17 K in compressed sulfur, *Nature* **390**, pp. 382–384.

[920] Kometani, S., Eremets, M. I., Shimizu, K., Kobayashi, M. and Amaya, K. (1997). Observation of pressure-induced superconductivity of sulfur, *J. Phys. Soc. Jpn* **66**, pp. 2564–2565.

[921] Hejny, C., Lundegaard, L. F., Falconi, S., McMahon, M. I. and Hanfland, M. (2005). Incommensurate sulfur above 100 GPa, *Phys. Rev. B* **71**, p. 020101.

[922] Degtyareva, O., Gregoryanz, E., Somayazulu, M., Mao, H. K. and Hemley, R. J. (2005). Crystal structure of superconducting phases of S and Se, *Phys. Rev. B* **71**, p. 214104.

[923] Shimizu, H., Nakamichi, Y. and Sasaki, S. (1991). Pressure induced phase transition in solid hydrogen sulfide at 11 GPa, *J. Chem. Phys.* **95**, pp. 2036–2040.

[924] Shimizu, H., Yamaguchi, H., Sasaki, S., Honda, H., Endo, S. and Kobayashi, M. (1995). Pressure-temperature phase diagram of hydrogen sulfide determined by Raman spectrocopy, *Phys. Rev. B* **51**, pp. 9391–9393.

[925] Sakashita, M., Yamawaki, H., Fujihisa, H., Aoki, K. Sasaki, S. and Shimizu, H. (1997). Pressure-induced molecular dissociation and metallization of hydrogen-bonded H_2S solid, *Phys. Rev. Lett.* **79**, pp. 1082–1085.

[926] Endo, S., Honda, A., Sasaki, S., Shimizu, H., Shimomura, O. and Kikegawa, T. (1996). High-pressure phase of solid hydrogen sulfide, *Phys. Rev. B* **54**, pp. R717–R719.

[927] Rousseau, R., Boero, M., Bernasconi, M., Parrinello, M. and Terakura, K. (2000). *Ab initio* simulation of phase transitions and dissociation of H_2S at high pressure, *Phys. Rev. Lett.* **85**, pp. 1254–1257.

[928] Rousseau, R., Boero, M., Bernasconi, M., Parrinello, M. and Terakura, K. (1999). Static structure and dynamic correlations in high pressure H_2S, *Phys. Rev. Lett.* **83**, pp. 2218–2221.

[929] Riggleman, B. M. and Drickamer, H. G. (1963). Approach to the metallic state as obtained from optical and electrical measurements, *J. Chem. Phys.* **38**, pp. 2721–2724.

[930] Pasternak, M., Farrell, J. N. and Taylor, R. D. (1987). Metallization and structural transformation of iodine under pressure: a microscopic view, *Phys. Rev. Lett.* **58**, pp. 575–578.

[931] Kenichi, T., Kyoko, S., Hiroshi, F. and Mitsuko, O. (2003). Modulated structure of solid iodine during its molecular dissociation under high pressure, *Nature* **423**, pp. 971–974.

[932] Takemura, K., Sato, K., Fujihisa, H. and Onoda, M. (2004). Structural phase transitions in iodine under high-pressure, *Z. Kristalogr.* **219**, pp. 749–754.

[933] Zeng, Q., He, Z., San, X., Ma, Y., Tian, F., Cui, T., Liu, B., Zou, G. and Mao, H. K. (2008). A new phase of solid iodine with different molecular covalent bonds, *Proc. Natl. Acad. Sci. USA* **105**, pp. 4999–5001.

[934] Mukose, K., Fukano, R., Miyagi, H. and Yamaguchi, K. (2002). First-principles studies of solid halogens under pressure: scaling rules for properties among I_2, Br_2 and Cl_2, *J. Phys.: Condens. Matter* **14**, pp. 10441–10444.

[935] Fujii, Y., Hase, K., Ohishi, Y., Fujihisa, H., Hamaya, N., Takemura, K., Shimomura, O., Kikegawa, T., Amemiya, Y. and Matsushita, T. (1989). Evidence for molecular dissociation in bromine near 80 GPa, *Phys. Rev. Lett.* **63**, pp. 536–539.

[936] Kumel, T., Hiraoka, T., Ohya, Y., Sasaki, S. and Shimizu, H. (2005). High pressure Raman study of bromine and iodine: soft phonon in the incommensurate phase, *Phy. Rev. Lett.* **94**, p. 065506.

[937] San-Miguel, A., Libotte, H., Gauthier, M., Aquilanti, G., Pascarelli. S. and Gaspard, J. P. (2007). New phase transition of solid bromine under high pressure, *Phys. Rev. Lett.* **99**, p. 015501.

[938] Li, P., Gao, G. and Ma, Y. (2012). Modulated structure and molecular dissociation of solid chlorine at high pressures, *J. Chem. Phys.* **137**, p. 064502.

[939] Onoda, S. and Shimizu, K. (2007). Pressure induced metallization of molecular crystal BI_3, *J. Phys. Soc. Jpn* **76** Suppl. A, pp. 33–34.

[940] Iwasaki, E., Shimizu, K., Amaya, K., Nakayama, A., Aoki, K. and Carlon, R. P. (2001). Metallization and superconductivity in hexaiodobenzene under high pressure, *Synth. Metals* **120**, pp. 1003–1004.

[941] Shirotani, I., Hayashi, J., Yakushi, K., Takeda, K., Yokota, T., Shimizu, K., Amaya, K., Nakayama, A. and Aoki, K. (2001). Pressure-induced insulator-to-metal transition and superconductivity in iodanil, $C_6I_4O_2$, *Physica B* **304**, pp. 6–11.

[942] Chen, A. L., Yu, P. Y. and Pasternak, M. P. (1991). Metallization and amorphization of the molecular crystals SnI_4 and GeI_4 under pressure, *Phys. Rev. B* **44**, pp. 2883–2886.

[943] Ashcroft, N. W. (2004). Hydrogen dominant metallic alloys: high temperature superconductors? *Phys. Rev. Lett.* **92**, p. 187002.

[944] Degtyareva, O., Martnez Canales, M., Bergara, A., Chen, X. J., Song, Y., Struzhkin, V. V., Mao, H. K. and Hemley, R. J. (2007). Crystal structure of SiH_4 at high pressure, *Phys. Rev. B* **76**, p. 064123.

[945] Feng, J., Grochala, W., Jaron, T., Hoffmann, R., Bergara, A. and Ashcroft, N. W. (2006). Structure and potential superconductivity in SiH_4 at high pressure: en route to "metallic hydrogen", *Phys. Rev. Lett.* **96**, p. 017006.

[946] Kim, D. Y., Scheicher, R. H., Lebègue, S., Prasongkit, J., Arnaud, B., Alouani, M. and Ahuja, R. (2008). Crystal structure of the pressure-induced metallic phase of SiH_4 from *ab initio* theory, *Proc. Natl. Acad. Sci. USA* **105**, pp. 16454–16459.

[947] Martinez-Canales, M., Oganov, A. R., Ma, Y., Yan, Y., Lyakhov, A. O. and Bergara, A. (2009). Novel structures and superconductivity of silane under pressure, *Phys. Rev. Lett.* **102**, p. 087005.
[948] Eremets, M. I., Trojan, I. A., Medvedev, S. A., Tse, J. S. and Yao, Y. (2008). Superconductivity in hydrogen dominant materials: silane, *Science* **319**, pp. 1506–1509.
[949] Martinez-Canales, M., Bergara, A., Feng, J. and Grochala, W. (2006). Pressure induced metallization of germane, *J. Phys. Chem. Solids* **67**, pp. 2095–2099.
[950] Gao, G., Oganov, A. R., Bergara, A., Martinez-Canales, M., Cui, T., Iitaka, T., Ma, Y. and Zou, G. (2008). Superconducting high pressure phase of germane, *Phys. Rev. Lett.* **101**, p. 107002.
[951] Gao, G., Oganov, A. R., Li, P., Li, Z., Wang, H., Cui, T., Ma, Y., Bergara, A., Lyakhov, A. O., Iitaka, T. and Zou, G. (2010). High pressure crystal structure and superconductivity of stannane (SnH_4), *Proc. Natl. Acad. Sci. USA* **107**, pp. 1317–1320.
[952] Jin, X., Meng, X., He, Z., Ma, Y., Liu, B., Cui, T., Zou, G. and Mao, H. K. (2010). Superconducting high-pressure phases of disilane, *Proc. Natl. Acad. Sci. USA* **107**, pp. 9969–9973.
[953] Citroni, M., Ceppatelli, M., Bini, R. and Schettino, V. (2007). Dimerization and polymerization of isoprene at high pressure, *J. Phys. Chem. B* **111**, pp. 3910–3917.
[954] Aoki, K., Usuba, S., Yoshida, M., Kakudate, Y., Tanaka, K. and Fujiwara, S. (1988). Raman study of the solid-state polymerization of acetylene at high pressure, *J. Chem. Phys.* **89**, pp. 529–534.
[955] Sakashita, M., Yamawaki, H. and Aoki, K. (1996). FT-IR study of solid state polymerization of acetylene under pressure, *J. Phys. Chem.* **100**, pp. 9943–9947.
[956] Gilman, J. J. (1996). Mechanochemistry, *Science* **274**, p. 65.
[957] McBride, J. M. (1983). The role of local stress in solid-state radical reactions, *Acc. Chem. Res.* **16**, pp. 304–312.
[958] Peachey, N. M. and Eckhardt, C. J. (1993). Energetics of organic solid-state reactions: lattice dynamics and chemical pressure in the 2,5-distyrylpyrazine photoreaction, *J. Phys. Chem.* **97**, pp. 10849-10856.
[959] Peachey, N. M. and Eckhardt, C. J. (1994). Energetics of organic solid-state reactions: electronic structure and photoreaction mechanism of the 2,5-distyrylpyrazine oligomer molecular crystal, *J. Phys. Chem.* **98**, pp. 685–691.
[960] Luty, T. and Fouret, F. (1989). On stability of molecular solids "under chemical pressure", *J. Chem. Phys.* **90**, pp. 5696–5703.
[961] Luty, T., Ordon, P. and Eckhardt, C. J. (2002). A model for mechanochemical transformations: application to molecular hardness, instabilities and shock initiation of reaction, *J. Chem. Phys.* **117**, pp. 1775–1785.
[962] Cohen, M. D. and Schmidt, G. M. J. (1964). Topochemistry. Part I. A survey, *J. Chem. Soc.*, pp. 1996–2000.

[963] Schmidt, G. M. J. (1971). Photodimerization in the solid state, *Pure Appl. Chem.* **27**, pp. 647–678.

[964] Balzaretti, N. M., Perottoni, C. A. and Herz da Jornada, J. A. (2003). High-pressure Raman and infrared spectroscopy of polyacetylene, *J. Raman Spectr.* **34**, pp. 259–263.

[965] LeSar, R. (1987). Calculated high-pressure properties of solid acetylene and possible polymerization paths, *J. Chem. Phys.* **86**, pp. 1485–1490.

[966] Trout, C. C. and Badding, J. V. (2000). Solid state polymerization of acetylene at high pressure and low temperature, *J. Phys. Chem. A* **104**, pp. 8142–8145.

[967] Baughman, R. (1974). Solid-state synthesis of large polymer single crystals, *J. Polym. Sci. Polym. Phys. Ed* **12**, pp. 1511–1535.

[968] Bernasconi, M., Chiarotti, G. L., Focher, P., Parrinello, M. and Tosatti, E. (1997). Solid-state polymerization of acetylene under pressure: *ab initio* simulation, *Phys. Rev. Lett.* **78**, pp. 2008–2011.

[969] Wieldraaijer, H., Schouten, J. A. and Trappeniers, N. J. (1983). Investigation of the phase diagrams of ethane, ethylene and methane at high pressures, *High Temp. High Press.* **15**, pp. 87–92.

[970] Citroni, M., Ceppatelli, M., Bini, R. and Schettino, V. (2002). IR study of the pressure induced solid state di- and polymerization in 1,3-butadiene, *High Pres. Res.* **22**, pp. 507–510.

[971] Citroni, M., Ceppatelli, M., Bini, R. and Schettino, V. (2003). Phase diagram and crystal phases of trans-1,3-butadiene probed by FTIR and Raman spectroscopy, *Chem. Phys. Lett.* **367**, pp. 186–192.

[972] Walling, C. and Peisach, J. (1958). Organic reactions under high pressure. IV. The dimerization of isoprene, *J. Am. Chem. Soc.* **80**, pp. 5819–5824.

[973] Henning, Th. and Salama, F. (1998). Carbon in the universe, *Science* **282**, pp. 2204–2210.

[974] Bridgman, P. W. (1914). Change of phase under pressure. I. The phase diagram of eleven substances with special reference to the melting curve, *Phys. Rev.* **3**, pp. 153–203.

[975] Cansell, F., Fabre, D. and Petitet, J. P. (1993). Phase transitions and chemical transformations of benzene up to 550 °C and 30 GPa, *J. Chem. Phys.* **99**, pp. 7300–7304.

[976] Thiéry, M. M. and Léger, J. M. (1988). High pressure solid phases of benzene. I. Raman and X-ray studies of C_6H_6 at 294 K and 25 GPa. *J. Chem. Phys.* **89**, pp. 4255–4271.

[977] Thiéry, M. M. and Rerat, C. (1996). High pressure solid phases of benzene. III. Molecular packing analysis of the crystalline structures of C_6H_6, *J. Chem. Phys.* **104**, pp. 9079–9089.

[978] Eijck, B. P. V., Spek, A. L., Mooij, W. T. M. and Kroon, J. (1998). Hypothetical crystal structures of benzene at 0 and 30 GPa, *Acta Cryst. B* **54**, pp. 291–299.

[979] Thiery, M. M., Besson, J. M. and Bribes, J. L. (1992). High pressure solid phases of benzene. II. Calculation of the vibration frequencies and evolution of the bonds in C_6H_6 and C_6D_6 up to 20 GPa, *J. Chem. Phys.* **96**, pp. 2633–2654.

[980] Raiteri, P., Martonak, R. and Parrinello, M. (2005). Exploring polymorphism: the case of benzene, *Angew. Chem. Int. Ed.* **44**, pp. 3769–3773.

[981] Wen, X. D., Hoffmann, R. and Ashcroft, N. W. (2011). Benzene under high pressure: a story of molecular crystals transforming to saturated networks, with a possible intermediate metallic phase, *J. Am. Chem. Soc.* **133**, pp. 9023–9035.

[982] Piermarini, G. J., Mighell, A. D., Weir, C. E. and Block, S. (1968). Crystal structure of benzene II at 25 kilobars, *Science* **165**, pp. 1250–1255.

[983] Block, S., Weir, C. E. and Piermarini, G. J. (1970). Polymorphism in benzene, naphtalene and anthracene at high pressure, *Science* **169**, pp. 586–587.

[984] Katriusak, A., Podsiado, M. and Budzianowski, A. (2010). Association CH$\cdots\pi$ and no van der Waals contacts at the lowest limits of crystalline benzene I and II stability regions, *Cryst. Growth Des.* **10**, pp. 3461–3465.

[985] Zhou, M., Wang, K., Men, Z., Gao, S., Li, Z. and Sun, C. (2012). Study of high-pressure Raman intensity behavior of aromatic hydrocarbons: benzene, biphenyl and naphtalene, *Spectrochim. Acta A* **97**, pp. 526–531.

[986] Pruzan, P., Chervin, J. C., Thiéry, M. M., Itié, J. P., Besson, J. M., Forgerit, J. P. and Revault, M. (1990). Transformation of benzene to a polymer after static pressurization to 30 GPa, *J. Chem. Phys.* **92**, pp. 6910–6915.

[987] Dunitz, J. D. (1995). Phase changes and chemical reactions in molecular crystals, *Acta Cryst. B* **51**, pp. 619–631.

[988] Engelke, R., Hay, P. J., Kleier, D. A. and Wadt, W. R. (1983). A theoretical study of possible benzene dimerization under high-pressure conditions, *J. Chem. Phys.* **79**, pp. 4367–4375.

[989] Engelke, R., Hay, P. J., Kleier, D. A. and Wadt, W. R. (1984). Theoretical study of dimeric forms of ground-state benzene molecules, *J. Am. Chem. Soc.* **106**, pp. 5439–5446.

[990] Engelke, R. (1986). Theoretical study of a symmetry-allowed dimerization of benzene, *J. Am. Chem. Soc.* **108**, pp. 5799–5803.

[991] Wen, X. D., Hand, L., Labet, V., Yang, T., Hoffmann, R., Ashcroft, N. W., Oganov, A. R. and Lyakhov, A. O. (2011). Graphane sheets and crystals under pressure, *Proc. Natl. Acad. Sci. USA* **108**, pp. 6833–6837.

[992] Cansell, F., Fabre, D., Petitet, J. P., Itie, J. P. and Fontaine, A. (1995). Study of high pressure reaction paths of C=C bonds and aromatic rings opening by X-ray absorption near-edge structure and Raman scattering, *J. Phys. Chem.* **99**, pp. 13109–13114.

[993] Farina, L., Syassen, K., Brillante, A., Della Valle, R. G. Venuti, E. and Karl, N. (2003). Pentacene at high pressure, *High Press. Res.* **23**, pp. 349–354.

[994] Bastron, V. C. and Drickamer, H. G. (1971). Solid state reactions in organic crystals at very high pressure, *J. Sol. State Chem.* **3**, pp. 550–563.

[995] Farina, L., Brillante, A., Della Valle, R. G., Venuti, E., Amboage, M. and Syassen, K. (2003). Pressure-induced phase transition in pentacene, *Chem. Phys. Lett.* **375**, pp. 490–494.

[996] Jones, P. F. and Nicol, M. (1965). Excimer fluorescence of crystalline anthracene and naphtalene produced by high pressure, *J. Chem. Phys.* **43**, pp. 3759–3760.

[997] Engelke, R. and Blais, N. C. (1994). Chemical dimerization of crystalline anthracene produced by transient high pressure, *J. Chem. Phys.* **101**, pp. 10961–10972.

[998] Tapilin, V. M., Bulgakov, N. N., Chupakhin, A. P. and Politov, A. A. (2008). On the mechanism of mechanochemical dimerization of anthracene. quantum-mechanical calculation of the elctronic structure of anthracene and its dimer, *J. Struct. Chem.* **49**, pp. 581–586.

[999] Dreger, Z. A., Lucas, H. and Gupta, Y. M. (2003). High-pressure effects on fluorescence of anthracene crystals, *J. Phys. Chem. B* **107**, pp. 9268–9274.

[1000] Grimme, S., Diedrich, C. and Korth, M. (2006). The importance of inter- and intramolecular van der Waals interactions in organic reactions: the dimerization of anthracene revisited, *Ang. Chem. Int. Ed.* **45**, pp. 625–629.

[1001] Zade, S. S., Zamoshchik, N., Reddy, A. R., Fridman-Marueli, G., Sheberla, D. and Bendikov, M. (2011). Products and mechanisms of acene dimerization. a computational study, *J. Am. Chem. Soc.* **133**, pp. 10803–10816.

[1002] Zhuravlev, K. K., Traikov, K., Dong, Z., Xie, S. and Song, Y. (2010). Raman and infrared spectroscopy of pyridine under high pressure, *Phys. Rev. B* **82**, p. 064116.

[1003] Podsiadlo, M., Jakobek, K. and Katrusiak, A. (2010). Density, freezing and molecular aggegation in pyridazine, pyridine and benzene, *CrystEngComm* **12**, pp. 2561–2567.

[1004] Ceppatelli, M., Santoro, M., Bini, R. and Schettino, V. (2003). High pressure reactivity of solid furane probed by infrared and Raman spectroscopy, *J. Chem Phys.* **118**, pp. 1499–1506.

[1005] Shimizu, H. and Matsunami, M. (1988). High-pressure Raman study of liquid and solid thiophene up to 100 Kbar and 300 K, *J. Raman Spectr.* **19**, pp. 199–201.

[1006] Pruzan, P., Chervin, J. C. and Forgerit, J. P. (1992). Chemical and phase transformations of thiophene at high pressures, *J. Chem. Phys.* **96**, pp. 761–767.

[1007] Gauthier, M., Chervin, J. C. and Pruzan, P. (1991). Pressure induced polymerization of cyclic molecules: a study of benzene and thiophene, in Hochheimer, H. and Etters, R. (Eds.), *Frontiers of high-pressure Research* (Plenum Press, New York).

[1008] Katz, A. I., Schiferl, D. and Mills, R. L. (1988). New phases and chemical reactions in solid CO under pressure, *J. Phys. Chem.* **88**, pp. 3176–3179.

[1009] Lipp, M., Evans, W. J., Garcia-Baonza, V. and Lorenzana, H. E. (1998). Carbon monoxide: spectroscopic characterization of the high-pressure polymerized phase, *J. Low Temp. Phys.* **111**, pp. 247–256.

[1010] Lipp, M. J., Evans, W. J., Baer, B. J. and Yoo, C. S. (2005). High-energy-density extended CO solid, *Nat. Mater.* **4**, pp. 211–215.

[1011] Bernard, S. J., Chiarotti, G. L., Scandolo, S. and Tosatti, E. (1998). Decomposition and polymerization of solid carbon monoxide under pressure, *Phys. Rev. Lett.* **81**, pp. 2092–2095.

[1012] Aoki, K. and Yamawaki, H. (1996). Phase transition and chemical reaction in cyanide, *Rev. High Press. Sci. Technol.* **5**, pp. 137–142.

[1013] Yamawaki, H., Sakashita, M. and Aoki, K. (1998). High pressure solid-state polymerization of molecules having a triple bond, *J. of NIMC* **6**, pp. 169–178.

[1014] Volker, T. (1960). Polymere blausaure, *Ang. Chem.* **72**, pp. 379–384.

[1015] Mamajanov, I. and Herzfeld, J. (2009). HCN polymers characterized by solid state NMR: chains and sheets formed in the neat liquid, *J. Chem. Phys.* **130**, p. 134503.

[1016] Chall, M., Winkler, B. and Milman, V. (1996). *Ab initio* calculation of the high-pressure behaviour of hydrogen cyanide, *J. Phys.: Condens. Matter* **8**, pp. 9049–9057.

[1017] Khazaei, M., Liang, Y., Bahramy, M. S., Pichierri, F., Esfarjani, K. and Kawazoe, Y. (2011). High-pressure phases of hydrogen cyanide: formation of hydrogenated carbon nitride polymers and layers and their electronic properties, *J. Phys.: Condens. Matter* **23**, p. 405503.

[1018] Cairns, T. L., Larchar, A. W. and McKusick, B. C. (1952). The trimerization of nitriles at high pressures, *J. Am. Chem. Soc.* **74**, pp. 5633–5636.

[1019] Yoo, C. S. and Nicol, M. (1986). Chemical and phase transformations of cyanogen at high pressure, *J. Phys. Chem.* **90**, pp. 6726–6731.

[1020] Yoo, C. S. and Nicol, M. (1986). Kinetics of a pressure-induced polymerization reaction of cyanogen, *J. Phys. Chem.* **90**, pp. 6732–6736.

[1021] Aoki, K., Kakudate, Y., Yoshida, M., Usuba, S. and Fujiwara, S. (1989). High-pressure Raman study of a one-dimensional hydrogen-bonded crystal of cyanoacetylene, *J. Chem. Phys.* **91**, pp. 2814–2817.

[1022] Aoki, K., Kakudate, Y., Yoshida, M., Usuba, S. and Fujiwara, S. (1989). Solid state polymerization of cyanoacetylene into conjugated linear chains under pressure, *J. Chem. Phys.* **91**, pp. 778–782.

[1023] Khazaei, M., Liang, Y., Venkataramanan, N. S. and Kawazoe, Y. (2012). Polymerization of cyanoacetylene under pressure: formation of carbon nitride polymers and bulk structures, *Phys. Rev. B* **85**, p. 054101.

[1024] Santoro, M., Ciabini, L., Bini, R. and Schettino, V. (2003). High-pressure polymerization of phenylacetylene and of the benzene and acetylene moieties, *J. Raman Spectr.* **34**, pp. 557–566.

[1025] Ceppatelli, M., Fontana, L. and Citroni, M. (2007). The high pressure reactivity of substituted acetylenes: a vibrational study on diphenylacetylene, *Phase Trans.* **80**, pp. 1085–1101.

[1026] Pérez Tejedor, F. J., Montoro, O. R., Rodrguez-Jiménez, O., Baonza, V.G. and Núnez Delgado, J. (2002). Polymerization at high pressure of dimethylacetylene: Raman microscopy, vibrational analysis, calculation of force constants and internal coordinates, in Lojkowski, W. (Ed.), *High-pressure Effects in Chemistry, Biology and Materials Science* (Scitec Publications, Zurich).

[1027] Mediavilla, C., Tortajada, J. and Baonza, V. G. (2009). Modeling high pressure reactivity in unsaturated systems: application to dimethylacetylene, *J. Comput. Chem.* **30**, pp. 415–422.

[1028] Gourdain, D., Chervin, J. C. and Pruzan, P. (1996). Transformations of styrene at high pressure, *J. Chem. Phys.* **105**, pp. 9040–9045.

[1029] Walley, E. (1960). Structure of the Bridgman's black carbon disulfide, *Can. J. Chem.* **38**, pp. 2105–2108.

[1030] Butcher, E. G., Alsop, M., Weston, J. A. and Gebbie, H. A. (1963). Formation and properties of the black form of carbon disulfide, *Nature* **199**, pp. 756–758.

[1031] Chan, W. S. and Jonscher, A. K. (1969). Structural and electrical properties of solid polymeric carbon disulphide, *Phys. Stat. Sol.* **32**, pp. 749–761.

[1032] Colman, J. and Trogler, W. (1995). Photopolymerization of carbon disulfide yields the high-pressure phase $(CS_2)x$, *J. Am. Chem. Soc.* **117**, pp. 11270–11277.

[1033] Akahama, Y., Minamoto, Y. and Kawamura, H. (2002). X-ray powder diffraction study of CS_2 at high pressures, *J. Phys.: Condens. Matter* **14**, pp. 10457–10460.

[1034] Bolduan, F., Hocheimer, H. D. and Jodl, H. J. (1986). High pressure Raman study of solid CS_2, *J. Chem. Phys.* **84**, pp. 6997–7004.

[1035] Schatschneider, B., Liang, J. J., Jezowski, S. and Tkatchenko, A. (2012). Phase transitions between cubic and monoclinic polymorphs of tetracyanoethylene crystal: the role of temperature and kinetics, *CrystEngComm.* **14**, pp. 4656–4663.

[1036] Chaplot, S. L. and Mukhopadyay, R. (1986). Monoclinic-amorphous-cubic phase transitions in tetracyanoethylene under high pressure, *Phys. Rev. B* **33**, pp. 5099–5101.

[1037] Mukhopadhyay, R., Deb, S. K., Das, A. and Chaplot, S. L. (2009). Phase transitions at high pressure in tetracyanoethylene, *Sol. State Comm.* **149**, pp. 1914–1918.

[1038] Yamawaki, H., Aoki, K., Kakudate, Y., Yoshida, M., Usuba, S. and Fujiwara, S. (1992). Infrared study of phase transition and chemical reaction in tetracyanoethylene under high pressure, *Chem. Phys. Lett.* **198**, pp. 183–187.

[1039] Rao, R., Sakuntala, T., Deb, S. K. and Mukhopadhyay, R. (2005). Pressure induced transitions in tetracyanoethylene: a Raman scattering study, *J. Phys.: Condens. Matter* **17**, pp. 2633–2644.

[1040] Mukhopadhyay, R. and Chaplot, S. L. (2002). Structural changes in tetracyanoethylene at high pressures: neutron diffraction study, *J. Phys. Condens. Matter* **14**, pp. 759–768.

[1041] Nesting, D. C. and Badding, J. V. (1996). High-pressure synthesis of sp^2-bonded carbon nitrides, *Chem. Mater.* **8**, pp. 1535–1539.

[1042] Ostwald, I. D. H. and Urquhart, A. I. (2011). Polymorphism and polymerization of acrylic and methacrylic acid at high pressure, *CrystEngComm* **13**, pp. 4503–4507.

[1043] Murli, C. and Song, Y. (2010). Pressure-induced polymerization of acrylic acid: a Raman spectroscopy study, *J. Phys. Chem. B* **114**, pp. 9744–9750.

[1044] Cansell, F., Petitet, J. P. and Fabre, D. (1989). Evidence of high pressure stability of solid methylmethacrylate by Raman spectroscopy, *Polym. Comm.* **30**, pp. 234–236.

[1045] Tabata, Y. and Suzuki, T. (1965). Effect of pressure on the radiation-induced solid state polymerization of acrylamide, *Macromol. Chem. Phys.* **81**, pp. 223–233.

[1046] Bradbury, M. G., Hamann, S. D. and Linton, M. (1970). Solid-phase polymerizations at high pressures, *Aust. J. Chem.* **23**, pp. 511–523.

[1047] Peiris, S. M. and Piermarini, G. J (Eds.) (2008). *Static Compression of Energetic Materials* (Springer-Verlag, Berlin Heidelberg).

[1048] Trevino, S. F., Prince, E. and Hubbard, C. R. (1980). Refinement of the structure of nitromethane, *J. Chem. Phys.* **73**, pp. 2996–3000.

[1049] Brasch, J. W. (1980). Irreversible reaction of nitromethane at elevated pressure and temperature, *J. Phys. Chem.* **84**, pp. 2084–2085.

[1050] Piermarini, C. J., Bloch, S. and Miller, P. J. (1989). Effects of pressure on the thermal decomposition kinetics and chemical reactivity of nitromethane, *J. Chem. Phys.* **93**, pp. 457–462.

[1051] Courtecuisse, S., Cansell, F., Fabre, D. and Petitet, J. P. (1995). Phase transitions and chemical transformations of nitromethane up to 350°C and 35 GPa, *J. Chem. Phys.* **102**, pp. 968–974.

[1052] Pruzan, P., Canny, B., Power, C. and Chervin, J. C. (2000). Infrared spectroscopy of nitromethane up to 50 GPa, in Zhang, S. and Zhu, B. (Eds.), *Proceedings of the XVII International Conference on Raman Spectroscopy (ICORS 2000)* (John Wiley and Sons, New York).

[1053] Rice, B. M. and Byrd, E. F. C. (2008). Theoretical chemical characterization of energetic materials, *J. Mater. Res.* **21**, pp. 2444–2452.

[1054] Byrd, E. F. C. and Rice, B. M. (2007). *Ab Initio* study of compressed 1,3,5,7-tetranitro-1,3,5,7-tetraazacyclooctane (HMX), cyclotrimethylenetrinitramine (RDX), 2,4,6,8,10,12-hexanitrohexaazaisowurzitane (CL-20), 2,4,6-trinitro-1,3,5-benzenetriamine (TATB), and pentaerythritol tetranitrate (PETN), *J. Phys. Chem. C* **111**, pp. 2787–2796.

[1055] Dreger, Z. A. (2012). Energetic materials under high pressures and temperatures: stability, polymorphism and decomposition of RDX, *J. Phys. Conf. Series* **377**, p. 012047.

[1056] Miller, P. J., Block, S. and Piermarini, G. J. (1991). Effect of pressure on the thermal decomposition kinetics, chemical reactivity and phase behaviour of RDX, *Combust. Flame* **83**, pp. 174–184.

[1057] McWilliams, R. S., Kadry, Y., Mahmood, M. F., Goncharov, A. F. and Ciezak-Jenkins, J. (2012). Structural and chemical properties on nitrogen-rich energetic material triaminoguanidinium 1-methyl-5-nitriminotetrazolate under pressure, *J. Chem. Phys.* **137**, p. 054501.

[1058] Fanetti, S., Ceppatelli, M., Citroni, M. and Bini, R. (2012). High-pressure photoinduced reactivity of CH_3OH and CD_3OH, *J. Phys. Chem. C* **116**, pp. 2108–2116.

[1059] Offen, H. W. (1966). Fluorescence spectra of several aromatic crystals under high pressures, *J. Chem. Phys.* **44**, pp. 699–703.

[1060] Jones, P. F. and Nicol, M. (1968). Excimer emission of naphthalene, anthracene and phenanthrene produced by very high pressures, *J. Chem. Phys.* **48**, pp. 5440–5447.

[1061] Wang, C. C. and Davis, L. I. (1975). Two-photon dissociation of water: a new OH source for spectroscopic studies, *J. Chem. Phys.* **62**, pp. 53–55.

[1062] Nisini, B. (2000). Water's role in making stars, *Science* **290**, pp. 1513–1514.

[1063] Watanabe, N., Nagaoka, A., Shiraki, T. and Kouchi, A. (2004). Hydrogenation of CO on pure solid CO and CO-H_2O mixed ices, *Astrophys. J.* **616**, pp. 638–642.

[1064] Wu, C. Y. R., Judge, D. L., Cheng, B. M., Shih, W. H., Yih, T. S. and Ip, W. H. (2002). Extreme ultraviolet photon-induced chemical reactions in the C_2H_2-H_2O mixed ices at 10 K, *Icarus* **156**, pp. 456–473.

[1065] Mao, W. L., Mao, H. K., Meng, Y., Eng, P. J., Hu, M. Y., Chow, P., Cai, Y. Q., Shu, J. and Hemley, R. J. (2006). X-ray-induced dissociation of H_2O and formation of an O_2-H_2 alloy at high pressure, *Science* **314**, pp. 636–638.

[1066] Sloan, E. D. Jr. (1998). *Clathrate Hydrates of Natural Gases*, Second edition (Marcel Dekker Inc., New York).

[1067] Ceppatelli, M., Bini, R. and Schettino, V. (2009). High-pressure reactivity of model hydrocarbons driven by near-uv photodissociation of water, *J. Phys. Chem. B* **113**, pp. 14640–14647.

Appendix A

Table A.1: Elastic constants (GPa) of molecular crystals. For trigonal, tetragonal and triclinic crystals some additional elastic constants are reported in the notes. Whenever necessary the pressure of the experiment is reported in the notes

Material	Structure	T(K)	c11	c22	c33	c44	c55	c66	c12	c13	c23	c15	c25	c35	c46	Method	Ref.
Neon	C	5	17			9.5			8.9							comput, Brillouin	1-3
Argon	C	10	42			22.9			24.2							comput, Brillouin	1, 4-6
Krypton	C	10	51.5			26.9			28.5							ultrasonic, comput, Brillouin	1, 7-10
Xenon	C	10	53			29			28							ultrasonic, comput, Brillouin	1, 11, 12
p-H_2	H	5.4	42		51	11			18	5						Brillouin, diffraction, ultrasonic	13-19
o-H_2	C	4	27.3			19.4			14.7							comput	20
o-D_2	H	4	82		102	23			29	9						diffraction	20, 22
p-D_2	C	4	45.8			30.5			26.1							comput	18-20
n-H_2 [a]	H	295	30.7		36.0	6.2			19.2	13.5						Brillouin	13
α-N_2	C	15	2.90			1.35			2.00							ultrasonic, diffraction	21
β-N_2	H		1.83		1.98	0.32			1.13	0.98						Brillouin, comput, ultrasonic	21, 23-25
α-CO	C	61	2.41		1.06				1.44							ultrasonic, diffraction	21
β-CO	H	68	1.9		2.1	0.36			1.15	0.95						Brillouin, ultrasonic	21, 23
γ-O_2	C	54.3	2.57			0.28			2.01							Brillouin, ultrasonic	26-28
CO_2	C	195	12.34			5.51			7.01							Brillouin, comput, diffraction	21, 29-32
H_2O ice I_h	H	257	13.93		15.01	3.01			7.08	5.7						Brillouin, ultrasonic	33-36
D_2O ice I_h	H	263	13.9		16.37	3.53			7.15	4.32						Brillouin	37
H_2O ice III	T	253	15.37		11.55	4.46			9.95	6.51						Brillouin	38, 39
H_2O ice V	M	238	21.4	19.3	21.1	7.5	3.7	7.5	12.2	9.5	11.8	0.17	-0.1	-0.3	-2.1	Brillouin	38
H_2O ice VI [b]	T	300	32.8		27.8	6.3		5.9	11.8	14.7						Brillouin	38, 40-44
H_2O ice VII [c]	C	300	38			21.5			19							Brillouin	40-42, 44
H_2O ice VIII	T	160	28.4		23.8	8.1		5.7	10.6	12.3						comput	45
H_2O ice I_c	C	250	9.2						8.6	2.3						comput	46
H_2S	C	RT	13.6			5.9			10.1							Brillouin	47-50
NH_3 III	C	RT	12.5			6.3			8.4							Brillouin	51, 52

continued . . .

. . . continued

Material	Structure	T(K)	c11	c22	c33	c44	c55	c66	c12	c13	c23	c15	c25	c35	c46	Method	Ref.
NH_3 IV [d]	C	RT	104			23.4			65							Brillouin	51
HF [e]	O	RT	15.1	28.5		9.74		2.56	12							Brillouin	53
HCl [f]	C	300	18			7.5			12.5							Brillouin	54, 55
HBr [f]	C	RT	17.3			8			12.3							Brillouin	56
SF_6	C	220	3.7			1.4			2.4							Brillouin, comput	57, 58
Urea	T	RT	11.73	11.73	53.98	6.19	6.19	10.63	10.72	9.16	9.16					ultrasonic, Brillouin	59-62
Thiourea	O	RT	10.24	25.95	15.03	2.22	6.08	0.57	2.23	5.67	4.43					Brillouin, ultrasonic	59, 63-66
Neopenthane	C	173	4.65			1.68			3.37							diffraction	67
Cyclohexane	C	279	2.58			0.36			2.25							ultrasonic, Brillouin	59[g], 68, 69
Cyclooctane	C	284	2.94			0.24			2.41							Brillouin	26
Benzene	O	250	6.14	6.56	5.83	1.97	3.78	1.53	3.52	4.01	3.9					ultrasonic, comput, Brillouin	59, 62, 70-72
Pyrazine-d_4	O	RT	22.7	9.3	5.1	2.6	2.9	1.5	5.5	1.4	2.45					Diffraction	59[g], 73
Naphtalene	M	298	7.8	9.9	11.9	3.3	2.1	4.15	4.45	3.4	2.3	-0.6	-2.7	2.9	-0.05	ultrasonic, comput, diffrac-tion	30, 59, 74-78
Anthracene	M	RT	8.52	11.7	15.22	2.72	2.42	3.99	6.72	5.9	3.75	-1.92	-1.7	-1.87	1.38	ultrasonic, Brillouin, comput	59[g],76, 79-84
Phenantrene	M	RT	10.1	1.98	13.5	0.29	2.6	0.81	1.9		-1.8	3.4	1.1	-6	1.7	comput	85
Acenaphtene	O	RT	11.05	10.22	9.40	4.25	3.69	4.87	-1.93	3.34	2.34					ultrasonic, diffrac-tion	59, 86
Acenaphtylene	O	0	7.67	7.61	10.23	0.88	0.96	1.02	6.89	3.83	4.51					comput	87
Tolan	M	RT	11.43	8.5	9.29	2.81	2.65	2.09	3.76	1.94	4.15	2	0.2	-1.84	0.45	ultrasonic	59, 88
Diphenyl	M	RT	5.95	6.97	14.6	1.83	2.26	4.11	4.05	2.88	6.11	0.4	0.94	2.02	-0.89	Brillouin, comput	59[g], 89-91
Diphenyl-d_{10}	M	RT	7.58	9.01	18.08	2.13	2.2	4.59	5	3.95	7.75	0.1	1	1.9	0.83	Brillouin	90
p-terphenyl	M	RT	8.18	9.75	26.4	3.36	2.7	5.13	5.5	4.33	8.1	0	-0.4	-3.16	1.49	Brillouin	59[g], 90, 92, 93
p-terphenyl-d_{14}	M	RT	8.20	9.72	26.64	2.94	2.48	5.29	5.5	4.56	8.1	0	-0.4	-3.42	1.59	Brillouin	90
o-terphenyl	O	RT	13.9	6.34	8.07	4.63	3.08	2.94	4.23	4.75	1.39					Brillouin	94
t-Stilbene	M	RT	10.86	9.09	10.5	3.48	4.46	2.28	5.27	4.02	5.21	2.31	0.2	3.03	0.02	ultrasonic	59, 88
t-azobenzene	M	RT	11.97	9.47	14.08	3.49	4.61	2.38	2.81	3.49	5.81	2.83	-2.1	-3	-0.3	ultrasonic	59
Durene	M	RT	9.08	10	10.03	1.84	2.17	7.27	7.75	2.57	3.19	0.11	1.04	-0.13	0.13	Brillouin	59, 95
Iododurene	O	RT	7.69	7.43	7.88	3.31	4.08	3.94	5.86	4.3	4.27					ultrasonic	59[g]
Dibenzyl	M	RT	6.04	6.74	9.8	2.47	2.81	3.57	3.63	4.49	4.97	-0.7	-0.3	-1.85	0.73	ultrasonic	59
Benzyl [h]	TR	RT	10.92		8.23	10.8		2.74	5.44	3.26						ultrasonic	59[g], 96, 97
s-$C_6H_3Cl_3$	O	RT	8.03	10.98	7.89	3.49	3.75	3.38	4.47	3.85	3.88					Brillouin	98, 99

continued . . .

... continued

Material	Structure	T(K)	c11	c22	c33	c44	c55	c66	c12	c13	c23	c15	c25	c35	c46	Method	Ref.
m-$C_6H_4(NO_2)_2$	O	RT	10.7	11.3	20.27	4.37	2.04	5.31	6.3	1.95	3.19					Brillouin	59, 61, 100, 101
m-nitroaniline	O	RT	13.91	10.47	22.07	4.64	12.17	4.26	6.27	9.73	14.07					ultrasonic	102
α-phenazine	M	0	16.82	15.45	18.81	3.93	7.12	12.23	13.64	13.51	12.67	-1.11	2.73	-3.33	5.03	comput	103
Phenothiazine	O	RT	12.8	6.08	3.3	0.72	0.3	2.89	4.38	0.01	-3.45					Brillouin, ultrasonic	104-107
Benzophenone	O	RT	10.7	10.56	7.12	2	1.55	3.59	5.86	5.29	3.3					ultrasonic	59, 108, 109
Diphenyl sulfone	M	RT	13.65	9.29	10.79	5.57	1.57	1.7	2.8	2.8	5.6	-1.5	1.3	0.1	1.32	ultrasonic	59
2,2'-biphenol	M	RT	8.24	9.74	8.63	4.15	2.8	3.18	6.1	5.04	6.13	0.6	0.1	-0.4	-1.26	ultrasonic	59
Oxalic acid dihydrate	M	RT	21.72	11.63	40.09	2.08	8.09	9.16	9.46	12.18	6.14	-0.39	0.96	-7.04	0.36	ultrasonic	59^g
Taurine	M	RT	31.78	46.92	13.99	7.22	9.6	10	16.31	13.47	14.95	-1.35	3.9	0.9	-0.08	ultrasonic	59^g
Succinonitrile	C	RT	5.07			0.66			3.54							ultrasonic	59^g
Succinic acid	M	RT	10.14	13.24	125.27	7.38	3.87	4.47	7.29	4.3	8	-0.54	1.9	-4.31	0.68	ultrasonic	110
PE	O	RT	7.99	9.92	315.92	3.19	1.62	3.62	3.28	1.13	2.14					comput	111-114
PVA^i	M	RT	21.79	11.95	288.47	10.78	1.64	4.28	7.72	2.98	3.30					comput	113
PVF	O	RT	23.60	10.64	238.24	4.40	6.43	2.15	1.92	3.98	2.19					comput	115
Nylon 6 α ^{j}	M	RT	12.44	5.20	312.33	2.33	0.95	3.65	2.64	2.00	0.75					comput	116
Nylon 6 γ ^{k}	M	RT	5.27	17.52	54.96	4.44	2.35	2.75	1.77	0.80	4.47					comput	116
Polypropylene	M	RT	7.78	11.55	42.44	4.02	3.10	3.99	3.91	3.72	3.99	0.90	-0.36	-0.57	-0.12	comput	117
Tartaric acid	M	RT	9.3	19.2	46.5	8.12	8.2	10.6	20.2	36.7	13.9	-12	-4	-0.4	1.4	ultrasonic	59^g, 118
Melamine	M	RT	15.29	39.72	36.26	10.17	10.78	7.33	8.93	11.44	14.31	-3.5	-1.9	-1.34	4.16	ultrasonic	59
L-glutamic acid hydrochloride	O	RT	16.6	20.1	37.63	9.22	8.04	8.37	10.02	10.88	8.65					ultrasonic	119
Succinimide	O	RT	10.54	13.18	10.58	5.48	5.29	3.06	4.36	6.49	7.89					ultrasonic	59
Urotropine	C	RT	16.51			5.03			4.76							ultrasonic	59
Malonic acid ^{m}	TR	RT	16.52	16.97	17.85	7.1	8.05	10.61	13.2	7.6	7.7	-4.7	-5	-3.2	-2.8	ultrasonic	59
Dimethyl malonic acid	T	RT	9.66		13.57	5.59		5.26	6.56	6.92						ultrasonic	59
Maleic acid	M	RT	10.37	26.92	54.91	12.64	0.4	2.04	2.5	1	1.01	0.2	0.9	-1.8	-0.1	ultrasonic	59
Hippuric acid	O	RT	13.16	10.38	23.89	6.9	1.5	4.18	8.36	6.9	9.73					ultrasonic	59, 120
Pivalic acid	C	RT	3.43			0.64			1.38							ultrasonic	59^g
Betaine hydrate	O	RT	18.72	15.57	17.23	6.23	3.74	5.36	8.93	7.49	8.47					ultrasonic	121
3-methyl 4-nitro-pyridine 1-oxide	O	RT	13.29	18.14	12.20	7.8	5.2	5.4	4.9	-2.6	10.6					ultrasonic	59^g
Citric acid	M	RT	20.07	23.22	22.04	8.52	9.67	9.4	6.04	8.98	10.19	2	-2.77	-4.07	-5.46	ultrasonic	59, 122
Citric acid hydrate	O	RT	17.44	33.57	23.6	2.76	7.36	3.75	6.92	6.65	5.67					ultrasonic	59, 122, 123
Nitrilotriacetic acid	M	RT	25.72	30.94	22.97	9.88	5.97	12.1	20.61	9.31	11.16	0.02	5.34	0.38	3.59	ultrasonic	124
Sulphamic acid	O	293	35.51	43.23	41.83	15.82	14.51	16.78	15.99	14.87	16.25					ultrasonic	125
1-rhamnose hydrate	M	RT	38.2	21.9	19.8	5.37	5.02	9.11	16	16.6	8.88	-0.3	1.22	-1.18	0.22	ultrasonic	59^g
2-methyl 1,3-cyclohexandione	O	RT	14.9	50.5	9.1	1.15	1.65	6.43	11.1	3.26	-1.15					Brillouin	126
Norbornylene	H	RT	4.01		4.01	0.91			3.59	3						Brillouin	59^g, 127

continued ...

...continued

Material	Structure	T(K)	c11	c22	c33	c44	c55	c66	c12	c13	c23	c15	c25	c35	c46	Method	Ref.
Phtalic acid	M	RT	21.85	24.09	17.03	9.52	8.32	13.66	9.94	11.13	9.54	-5.86	-3.55	-6.14	-8.6	ultrasonic	59^{g}, 128
Cis -terpine hydrate	O	RT	12.5	9.9	15.3	2.43	2.23	3.46	3.8	6.2	4.1					ultrasonic	59^{g}
P-dichloro benzophenone	M	RT	15.85	14.58	8.92	2.45	1.72	2.94	3.75	3.7	6.02	-1.15	-0.9	-0.64	0.23	ultrasonic	59^{g}
(-)-2-α-methyl-benzyl-amino-5-nitropyridine	M	RT	10.26	12.8	12.66	5.99	3.13	3.08	8.24	9.24	6.32	3.07	5.61	3.89	-0.86	ultrasonic	59^{g}, 129
Chrysazine [n]	T	RT	14		20.4	8.4		9.2	-6.3	1.6						ultrasonic	59^{g}
Benzalazine	O	RT	14.28	7.99	6.36	2.4	1.06	3.24	5.11	3.37	4.94					diffraction, ultrasonic	59^{g}, 130, 131
4-methyl-benzophenone [o]	TRG	RT	8.94		7.48	2.48		2.87	3.19	4.72	0.59					ultrasonic	132
4-methyl-benzophenone	M	RT	9.41	9.26	7.69	5.19	2.83	2.57	5.1	4.26	6.68	-0.7	0.3	0.2	0.66	ultrasonic	132
N-isopropyl carbazole	O	RT	10.03	7.15	8.08	3.56	1.14	2.87	5.1	3.4	5.6					ultrasonic	59^{g}
Carbazole 1,3,5-trinitrobenzene	O	RT	12.13	9.52	14.67	3.92	1.4	2.16	2.58	2.97	7.41					ultrasonic	59^{g}
Anthracene - tetracyanobenzene	M	RT	22.4	14.67	10.91	0.31	4.8	1.59	6.29	5.1	-0.64	-4.81	-1.34	1.23	0.25	ultrasonic	59^{g}
1,3,5-triphenylbenzene	O	RT	7.2	13.53	14.33	6.32	0.97	1.85	4.3	4.21	4.68					ultrasonic, diffraction	59^{g}, 133, 134
Tetraphenil silicon	T	RT	10.76		6.7	3.89		2.43	5.4	5.35						ultrasonic	59
Br_2 [p]	O	250	417	418.6	232.6	140.6	146.1	-15.1	344.8	237.9	235.6					comput	135
I_2	O	0 (P=0)	20.55	30.75	65.78	38.43	3.83	8.77	5.67	-1.62	29.56					comput	136
Pentaerithritol [q]	T	RT	38.98		13.43	4.32		9.77	30.39	6.94						diffraction, ultrasonic Brillouin	59, 61, 137-142
Resorcinol	O	298	8.6	28.8	19.5	3.26	4.35	4	9.5	7.5	19.1						59, 61, 143
β-HMX	M	RT	20.58	19.69	18.24	9.92	7.69	10.67	9.65	9.75	12.93	-0.61	4.89	1.57	4.42	ISTS, Brillouin, comput	144-147
α-HMX	O	0	30.6	23.3	31.4	0.8	3.3	3.3	5.7	13.8	6					comput	147
δ-HMX	H	0	14.5		18	4.4		2.3	10.4	10.4						comput	147
RDX	O	RT	25.6	21.3	19	5.38	4.27	7.27	8.67	5.72	6.4						59, 148-150, 153
PETN	T	RT	16.92		11.62	4.45		3.8	5.14	7.8						Brillouin, ultrasonic	59, 151-154
ϵ CL-20	M	RT	7.70	28.29	28.05	12.64	3.86	4.73	5.69	9.21	-1.22	1.23	1.01	3.07	0.74	Brillouin, comput	155, 156
TNAD [r]	TR	RT	17.2	12.5	18.6	2.5	1.7	6	2.3	1.4	4.8	0.8	0.2	2.5	-0.2	comput	157
Hexamethylen tetramine	C	RT	16.43			5.15			4.33							ultrasonic, diffraction	61, 158, 159

continued ...

…continued

Material	Structure	T(K)	c11	c22	c33	c44	c55	c66	c12	c13	c23	c15	c25	c35	c46	Method	Ref.
Sulphur	O	298	24	20.5	48.3	0.43	0.87	0.76	13.3	17.1		15.9					22, 160, 161
Adamantane	C	RT	6.15			3.4			3.02							ultrasonic	59[g], 162, 163
1-cyanoadamantane	C	294	7.79			1.88			4.47							Brillouin, diffraction	164, 165
Tetracyanoethylene	C	RT	20.35			4.83			1.18							Brillouin	59[g]
TCNQ	M		18.39	12.79	14.24	3.14	11.06	0.28	4.16	6.96	11.83	-2.65	2.11	2.33	0.37	comput	166
N_2O	C	300	13			4.6			8							Brillouin	31
Dianin clathrand [s]	TRG	RT	8.21		10.61	3.33			2.98	5.26			0.6			Brillouin	167, 168
Dianin clathrate I [t]	TRG	RT	8.61		8.46	3.18			3.51	4.29			-0.09			Brillouin	167
Dianin clathrate II [u]	TRG	RT	6.86		7.97	2.94			2.39	2.1			0.24			Brillouin	167
HPTB [v]	TR	0	3.83	3.53	4.77	2.1	1.42	2.14	0.56	-0.34	1.48	-0.34	-0.12	0.38	0.26	comput	169
C_{60}	C(sc)	RT a	14.9			6.6			8.8							ultrasonic, Brillouin, comput	170-176
C_{60}	C(fcc)	275	20.8			10.2			10.1							ultrasonic, comput	173
Lysozyme [w]	T	RT	12.44		12.97	2.97		2.63	7.03	8.36						ultrasonic, Brillouin	177-183
Lysozyme [x]	T	RT	5.72		5.68	1.32		1.16	3.12	3.71						ultrasonic	180
Aspirin I	M	RT	11.55	11.36	10.68	3.67	2.68	4.41	7.37	4.98	5.44	-0.49	0.28	-0.42	-0.08	ultrasonic, Brillouin, comput	184-187
Aspirin II	M	RT	11.93	14.25	14.86	3.95	4.35	2.93	9.75	7.08	6.79	1.57	-0.86	2.2	0.86	comput	187
Primidone B	O	RT	22.9	23.3	10	9.7	3.2	13.3	7.1	3.3	5.4					comput	188
Paracetamol I	M	RT	21.2	14.4	21.6	5.5	5.9	8.9	12.3	10.5	11	-0.4	1.4	2.8	2.3	comput	189
Paracetamol II	O	RT	53.2	26.3	12	3.5	0.7	11.3	21.2	3.6	8.3					comput	189
Carbamazepine	M	RT	10.89	11.47	11.32	3.68	0.85	2.89	4.57	6.45	1.51	0.78	0.23	0.04	-0.08	Brillouin	190
BRL61063-A [y]	TR	RT	22.44	18.60	20.81	6.94	3.98	8.51	12.91	9.28	11.62	3.23	0.79	-0.18	-1.69	comput	191
BRL61063-B	M	RT	17.24	21.25	13.68	8.14	6.35	5.34	13.26	10.21	11.52	-3.28	0.85	1.90	1.28	comput	191
BRL61063-B[y]	TR	RT	15.58	22.22	22.38	7.66	4.55	6.73	1.77	8.08	12.92	1.24	-0.40	0.64	-0.14	comput	191
CD_4	C	85.6	2.06			0.94			1.54							ultrasonic	68, 95, 192, 194
Methane	C	90.4	1.96			0.92			1.45							Brillouin	59[g], 193-195
Methane hydrate I	C (0.2 GPa)	296	12.5			3.6			6.7							Brillouin	196
CCl_4 [z]	TRG	247.7	4.54		4.33	9.5			2.92	2.73						Brillouin	197
CCl_4	C	244	3.57			1.11			3.16							Brillouin	198
CBr_4	C	333	4.22			1.61			3.36							Brillouin	199
CBr_4	M		16.1	14.74	13.39	4.04	4.59	2.67	3.85	5.02	6.26	-1.47	-1.35	2.08	-1.32	comput	200
CBr_4-HPTB clathrate[aa]	TRG	RT	12		16.9	2.9			1.1	3.9			2.1			Brillouin	201

C: cubic; M: monoclinic; O: orthorhombic; T: tetragonal; TR: triclinic; TRG: trigonal;

RT: room temperature;

[a]: elastic constants at 24 GPa;

[b]: elastic constants at 1.23 GPa;

[c]: elastic constants at 2.25 GPa;

[d]: estimated elastic constants at 16 GPa;

[e]: ratios of adiabatic constants to density at 24 GPa;

[f]: values of elastic constants at 2 GPa;

[g]: the constants of this crystal reported in ref. 85 are taken from Lanolt-Börnstein III 11, III 18 and III 29a (1979, 1984, 1992);

[h]: additional constant c_{14}=-0.51;

[i]: additional constants c_{16}=3.76, c_{26}=1.08, c_{36}=-0.29, c_{45}=0.12;

[j]: additional constants c_{16}=4.40, c_{26}=0.82, c_{36}=0.06, c_{45}=-0.39;

[k]: additional constants c_{16}=-0.96, c_{26}=0.67, c_{36}=-0.44, c_{45}=-0.63;

[m]: additional constants c_{14}=3.3, c_{16}=5.0, c_{24}=-5.6, c_{26}=6.4, c_{34}=-4.8, c_{36}=5.1, c_{45}=-4.7, c_{56}=-2.1;

[n]: c_{16}=0.10;

[o]: c_{14}=0.45;

[p]: elastic constants at 75 GPa;

[q]: c_{16}=0.8;

[r]: additional constants: c_{14}=0.1, c_{16}=1.0, c_{24}=2.0, c_{26}=-1.7, c_{34}=0.9, c_{36}=-3.2, c_{45}=-0.4, c_{16}=1.5;

[s]: c_{14}=1.08;

[t]: c_{14}=1.03;

[u]: c_{14}=0.93;

[v]: additional constants c_{14}=0.63, c_{16}=0.20, c_{24}=-0.15, c_{26}=-0.92, c_{34}=-0.32, c_{36}=0.15, c_{45}=-0.06, c_{56}=-0.34;

[w]: elastic constants of lysozyme dehydrated at 42 % hydration;

[x]: elastic constants of lysozyme at 98 % relative humidity;

[y]: the report on the elastic constants for the triclinic polymorphs is not complete;

[z]: c_{14}=-0.12;

[aa]: c_{14}=1.4.

References to Table A.1

[1] Gupta, S. and Goyal, S. C. (2004). Effect of temperature on elastic properties of rare gas solids, *Physica B* **352**, pp. 24–35.

[2] Shimizu, H., Imaeda, H., Kume, T. and Sasaki, S. (2005). High pressure elastic properties of liquid and solid neon to 7 GPa, *Phys. Rev. B* **71**, p. 014108.

[3] McLaren, R. A., Kiefte, H., Landheer, D. and Stoicheff, B. P. (1975). Elastic constants of neon single crystals determined by Brillouin scattering, *Phys. Rev. B* **11**, pp. 1705–1717.

[4] Gsänger, M., Egger, H. and Lüscher, E. (1968). Determination of elastic constants of argon, *Phys. Letters A* **27**, pp. 695–696.

[5] Shimizu, H., Hideyuki, T., Kume, T. and Sasaki, S. (2001). High-pressure elastic properties of solid argon to 70 GPa, *Phys. Rev. Lett.* **81**, pp. 4568–4571.

[6] Polian, A. and Grimsditch, M. (1986). Brillouin scattering from argon at high pressure, *Physica B+C* **139–140**, pp. 187–188.

[7] Korpiun, P., Burmeister, A. and Lüscher, E. (1972). The elastic constants of solid krypton determined from measurements of isothermal compressibility and ultrasound velocities, *J. Phys. Chem Solids* **33**, pp. 1411–1421.

[8] Shimizu, H., Saitoh, N. and Sasaki, S. (1998). High pressure elastic properties of liquid and solid krypton to 8 GPa, *Phys. Rev. B* **57**, pp. 230–233.

[9] Tse, J. S., Shpakov, V. P. and Belosludov, V. R. (1998). High-pressure elastic constants of solid krypton from quasi-harmonic lattice-dynamics calculations, *Phys. Rev. B* **58**, pp. 2365–2368.

[10] Landheer, D., Jackson, H. E., McLaren, R. A. and Stoicheff, B. P. (1976). Elastic constants of krypton single crystals determined by Brillouin scattering, *Phys. Rev. B* **13**, pp. 888–895.

[11] Sasaki, S., Wada, N., Kume, T. and Shimizu, H. (2008). High-pressure Brillouin study on solid xenon, *J. Phys.: Conf. Series* **121**, p. 042013.

[12] Sasaki, S., Wada, N., Kume, T. and Shimizu, H. (2009). High-pressure Brillouin study of the elastic properties of rare gas solid xenon at pressures up to 45 GPa, *J. Raman Spectr.* **40**, pp. 121–127.

[13] Zha, C. S., Duffy, T. S., Mao, H. K. and Hemley, R. J. (1993). Elasticity of hydrogen to 24 GPa from single-crystal Brillouin scattering and synchroton X-ray diffraction, *Phys. Rev. B* **48**, pp. 9246–9255.

[14] Nielsen, M. (1973). Phonons in solid hydrogen and deuterium studied by inelastic coherent neutron scattering, *Phys. Rev. B* **7**, pp. 1626–1635.

[15] Wanner, R. and Meyer, H. (1973). Velocity of sound in solid hexagonal close-packed H_2 and D_2, *J. Low Temp. Phys* **11**, pp. 715–744.

[16] Thomas, P. J., Rand, S. C. and Stoicheff, B. P. (1978). Elastic constants of parahydrrogen determined by Brillouin scattering, *Can. J. Phys.* **56**, pp. 1494–1501.

[17] Goldman, V. V. (1977). Effect of internal strains on the elastic constants of hcp hydrogen and deuterium, *J. Low Temp. Phys.* **26**, pp. 203–209.

[18] Goldman, V. V. (1980). The phonon absorption line shape of solid molecular H_2 and D_2 as a function of density, *J. Low Temp. Phys.* **38**, pp. 149–165.

[19] Goldman, V. V. (1979). Density dependence of the elastic constants of hcp hydrogen and deuterium, *J. Low Temp. Phys.* **368**, pp. 521–526.

[20] Manzhelii, V. G. and Strzhemechny, M. A. (1996). Quantum molecular crystals, in Manzhelii, V. G. and Freiman, Y. A. (Eds.), *Physics of Cryocrystals* (AIP Press, Woodbury).

[21] Erenburg, A. I. and Freiman, Y. A. (1996). Heat capacity and compressibility, in Manzhelii, V. G. and Freiman, Y. A. (Eds.), *Physics of Cryocrystals* (AIP Press, Woodbury).

[22] Hearmon, R. F. S. (1956). The elastic constants of anisotropic materials II, *Adv. Phys.* **5**, pp. 323–382.

[23] Askarpour, V., Klefte, H. and Clouter, M. J. (1988). Brillouin scattering near the $\alpha - \beta$ phase transition of N_2 and CO, *Can. J. Chem.* **66**, pp. 541–548.

[24] Kiefte, H. and Clouter, M. J. (1976). Brillouin scattering in single crystal of β-nitrogen: determination of the elastic constants in the triple point, *J. Chem. Phys.* **64**, pp. 1816–1819.

[25] Goldman, V. V. and Klein, M. L. (1976). Self-consistent phonon calculation of the elastic constants of the phase of solid N_2, *J. Chem. Phys.* **64**, pp. 5121–5125.

[26] Kiefte, H., Askarpour, V., Zuk, J. and Clouter, M. J. (1992). Comparative Brillouin scattering study of γ-oxygen and cyclooctane. I. Rotation-translation coupling effects, *J. Phys. Chem.* **96**, pp. 1449–1455.

[27] Clouter, M. J., Kiefte, H. and Cho, C. W. (1974). Brillouin scattering in crystals of γ-oxygen, *Sol. State Comm.* **14**, pp. 579–581.

[28] Kiefte, H. and Clouter, M. J. (1975). Brillouin scattering in single crystal of γ-oxygen: determination of the elastic constants at the triple point, *J. Chem. Phys.* **62**, pp. 4780–4786.

[29] Shimizu, H., Kitagawa, T. and Sasaki, S. (1993). Acoustic velocities, refractive index and elastic constants of liquid and solid CO_2 at high pressures up to 6 GPa, *Phys. Rev. B* **47**, pp. 567–570.

[30] Walmsley, S. H. (1968). Lattice vibrations and elastic constants of molecular crystals in the pair potential approximation, *J. Chem. Phys.* **48**, pp. 1438–1444.

[31] Shimizu, H., Sakoh, H. and Sasaki, S. (1998). High-pressure Brillouin scattering and elastic properties of liquid and solid N_2O and CO_2, *J. Chem. Phys.* **94**, pp. 670–673.

[32] Pavlides, P., Pugh, D. and Roberts, K. J. (1991). Elastic tensor atom-atom potential calculation for molecular crystals: carbon dioxide (CO_2), *J. Phys. D Appl. Phys.* **24**, pp. 100–104.

[33] Gammom, P. H., Klefte, H., Clouter, M. J. and Denner, W. W. (1983). Elastic constants of artificial and natural ice samples from Brillouin spectroscopy, *J. Glaciol.* **29**, pp. 433–460.

[34] Proctor, T. M. (1966). Low-temperature speed of sound in single-crystal ice, *J. Acoust. Soc. Am.* **39**, pp. 972–977.

[35] Gagnon, R. E., Klefte, H., Clouter, M. J. and Whalley, E. (1988). Pressure dependence of the elastic constants of ice I_h to 2.8 Kbar by Brillouin spectroscopy, *J. Chem. Phys.* **89**, pp. 4522–4528.

[36] Gagnon, R. E., Klefte, H., Clouter, M. J. and Whalley, E. (1987). Elastic constants of ice I_h, up to 2.8 Kbar, by Brillouin spectroscopy, *J. Physique C1* **48**, pp. 23–28.

[37] Mitzdorf, V. and Helmreich, D. (1971). Elastic constants of D_2O ice and variation of intermolecular forces on deuteration, *J. Acoust. Soc. Am.* **49**, pp. 723–728.

[38] Tulk, C. A., Gagnon, R. E., Klefte, H. and Clouter, M. J. (1997). Elastic constants of ice III, V and VI by Brillouin spectroscopy, *J. Phys. Chem. B* **101**, pp. 6154–6157.

[39] Tulk, C. A., Gagnon, R. E., Klefte, H. and Clouter, M. J. (1994). Elastic constants of ice III by Brillouin spectroscopy. *J. Chem. Phys.* **101**, pp. 2350–2354.

[40] Shimizu, H. (1996). Elastic properties of dense H_2O-ices studied by Brillouin spectroscopy, *Physica B* **219–220**, pp. 559–561.

[41] Shimizu, H., Ohnishi, M., Sasaki, S. and Ishibashi, Y. (1995). Cauchy relation in dense H_2O ice VII, *Phys. Rev. Lett.* **74**, pp. 2820–2823.

[42] Shimizu, H., Nabetani, T., Nishiba, T. and Sasaki, S. (1996). High-pressure elastic properties of the VI and VII phases of ice in dense H_2O and D_2O, *Phys. Rev. B* **53**, pp. 6107–6110.

[43] Tulk, C. A., Gagnon, R. E., Klefte, H. and Clouter, M. J. (1996). Elastic constants of ice VI by Brillouin spectroscopy, *J. Chem. Phys.* **104**, pp. 7855–7859.

[44] Baer, B. J., Brown, J. M., Zaug, J. M., Schiferl, D. and Chronister, E. L. (1998). Impulsive stimulated scattering in ice VI and ice VII, *J. Chem. Phys.* **108**, pp. 4540–4544.

[45] Tse, J. S., Shpakov, V. P. and Belosludov, V. R. (1999). Vibrational spectrum, elastic constants and mechanical instability of ice VIII, *J. Phys. Chem.* **111**, pp. 11111–11116.

[46] Shpakov, V. P., Tse, J. S., Belosludov, V. R. and Belosludov, R. V. (1997). Elastic moduli and instability in molecular crystals, *J. Phys. Condens. Matter* **9**, pp. 5853–5865.

[47] Shimizu, H. and Sasaki, S. (1992). High-pressure Brillouin studies of elastic properties of single-crystal H_2S grown in a diamond cell, *Science* **257**, pp. 514–516.

[48] Murase, S., Yanagisawa, M., Sasaki, S., Kume, T. and Shimizu, H. (2002). Development of low-temperature and high-pressure Brillouin scattering spectroscopy and its application to solid form I of hydrogen sulphide, *J. Phys. Condens. Matter* **14**, pp. 11537–11541.

[49] Shimizu, H. and Sasaki, S. (1992). High-pressure Raman and Brillouin scattering in hydrogen bonded crystalline H_2S, *Mol. Cryst. Liq. Cryst.* **218**, pp. 97–102.

[50] Shimizu, H. and Sasaki, S. (1992). New development of high-pressure Brillouin study of elastic properties of molecular solids, *Jpn J. Appl. Phys.* **32**, pp. 355–357.

[51] Gauthier, M., Pruzan, P., Chervin, J. C. and Polian, A. (1988). Brillouin study of liquid and solid ammonia up to 20 GPa, *Sol. State. Commun.* **68**, pp. 149–153.

[52] Kume, T., Daimon, M., Sasaki, S. and Shimizu, H. (1998). High-pressure elastic properties of liquid and solid ammonia, *Phys. Rev. B* **57**, pp. 13347–13350.

[53] Lee, S. A., Pinnick, D. A., Lindsay, S. M. and Hanson, R. C. (1986). Elastic and photoelastic anisotropy of solid HF at high pressure, *Phys. Rev. B* **34**, pp. 2799–2806.

[54] Sasaki, S., Kamabuchi, K., Kume, T. and Shimizu, H.(1999). High-pressure Brillouin study on hydrogen chloride up to 8 GPa, *Physica B* **263**, pp. 666–669.

[55] Shimizu, H., Kamabuchi, K., Kume, T. and Sasaki, S. (1999). High-pressure elastic properties of the orientationally disordered and hydrogen bonded phase of solid HCl, *Phys. Rev. B* **59**, pp. 11727–11732.

[56] Shimizu, H., Kanazawa, M., Kume, T. and Sasaki, S. (1999). High pressure Brillouin study of solid HBr at pressures up to 7 GPa, *J. Chem. Phys.* **111**, pp. 10617–10621.

[57] Kiefte, H., Penney, R. and Clouter, M. J. (1988). Brillouin scattering studies of the SF_6 crystal, *J. Chem. Phys.* **88**, pp. 5846–5849.

[58] Dove, M. T. (1988). The elastic constants of the disordered phase of SF_6: a computer simulation calculation, *Chem. Phys. Lett.* **150**, pp. 303–306.

[59] Haussühl, S. (2001). Elastic and thermoelastic properties of selected organic crystals: acenaphtene, trans-azobenzene, benzophenone, tolane, trans-stilbene, dibenzyl, dephenil sulphone, 2,2'-biphenol, urea, melamine, hexogen, succinimide, penthaerythritol, urotropine, malonic acid, dimethyl malonic acid, maleic acid, hippuric acid, aluminium acetylacetonate, iron acetylacetonate, and tetraphenyl silicon, *Z. Kristallogr.* **216**, pp. 339–353.

[60] Yoshihara, A. and Bernstein, E. R. (1982). Brillouin and Rayleigh studies of urea single crystals, *J. Chem. Phys.* **77**, pp. 5319–5326.

[61] Day, G. M., Price, S. L. and Leslie, M. (2001). Elastic constants calculations for molecular organic crystals, *Cryst. Growth Des.* **1**, pp. 13–27.

[62] Pavlides, P., Pugh, D. and Roberts, K. J. (1991). Elastic-tensor atom-atom potential calculations for molecular crystals: C_6H_6 and $CO(NH_2)_2$, *Acta Cryst. A* **47**, pp. 846–850.

[63] Haussühl, S. and Pähl, M. (1986). Thermoelastic and third-order elastic constants of orthorhombic thiourea $CS(NH_2)_2$, *Zeit. Krist.* **176**, pp. 147–157.

[64] Jakubowski, B. and Ecolivet, C. (1980). Thermal expansion and elastic properties of thiourea crystal at room temperature, *Mol. Cryst. Liq. Cryst.* **62**, pp. 33–39.

[65] Benoit, J. P. and Chapelle, J. P. (1974). Study of the elastic stiffness constants of thiourea by Brillouin scattering, *Sol. State Commun.* **14**, pp. 883–885.

[66] An, C. X., Benoit, J. P., Hauret, G. and Chapelle, J. P. (1979). New results in the study of Brillouin scattering in thiourea, *Sol. State Commun.* **31**, pp. 581–583.

[67] Debeau, M., Depondt, Ph., Hennion, B. and Reichardt, W. (1993). Phonons in orientationally disordered neopentane $C(CD_3)_4$, *J. Phys. I France* **3**, pp. 1617–1627.

[68] Marx, S. V. and Simmons, R. O. (1984). Ultrasonic sound velocities and elastic constants of liquid and crystalline CD_4 and C_6H_{12}, *J. Chem. Phys.* **81**, pp. 944–950.

[69] Ahmad, S. F., Kiefte, K. and Clouter, M. J. (1978). Brillouin scattering in single crystals of plastic cyclohexane, *J. Chem. Phys.* **69**, pp. 5468–5472.

[70] Heseltine, J. C. W., Elliott, D. W. and Wilson, O. B. (1964). Elastic constants of single-crystal benzene, *J. Chem. Phys.* **40**, pp. 2584–2587.

[71] Catti, M. (1985). Calculation of elastic constants by the method of crystal static deformation, *Acta Cryst. A* **41**, pp. 494–500.

[72] Brunel, L. C. (1978). Brillouin scattering from benzene single crystal, *Chem. Phys. Lett.* **54**, pp. 488–489.

[73] Reynolds, P. A. (1973). Lattice dynamics of the pyrazine crystal studied by coherent inelastic neutron scattering, *J. Chem. Phys.* **59**, pp. 2777–2786.

[74] Alexandrov, K. S., Belikova, G. S., Ryzhenkov, A. P., Teslenkov, V. F. and Kitaigorodskii, A. I. (1963). The elastic constants of molecular crystals. The elastic constants of naphtalene. *Sov. Phys. Cryst.* **8**, pp. 164–166.

[75] El Hamamsyb, M., Elnahwy, S., Damask A. C., Taub, H. and Ul, W. B. (1977). Pressure dependence of the lattice parameters of naphthalene up to 5.5 Kbar and a re-evaluation of the lattice constants, *J. Chem. Phys.* **67**, pp. 5501–5504.

[76] Pawley, G. S. (1967). A model for the lattice dynamics of naphthalene and anthracene, *Phys. Stat. Sol.* **20**, pp. 347–360.

[77] Hovi, V. and Mäntysalo, E. (1959). Ultrasonic investigation of the elastic constants of solid carbon dioxide, *Ann. Acad. Sci. Fenn. Ser. A VI Phys.* **24**, pp. 1–11.

[78] Pertsin, A. J. (1985). Thermodynamic properties of solid naphtalene: cell model and quasiharmonic calculations, *J. Chem. Phys. Solids* **46**, pp. 1019–1024.

[79] Afanaseva, G. K., Salexandrov, K. and Kitaigorodskii, A. I. (1967). Elastic constants of anthracene, *Phys. Stat. Sol.* **24**, pp. K61–K63.

[80] Luty, T. and Eckhardt, C. J. (1985). Piezomodulated Raman spectroscopy of molecular crystals: an experimental method for study of the anharmonic properties of solids, *J. Chem. Phys.* **82**, pp. 1515–1521.

[81] Huntington, H. B., Gangoli, S. G. and Mills, J. L. (1969). Ultrasonic measurements of elastic constants of anthracene, *J. Chem. Phys.* **50**, pp. 3844–3849.

[82] Danno, T. and Inokuchi, H. (1968). Dynamic mechanical behaviour of organic molecular crystals. II. Elastic constants of single crystal anthracene, *Bull. Chem. Soc. Japan* **41**, p. 1783.

[83] Dye, R. C. and Eckhardt, C. J. (1989). A complete set of elastic constants of crystalline anthracene by Brillouin scattering, *J. Chem. Phys.* **90**, pp. 2090–2096.

[84] Fukuhara, M., Matsui, A. H. and Takeshima, M. (2000). Low-temperature elastic anomalies in an anthracene single crystal, *Chem. Phys.* **258**, pp. 97–106.

[85] Kulver, R., Brose, K. H. and Eckhardt, C. J. (1987). Calculation of the elastic constants of phenanthrene crystal, *Chem. Phys.* **115**, pp. 239–242.

[86] Chatterjee, S. and Chakraborty, S. (1981). Elastic constants of acenaphthene determined from studies of thermal diffuse scattering of X-rays, *Acta Cryst. A* **37**, pp. 645–649.

[87] He, H. and Welberry, T. R. (1988). Calculation of elastic constants of acenaphthylene, $C_{12}H_8$, using semiempirical atom-atom potentials, *J. Chem. Phys. Solids* **49**, pp. 421–424.

[88] Kitaigorodsky, A. I. (1973). *Molecular Crystals and Molecules* (Academic Press, New York and London).

[89] Luty, T. (1972). Lattice dynamics of biphenyl, *Mol. Cryst. Liq. Cryst.* **17**, pp. 327–354.

[90] Ecolivet, C. and Sanquer, M. (1980). Brillouin scattering in polyphenyls. I. Room temperature study of p-terphenyl (H14, D14) and biphenyl (D10), *J. Chem. Phys.* **72**, pp. 4145–4152.

[91] Ecolivet, C., Sanquer, M., Pellegrin, J. and De Witte, J. (1983). Brillouin scattering in polyphenyls. II. Through the incommensurate phases of biphenyl, *J. Chem. Phys.* **78**, pp. 6317–6324.

[92] Ecolivet, C., Toudic, B. and Sanquer, M. (1984). Acoustic anomalies at the improper ferroelastic phase transition of p-terphenyl, *Ferroelectrics* **54**, pp. 273–276.

[93] Ecolivet, C., Toudic, B. and Sanquer, M. (1984). Brillouin scattering in p-polyphenyls. III. The improper ferroelastic phase transition of p-terphenyl, *J. Chem. Phys.* **81**, pp. 599–606.

[94] Kondo, S., Tamura, K. and Naoki, M. (2000). Brillouin scattering study of o-terphenyl single crystal, *Mol. Cryst. Liq. Cryst.* **338**, pp. 223–241.

[95] Sanquer, M. and Ecolivet, C. (1987). Elastic constants in molecular crystals: experiments and intermolecular potentials, in Lascombe, J. (Ed.), *Dynamics of Molecular Crystals*, (Elsevier, Amsterdam).

[96] Dows, D. A. and Chiang, S. R. (1976). Brillouin scattering in molecular crystals: benzil, in Grossmann, M., Elkomoss, G. and Ringeissen J. (Eds.), *Molecular Spectroscopy of Dense Phases*, (Elsevier, Amsterdam).

[97] Haussühl, S. (1967). Elastische und thermoelastische konstanten von benzil $C_6H_5COCOC_6H_5$, gemessen mit dem schaefer-bergmann-verfahren, *Acta Cryst.* **23**, pp. 666–667.

[98] Swanson, D. L. and Dows, D. A. (1973). Brillouin scattering from 1,3,5-trichlorobenzene single crystal, *Chem. Phys. Lett.* **23**, pp. 430–431.

[99] Swanson, D., Brunel, L. C. and Dows, D. A. (1975). Brillouin scattering by molecular crystals: s-trichlorobenzene, *J. Chem. Phys.* **63**, pp. 3863–3869.

[100] Mitsui, T. and Iio, K. (1980). Brillouin scattering in organic molecular crystal: meta-dinitrobenzene, *Jpn. J. Appl. Phys.* **19**, pp. 2511–2512.

[101] Sarkar, S. B. and Talapatra, S. K. (1987). Elastic constants of m-dinitrobenzene from diffuse reflection of X-rays, *Acta Cryst.* **43**, p. C96.

[102] Bain, A. M., El-Korashi, N., Gilmour, S., Pethrick, R. A. and Sherwood, J. N. (1992). Ultrasonic and piezoelectric investigations on single crystals 3-nitroaniline, *Phil. Mag. B* **66**, pp. 293–305.

[103] Michalski, D. and Echkardt, C. J. (1997). Computational determination of the elastic properties of the α-phenazine crystal, *J. Phys. Chem. B* **101**, pp. 9690–9694.

[104] Sartwell, J. and Eckhardt, C. J. (1993). Brillouin scattering study of the elastic constants of phenothiazine through the phase transition, *Phys. Rev. B* **48**, pp. 12438–12448.

[105] Morita, N., Nakayama, H. and Ishii, K. (1991). Elastic constants of phenothiazine crystal around its ferroelastic phase-transition temperature, *J. Phys. Soc. Jpn* **60**, pp. 2684–2690.

[106] Sartwell, J. and Eckhardt, C. J. (1995). Reply to comment on "Brillouin-scattering study of the elastic constants of phenothiazine through the phase transition", *Phys. Rev. B* **52**, pp. 10635–10636.

[107] Ecolivet, C., Sanquer, M., Ishii, K. and Nakayama, H. (1991). Ferroelasticity in phenothiazine: a Brillouin-scattering study, *Phys. Rev. B* **44**, pp. 4185–4191.

[108] Alexandrov, K. S. and Ryzihova, T. V. (1961). *Sov. Phys. Cryst.* **6**, pp. 228–252.

[109] Pethrick, R. A., Sherwood, J. N. and Choon, S. (1992). Ultrasonic studies of benzophenone single crystals. II elastic constants measurements on a perfect single crystal, *Phil. Mag. A* **65**, pp. 1033–1047.

[110] Haussühl, S. (1995). Beta-succinic acid, a crystal of extreme elastic anisotropy, *Z. Kristallogr.* **210**, pp. 903–904.

[111] Odajima, A. and Maeda, T. (1967). Calculation of the elastic constants and the lattice energy of the polyethylene crystal, *J. Pol. Sci. C Polymer Symposia* **15**, pp. 55–74.

[112] Karasawa, N., Dasgupta, S. and Goddard III, W. A. (1991). Mechanical properties and force field parameters for polyethylene crystal, *J. Phys. Chem.* **95**, pp. 2260–2272.

[113] Tashiro, K., Kobayashi, M. and Tadokoro, H. (1978). Calculation of three-dimensional elastic constants of polymer crystals. 2. Application to orthorhombic polyethylene and poly(vinyl alcohol), *Macromolecules* **11**, pp. 914–918.

[114] Sorensen, R. A., Liau, W. B., Kesner, L. and Boyd, R. H. (1988). Prediction of polymer crystal structures and properties: polyethylene and poly(oxymethylene), *Macromolecules* **21**, pp. 200–208.

[115] Tashiro, K., Kobayashi, M., Tadokoro, H. and Fukada, E. (1980). Calculation of elastic and piezoelectric constants of polymer crystals by a

point charge model: application to poly(vinylidene fluoride) form I, *Macromolecules* **13**, pp. 691–698.

[116] Tashiro, K. and Tadokoro, H. (1981). Calculation of three-dimensional elastic constants of polymer crystals. 3. α and γ forms of nylon 6, *Macromolecules* **14**, pp. 781–785.

[117] Tashiro, K., Kobayashi, M. and Tadokoro, H. (1992). Vibrational spectra and theoretical three-dimensional elastic constants of isotactic polypropylene crystal: an important role of anharmonic vibrations, *Polymer J.* **24**, pp. 899–916.

[118] Küppers, H. and Pies, M. (1984). The elastic constants of tartaric acid, *Acta Cryst. A* **40** supplement, p. C147.

[119] Aleksandrov, K. S. and Haussühl, S. (1975). Elastic and thermoelastic properties of L-glutamic acid hydrochloride, *Z. Kristallogr.* **142**, pp. 328–331.

[120] Alex, A. V. and Philip, J. P. (2004). Ultrasonic measurement of the elastic properties of benzoyl glycine single crystal, *PRAMANA J. Phys.* **62**, pp. 87–94.

[121] Haussühl, S. (1989). Elastic and thermoelastic properties of twelve adducts of betaine, $(CH_3)_3NCH_2COO$, with H_2O, HCl, HBr, HI, HNO_3, H_2SO_4, H_3PO_3, H_3PO_4, 1,4-toluene sulfonic acid, $MnCl_2$ and KBr, *Z. Kristallogr.* **188**, pp. 311–320.

[122] Haussühl, S. and Wang, J. (1999). Elastic constants of citric acid, citric acid hydrate, trilithium citrate tetrahydrate, trisodium citrate pentahydrate and tripotassium citrate hydrate, *Z. Kristallogr.* **214**, pp. 85–89.

[123] Khan, M. S., Narashimhamurty, T. S. and Ramana, Y. V. (1983). Elastic constants of citric acid monohydrate, *Sol. State Commun.* **48**, pp. 169–172.

[124] Haussühl, S. and Richter, U. (1996). Elastic properties and piezoelectric constants of nitriloacetic acid, *Z. Kristallogr.* **211**, pp. 101–105.

[125] Haussühl, E. and Haussühl, S. (1995). Elastic properties of sulfamic acid and sulfamates of Li, Na, K, Rb, Cs, Tl, NH_4, $C(NH_2)_3$ and $(CH_3)_3NCH_2COOH$, *Z. Kristallogr.* **210**, pp. 269–275.

[126] Wasicki, J., Czarnecki, J., Katrusiak, A., Ecolivet, C. and Bertault, M. (1995). The phase transition in 2-methyl-1,3-cyclohexanedione crystal, *J. Phys.: Condens. Matter* **7**, pp. 7489–7500.

[127] Folland, R., Jackson, D. A. and Rajagopal, S. (1975). Light scattering in organic plastic crystals. I. Brillouin scattering in the hexagonal plastic crystal norbornylene, *Mol. Phys.* **30**, pp. 1053–1061.

[128] Haussühl, S. (1991). Physical properties of phtalic acid and of eight salts of phtalic acid with monovalent cations, *Z. Kristallogr.* **196**, pp. 47–60.

[129] Gilmour, S., Pethrick, R. A., Pugh, D. and Sherwood, J. N. (1993). Ultrasonic, piezoelectric and dielectric investigations of (-)-2-α-methylbenzylamino-5-nitropyridine, *Phil. Mag. B* **67**, pp. 855–868.

[130] Joshi, S. K. and Kashyap, B. M. S. (1964). Determination of the elastic constants of benzalazine from thermal diffuse scattering of X-rays, *Acta Cryst.* **17**, pp. 629–632.

[131] Haussühl, S. (1965). Elastische und thermoelastische konstanten von benzalazin, gemessen mit dem schaefer-bergmann-verfahren, *Acta Cryst.* **18**, pp. 980–981.

[132] Klapper, H., Kutzke, H. and Haussühl, S. (2000). Physical properties of stable monoclinic and metastable trigonal 4-methylbenzophenone, *Z. Kristallogr.* **215**, pp. 187–189.

[133] Chandra, S. and Hemkar, M. P. (1973). Evaluation of elastic constants of 1,3,5-triphenylbenzene from thermal diffuse scattering of x-rays, *Acta Cryst. A* **29**, pp. 25–28.

[134] Haussühl, S. (1974). Elastic and thermoelastic properties of 1,3,5-triphenylbenzene derived from ultrasonic wave velocities, *Acta Cryst. A* **30**, pp. 106–107.

[135] Duan, D., Liu, Y., Ma, Y., Liu, Z., Cui, T., Liu, B. and Zou, G. (2007). *Ab initio* study of solid bromine under high pressure, *Phys. Rev. B* **76**, p. 104113.

[136] San, X., Wang, L., Ma, Y., Liu, Z., Cui, T., Liu, B. and Zou, G. (2008). Theoretical calculations of the phase transitions and optical properties of solid iodine under high pressure, *J. Phys.: Condens. Matter* **20**, p. 175225.

[137] Matsuura, H. and Miyazawa, T. (1974). Brillouin scattering and elastic constants of tetragonal pentaerythritol, *Bull. Chem. Soc. Jpn* **47**, pp. 1143–1147.

[138] Nomura, H., Higuchi, K., Kato, S. and Miyahara, Y. (1972). Determination of elastic constants of pentaerythritol crystal by ultrasonic method, *Jpn J. Appl. Phys.* **11**, pp. 304–308.

[139] Shrivastava, R. C. (1962). A note on the elastic properties of pentaerythritol, *Acta Cryst.* **15**, p. 1306.

[140] Shrivastava, R. C. and Chakraborty, S. C. (1962). Determination of the elastic constants of pentaerythritol by X-ray diffraction method, *J. Phys. Soc. Jpn* **17**, pp. 1767–1770.

[141] Ramamoorthy, P., Rajaram, R. K. and Krishnamurthy, N. (2001). Pressure dependence of the electromechanical and vibrational properties of pentaerythritol, *Cryst. Res. Technol.* **36**, pp. 169–182.

[142] Ramamoorthy, P. and Krishnamurthy, N. (1997). Elastic, piezoelectric and dielectric constants of pentaerythritol, *Cryst. Res. Technol.* **32**, pp. 525–536.

[143] Kopstik, V. A. (1960). The dielectric, piezoelectric and elastic parameters of resorcinol monocrystals, *Soviet physics. Crystallography* **4**, pp. 197–200.

[144] Sun, B., Winey, J. M., Gupta, Y. M. and Hooks, D. E. (2009). Determination of second-order elastic constants of cyclotetramethylene tetramine (β-HMX) using impulsive stimulated thermal scattering, *J. Appl. Phys.* **106**, p. 053505.

[145] Stevens, L. L. and Eckhartdt, C. J. (2005). The elastic constants and related properties of β-HMX determined by Brillouin scattering, *J. Chem. Phys.* **122**, p. 174701.

[146] Sewell, T. D., Menikoff, R., Bedrov, D. and Smith, G. D. (2003). A molecular dynamics simulation study of elastic properties of HMX, *J. Chem. Phys* **119**, pp. 7417–7426.

[147] Zamri, A. R. and De, S. (2010). Deformation distribution maps of β-HMX molecular crystal, *J. Phys. D Appl. Phys.* **43**, p. 035404.

[148] Schwarz, R. B., Hooks, D. E., Dick, J. J., Archuleta, J. I. and Martinez, A. R. (2005). Resonant ultrasound spectroscopy measurement of the elastic constants of cyclotrimetylene trinitramine, *J. Appl. Phys.* **98**, p. 056106.

[149] Haycraft, J. J., Stevens, L. L. and Eckhardt, C. J. (2006). The elastic constants and related properties of the energetic material cyclotrimethylene trinitramine (RDX) determined by Brillouin scattering, *J. Chem. Phys.* **124**, p. 024712.

[150] Sewell, T. D. and Bennet, C. M. (2000). Monte Carlo calculations of the elastic moduli and pressure-volume-temperature equation of state of hexahydro-1,3,5-trinitro-1,3,5-triazine, *J. Appl. Phys.* **88**, pp. 88–95.

[151] Stevens, L. L., Hooks, D. E. and Migliori, A. (2010). A comparative evaluation of elasticity in pentaerythritol tetranitrate using Brillouin scattering and resonant ultrasound spectroscopy, *J. Appl. Phys.* **108**, p. 053512.

[152] Winey, J. M. and Gupta, Y. M. (2001). Second-order elastic constants for pentaerythritol tetranitrate single crystals, *J. Appl. Phys.* **90**, pp. 1669–1671.

[153] Sun, B., Winey, J. M., Hemmi, N., Dreger, Z. A., Zimmerman, K. A., Gupta, Y. M., Torchinski, D. H. and Nelson, K. A. (2008). Second-order elastic constants of pentaerytrhritol tetranitrate and cyclotrimethylene trinitramine using impulsive stimulated thermal scattering, *J. Appl. Phys.* **104**, p. 073517.

[154] Zaoui, A. and Sekkal, W. (2001). Molecular dynamics study of mechanical and thermodynamic properties of pentaerythritol tetranitrate, *Sol. State Commun.* **118**, pp. 345–350.

[155] Haycraft, J. J. (2009). The elastic constants and related properties of the epsilon polymorph of the energetic material CL-20 determined by Brillouin scattering, *J. Chem. Phys.* **131**, p. 214501.

[156] Xu, X. J., Xiao, H. M., Xiao, J. J., Zhu, W., Huang, H. and Li, J. S. (2006). Molecular dynamics simulations for pure ϵ-CL-20 and ϵ-CL-20-based PBXs, *J. Phys. Chem. B* **110**, pp. 7203–7207.

[157] Qiu, L., Xiao, H. M., Zhu, W. H., Xiao, J. J. and Zhu, W. (2006). *Ab initio* and molecular dynamics studies of crystalline TNAD (trans-1,4,5,8-tetranitro-1,4,5,8-tatrazadecalin), *J. Phys. Chem. B* **110**, pp. 10651–10661.

[158] Haussühl, S. (1958). Elastiche konstanten von hexamethylentetramin, *Acta Cryst.* **11**, pp. 58–59.

[159] Ramachandran, G. N. and Wooster, W. A. (1951). Determination of elastic constants from diffuse reflections of X-rays. II. Application to some cubic crystals, *Acta Cryst.* **4**, pp. 431–440.

[160] Rao, R. V. G. S. (1950). Elastic constants of orthorhombic sulphur, *Proc. Mat. Sci.* **32**, pp. 275–278.

[161] Sounders, G. A., Yogurtcu, Y. K., McDonald, J. E. and Pawley, G. S. (1986). The elastic behaviour of orthorhombic sulphur under pressure, *Proc. Roy. Soc. London A* **407**, pp. 325–342.

[162] Damien, J. C. (1975). Plastic phase of adamantane ($C_{10}H_{16}$): measurement of the elastic constants study of the temperature dependence, *Sol. State Commun.* **16**, pp. 1271–1277.

[163] Damien, J. C. and Deprez, G. (1976). Light scattering study of dispersion and attenuation of hypersound in adamantane, *Sol. State Commun.* **20**, pp. 161–167.

[164] Bonnet, J. P., Boissier, M., Pelous, J., Vacher, R., Descamps, M. and Sauvajol, J. L. (1985). Glassy and plastic crystals of cyanoadamantane: a Brillouin scattering investigation, *J. Physique Lett.* **46**, pp. 617–621.

[165] Sauvajol, J. L., Lefebvre, J., Amoureux, J. P. and Bee, M. (1982). Dynamical properties of 1-cyanoadamantane in the disordered phase, *J. Phys. C: Sol. State Phys.* **15**, pp. 6523–6532.

[166] Brose, K. H. and Eckhardt, C. J. (1986). Calculation of elastic constants of TCNQ crystals using atom-atom potentials, *Chem. Phys. Lett.* **125**, pp. 235–240.

[167] Sandstedt, C. A., Michalski, D. and Eckhardt, C. J. (2000). Quantitative measurement of guest-host interaction in supramolecular systems: a comparative Brillouin scattering study of the dianin's compound clathrand and two of its isostructural clathrates, *J. Chem. Phys.* **112**, pp. 7606–7614.

[168] Selbo, J. G., Haycraft, J. J. and Eckhardt, C. J. (2003). Elastic and thermodynamic properties of dianin inclusion compounds and their host-guest interactions, *J. Phys. Chem. B* **107**, pp. 11163–11169.

[169] Michalski, D., Swanson, D. R. and Eckhardt, C. J. (1996). Elasticity, bulk modulus, and mode Grüneisen parameters of the HPTB molecular crystal:

computational investigation of a clathrate precursor, *J. Phys. Chem.* **100**, pp. 9506–9511.

[170] Kobelev, N. P., Nikolaev, R. K., Soifer, Y. M. and Khasanov, S. S. (1997). The elastic stiffness matrix of single-crystal C_{60}, *Chem. Phys. Lett.* **276**, pp. 263–265.

[171] Schranz, W., Fuith, A., Dolinar, P., Varhanek, H., Haluska, M. and Kuzmany, H. (1993). Low frequency elastic properties of the structural and freezing transitions in single-crystal C_{60}, *Phys. Rev. Lett.* **71**, pp. 1561–1564.

[172] Burgos, E., Halac, E. and Bonadeo, H. (1994). Intermolecular forces and phase transitions in solid C_{60}, *Phys. Rev. B* **49**, pp. 15544–15549.

[173] Haluska, M., Havlik, D., Kirlinger, G. and Schranz, W. (1999). Acoustic phonon dispersion in single crystal C_{60}, *J. Phys.: Condens. Matter* **11**, pp. 1009–1014.

[174] Kobelev, N. P., Nikolaev, R. K., Sidorov, N. S. and Soifer, Y. M. (2002). The temperature dependence of the elastic moduli of solid C_{60}. *Phys. Sol. State* **44**, pp. 429–431.

[175] Levin, V. M., Blank, V. D., Prokhorov, V. M., Soifer, J. M. and Kobelev, N. P. (2000). Elastic moduli of solid C_{60}: measurement and relationship with nanostructure, *J. Phys. Chem. Solids* **61**, pp. 1017–1024.

[176] Kobelev, N. P., Nikolaev, R. K., Soifer, Y. M. and Khasanov, S. S. (1998). Elastic moduli of single crystal C_{60}, *Phys. Sol. State* **40**, pp. 154–156.

[177] Hinsen, K. (2008). Structural flexibility in proteins: impact of the crystal environment, *Bioinformatics* **24**, pp. 521–528.

[178] Koizumi, H., Tachibana, M. and Kojima, K. (2006). Observation of all the components of elastic constants using tetragonal hen egg-white lysozyme crystals dehydrated at 42% relative humidity, *Phys. Rev. E* **73**, p. 041910.

[179] Speziale, S., Jiang, F., Caylor C. L., Kriminski, S., Zha, C. S., Thorne, R. E. and Duffy, T. S. (2003). Sound velocity and elasticity of tetragonal lysozyme crystals by Brillouin scattering, *Biophys. J.* **85**, pp. 3202–3213.

[180] Fourme, R., Kahn, R., Mezouar, M., Girard, E., Hoerentrup, C., Prange, T. and Ascone, I. (2001). High-pressure protein crystallography (HPPX): instrumentation, methodology and results on lysozyme crystals, *J. Synchro. Rad.* **8**, pp. 1149–1156.

[181] Zamiri, A. and De, S. (2010). Modeling the mechanical response of tetragonal lysozyme crystals, *Langmuir* **26**, pp. 4251–4257.

[182] Koizumi, H., Kawamoto, H., Tachibana, M. and Kojiwa, K. J. (2008). Effect of intracrystalline water on micro-Vickers hardness in tetragonal hen egg-white lysozyme single crystals, *J. Phys. D Appl. Phys.* **41**, p. 074019.

[183] Morozova, T. Y. and Morozov, V. N. (1982). Viscoelasticity of protein crystal as a probe of the mechanical properties of protein molecule: hen egg-white lysozyme, *J. Mol. Biol.* **157**, pp. 173–179.

[184] Bauer, J. D., Hausshül, E., Winkler, B. and Arbeck, D. (2010). Elastic properties, thermal expansion and polymorphism of acetylsalicylic acid, *Cryst. Growth Des.* **10**, pp. 3132–3140.

[185] Kim, Y., Machida, K., Taga, T. and Osaki, K. (1985). Structure redermination and packing analysis of aspirin crystal, *Chem. Pharm. Bull.* **33**, pp. 2641–2647.

[186] Ko, J. H., Lee, K. S., Ike, Y. and Kojima, S. (2008). Elastic properties of aspirin in its crystalline and glassy phases studied by micro-Brillouin scattering, *Chem. Phys. Lett.* **465**, pp. 36–39.

[187] Ko, J. H., Kim, T. H., Lee, K. S. and Kojima, S. (2011). Acoustic properties of aspirin in its various phases and transformation stages studied by Brillouin scattering, *J. Non Cryst. Sol.* **357**, pp. 547–551.

[188] Payne, R. S., Roberts, R. J., Rowe, R. C. McPartlin, M. and Bashal, A. (1996). The mechanical properties of two forms of primidone predicted from their crystal structure, *Int. J. Pharm.* **145**, pp. 165–173.

[189] Day, G. M., Beyer, T. and Price, S. L. (2001). The prediction, morphology, and mechanical properties of the polymorphs of paracetamol, *J. Am. Chem. Soc.* **123**, pp. 5086–5094.

[190] Mohapatra, H. and Eckhardt, C. J. (2008). Elastic constants and related mechanical properties of monoclinic polymorph of the carbamazepine molecular crystal, *J. Phys. Chem. B* **112**, pp. 2293–2298.

[191] Coombes, D. S., Catlow, C. R. A., Gale, J. D., Hardy, M. J. and Saunders, M. R. (2002). Theoretical and experimental investigation of the morphology of pharmaceutical crystals, *J. Pharm. Sci.* **91**, pp. 1652–1658.

[192] Singh, R. K. and Prabhakar, N. V. K. (1988). Temperature dependence of the phonon dynamics of CD_4 in plastic phase, *Nuovo Cimento D* **10**, pp. 989–998.

[193] Sasaki, S., Nakashima, N. and Shimizu, H. (1996). High-pressure Brillouin study on the orientationally disordered phase of methane, *Physica B* **219**, pp. 380–382.

[194] Rand, S. C. and Stoicheff, B. P. (1982). Elastic and photo-elastic constants of CH_4 and CD_4 obtained by Brillouin scattering, *Can. J. Phys.* **60**, pp. 287–298.

[195] Shimizu, H., Nakashima, N. and Sasaki, S.(1996). High-pressure Brillouin scattering and elastic properties of liquid and solid methane, *Phys. Rev. B* **53**, pp. 111–115.

[196] Shimizu, H., Kumazaki, T., Kume, T. and Sasaki, S. (2002). Elasticity of single-crystal methane hydrate at high pressure, *Phys. Rev. B* **65**, p. 212102.

[197] Zuk, J., Kiefte, H. and Clouter, M. J. (1991). Elastic constants of orientationally disordered phase Ib of CCl_4 by Brillouin spectroscopy, *J. Chem. Phys.* **95**, pp. 1950–1953.

[198] Zuk, J., Kiefte, H. and Clouter, M. J. (1990). Elastic constants of the metastable disordered phase Ia of CCl_4 by Brillouin scattering, *J. Chem. Phys.* **92**, pp. 917–922.

[199] Zuk, J., Brake, D. M., Kiefte, H. and Clouter, M. J. (1989). Elastic constants of disordered phase of CBr_4 by Brillouin scattering, *J. Chem. Phys.* **91**, pp. 5285–5290.

[200] Zielinski, P., Fouret, R., Foulon, M. and More, M. (1990). The structure and dynamics of the ordered solid phase of CBr_4, a rigid molecule model, *J. Chem. Phys.* **93**, pp. 1948–1954.

[201] Michalski, D., Mroz, B., Kiefte, H., White, M. A. and Clouter, M. J. (1993). Elastic behaviour of carbon tetrabromide clathrate with hexakis(phenylthio)benzene as determined by Brillouin spectroscopy, *J. Phys. Chem.* **97**, pp. 12949–12953.

Index

$(NSF)_3$, 181

absorption edge, 148, 151, 153, 185, 188
acetaminophen, 6, 41, 55, 56, 59
acetonitrile, 235
acetylene, xi, 175, 177, 181, 218, 220, 221, 237, 238, 254
acrylic acid, 240
activation, xi, 12, 19, 253, 256
  energy, 165, 176
  volume, 167, 168, 170, 172, 173, 223, 224, 236, 245
amorphization, 55, 72, 173, 191, 207, 219, 231, 249
amplitude of motion, 220, 228
anharmonicity, 1, 5, 37, 39, 47
anisotropy, 13, 31, 35–37, 39, 42, 45, 48, 51, 55, 242
anthracene, 36, 187, 250
archaeometric, 18
argon, 2, 11, 70, 118, 124, 145, 158
aromatics, 72, 225, 232, 248
Arrhenius equation, 165
aspirin, 41
Avrami, 175, 176
azide, 196, 200
azulene, 187

belt apparatus, 85, 88
benzene, xii, 6, 36, 48, 55, 72, 76, 146, 149, 153, 179, 181, 187, 191, 219, 225, 226, 237, 248, 250
  phase diagram, 225
BI_3, 216
biopolymers, 39, 61
biosciences, 10
Brillouin scattering, 32, 38, 131, 143, 144
BRL61063, 41, 42
bromine, xi, 11, 50, 216
Buckingham, 75
bulk modulus, 75, 77, 118, 129, 131, 164, 198, 204, 227
butadiene, xii, 172, 173, 191, 211, 218, 223, 224, 230, 248, 252

carbamazepine, 36, 41
carbon dioxide, vii, xi, 11, 35, 123, 128, 147, 203, 209, 213, 244, 246, 251, 258, 259
carbon disulphide, 147, 239
carbon monoxide, 35, 218, 233
Cauchy relations, 30
Challenger voyage, 21
chemical
  equilibria, 13, 15, 22, 159, 162
  potential, 1, 161
  pressure, 184, 219
chlorpropamide, 59
Christoffel equation, 31, 46
citric acid, 37
CL-20, 45
clathrate, 37

hydrate, 35, 158, 196, 257, 258
coesite CO_2, 204
Coherent Raman scattering, 146
cold working, 17, 18
compressibility, 9, 38, 41, 55, 56, 124, 129, 162, 186, 207, 238, 242
coefficient of activation, 167
linear, 41, 45, 59, 61
of liquids, 20
compression
anisotropic, 70, 176
hydrostatic, 84, 89
isothermal, 181, 183, 191, 205, 225, 228
linear, 59, 61, 62
media, 84, 88, 116, 125, 138
static, 79, 241, 244
strain, 72
conical support, 106
cooperative processes, 57, 164, 181, 183, 191, 192
copper, 17–19, 105, 138, 156, 158
cristobalite CO_2, 204
cryogenic loading, 113, 114
crystal engineering, 13
CsI, 6, 214
cubic gauche N_2, 197
cyanoacetylene, 233, 236
cyanogen, 233, 236
cyclohexanedione, 59
cyclopentanedione, 59

density functional theory, 2, 131, 201, 213
diacetylene, 177, 178, 236
diamond, 10, 62, 93, 102, 195, 203
Compton scattering, 149
fluorescence, 107
IR spectrum, 103
nano-polycrystalline, 88
tips, 104, 111
diamond anvil cell, ix, 19, 79, 93, 96, 140
Bassett, 96
hydrothermal, 99
Mao-Bell, 96
membrane-type, 100
Merrill-Bassett, 98
NBS, 94, 96
Syassen-Holzapfel, 98
dianin, 37
diffusion processes, 160, 172–175, 177, 181, 220, 236
dimedone, 60
dimethylacetylene, 238
dinitrobenzene, 36
diphenyl, 35
diphenylacetylene, 233, 237
dipole-dipole interactions, 36, 59
disilane, 217
dissociation energy
C-C, 217
N-N, 196

early metallurgy, 17
elastic
anisotropy, 31, 36, 70
constants, 28, 30–35, 39, 47, 70, 144, 184, 198
limit, 32
moduli, 27, 54
regime, 25
strain, 32
stress tensor, 183
wave velocity, 131
electronic structure, 2, 4, 10, 151, 160, 185, 222, 245
electrostriction, 21, 170, 192
energetic materials, 33, 44, 196, 233, 241, 244
environmental effects, 159, 163, 170, 177
equation of state, 23, 24, 70, 130
benzene, 76
Birch, 74
Birch–Murnagan, 73
methane, 5
Murnaghan, 71
Poirier–Tarantola, 74
Tait, 21, 70
Vinet, 75
equilibrium constant, 161, 165

ethanol, 37, 189, 246
  deuterated, 248
  methanol mixture, 89, 117
ethylene, xii, 55, 175, 178, 181, 211, 218, 222, 248, 254
excimer, 148, 189, 232, 250, 251

food treatment, 10, 21
formic acid, 208
free energy, 24, 32, 71, 72, 74, 246
  CsI, 6
  Gibbs, 164
  Helmholtz, 47, 70
  profile, 193
frictional effects, 83
fullerene, 46, 61, 153
furane, xii, 187, 231, 250

gas loading, 113, 114
gasket, 86, 88, 89, 94, 101, 109
  materials, 111
GeH_4, 217
GeI_4, 216
girdle device, 88
glycine, 61, 68
Grüneisen parameters, 47, 141

H_2S, xi, 51, 145, 215
HBr, 35
HCl, 34, 51
HCN, 177, 235
heteroaromatics, 231
hexaiodobenzene, 216
HMX, 45
HOMO-LUMO gap, 5, 185, 191, 222
HTPB, 37
Hugoniot, 130
  equations, 130
hydrogen, 114, 155, 246
  bonding, 2, 35, 36, 41, 42, 51, 56, 59, 62, 68, 146, 149, 170, 179, 209, 232, 236
  ionization, 211
  melting line, 125, 145
  metallization, xi, 11, 214
  ortho-para conversion, 158
  transfer, 172
hydrostatic
  compression, 65, 74, 101, 113, 116, 117, 136
  limit, 116, 118

ice, 35, 51, 156
indole, xii, 72, 149, 187, 232, 251
infrared spectroscopy, 107, 141
inorganic chemistry, 12
interaction potential, 2, 47, 76
internal energy, ix, 1, 74
inverse relationship, 52
iodine, xi, 11, 50, 215
ionization, 9, 196, 211, 213
isoprene, 218, 225, 230, 256

krypton, 51

large volume cells, 85, 116
laser chemistry, 13
laser heating, 101, 119, 121, 123, 198, 204, 212, 213
lateral shift, 186, 250
lattice
  dynamics, 32
  strain, 65, 68
Le Chatelier principle, 15, 16, 192
Lennard-Jones, 2, 74
linear growth, 220, 222, 224
loading techniques, 113
lysozyme, 39

malonic acid, 37
mechanochemistry, 20
metallization, 11, 142, 214
methacrylic acid, 240
methane, 5, 6, 35, 51, 72
  elastic constants, 50
  hydrate, 35
methanol, 158, 248
minerals, 10, 16, 124, 155
moissanite, 108, 158
molecular crystals, viii–x, xii, 2, 3, 5, 6, 12, 25, 31–33, 40, 47, 55, 59, 76, 193

molecular dynamics, 33, 55, 181, 191
  acetylene, 221
  benzene, 226, 228
  butadiene, 224
  CO, 234
  cyanoacetylene, 237
  ethylene, 222
  H_2S, 215
  nitromethane, 243
  propene, 224
Monte Carlo, 33
multi-anvil devices, 89, 101

naphthalene, 36, 147
neutron
  diffraction, 89, 155, 243
  scattering, 108, 115, 152, 153
NH_3, vii, 35, 51, 213
nitriles, 181, 235
nitrogen, x, 11, 35, 114, 118, 123, 138, 147, 166, 186, 196, 209, 233
  hydrate, 257
  non-molecular, 197
  oxygen mixture, 211
  phase diagram, 197
nitrogen dioxide, 211
nitromethane, 6, 26, 62, 71, 147, 191, 242, 243
nitrosonium nitrate, 211, 213
nitrous oxide, 35, 212, 213
NMR, 157
non-linear spectroscopy, 145
nucleation, 146, 159, 172–174, 176, 180, 219
  rate, 174
nylon, 38, 39

opposed anvils, 19, 86, 93
organic chemistry, 12
oxygen, x, xi, 153, 186, 201, 209
  magnetic ordering, 157
  metallization, 11, 214
  phase diagram, 201

p-benzoquinone, 60, 241
Paris-Edinburgh cell, 89, 115, 155
pentacene, 187, 230
pentaerythritol, 36, 60
PETN, 45
pharmaceuticals, 35, 40, 41
phenacetin, 58
phenylacetylene, 237
phonon, xi, 1, 47
  -phonon coupling, 46
  assistance, 181, 219, 223, 228
  density of states, 154, 155
  dispersion, 152
  dispersion curves, 32
  energy spectrum, 154
  softening, 181
photo-activation, xi, 12, 108, 160, 188, 189, 195, 196, 200, 211, 217, 219, 230, 239, 241, 246, 248, 250, 252, 254, 256
photo-crystallography, 13
photochemistry, 139, 146, 217, 218, 245
photodissociation, 213, 256
piezochromism, 185
piston-cylinder, 19, 80, 82, 85, 96, 98, 128
pneumatic chemistry, 16
poly-vinyl-alcohol, 38
poly-vinylidene-fluoride, 38
polyacetylene, xi, 175, 177, 220
polyacrilonitrile, 237
polyamorphism, 156
polyaromatics, 225, 230
polybutadiene, xii, 224, 252
polycrystals, 53
polyethylene, xii, 38, 39, 175, 178, 222, 254
polyisoprene, 225
polymeric
  carbon dioxide, 204
  carbon monoxide, 233
  nitriles, 235
  nitrogen, 197, 199
  sulphur, 209
polymerization, x, xi, 22, 55, 80, 100, 173, 175, 176, 181, 186, 196, 209, 217

polymers, 33, 35, 38, 39
polymorphism, 10, 35, 40, 45, 61, 177, 203, 244
polynitrogen, 196, 200
polypropylene, 38, 39
polystyrene, 239
pressure
  calibration, 101, 135, 139
  gauge, 82, 89, 128, 129, 137
  intensifiers, 85
  measurement, 128
  units, 80
primidone, 41, 44
propagation, 159, 160, 172, 179, 219
propene, 182, 191, 218, 224, 231, 256, 258
proteins, 39
pyridine, 149, 182, 187, 189, 231, 232, 251

quadrupolar interactions, 197

radial diffraction, 69
Raman spectroscopy, 143
  pressure determination, 138
  pulsed, 145
rare gases, 11, 117, 124, 150
RDX, 45, 245
reaction
  cavity, 178, 184, 185, 219
  coordinate, vii, 169, 170
  hydrothermal, 99
  irreversible, xi, 217
  kinetics, 100, 170, 172, 224
  mechanism, vii, 13, 171, 175, 218, 227, 231, 235, 246
  monitoring, 79, 83
  rate, x, 22, 159, 165, 166, 169, 173
  reversible, 196
  solid state, viii, 12, 139, 160, 172–174, 178, 219
  threshold, 179, 181, 196, 218, 221, 226, 233, 245
  volume, 162, 166, 171, 177, 196, 217
red shift, 5, 148, 187, 191, 230, 249
  R1 and R2 ruby lines, 134
resistive heating, 82, 85, 119, 121
resorcinol, 36
Reuss bound, 53
ruby
  absorption spectrum, 132
  energy levels, 132
  fluorescence, 101, 117, 118, 123, 131, 133, 135, 218, 233
  pressure scales, 136
  structure, 131

sapphire, 108, 146, 156
serine, 61
shear
  deformation, 26, 45, 98, 200, 239, 241
  elastic constants, 37, 43
  modulus, 53
  strain, 219, 245
  strength, 102, 109
  stress, 27, 116
shock compression, 130, 242
silane, 216
Sm fluorescence, 137
SnI_4, 216
sodium oxalate, 61
solid state chemistry, 12
stannane, 217
steric hindrance, 164, 170, 179, 224, 239, 256
stishovite CO_2, 204, 205
strain, 26, 185
  ellipsoid, 65
  Eulerian, 72, 73
  Henky, 72, 74
  Lagrangian, 65, 66, 68, 72
  tensor, 26, 63
  vector, 27
stress, 25, 26, 64, 185
  -strain curve , 25
  deviatoric, 68, 118, 134
  tensor, 26
  vector, 27
styrene, 233, 238
succinic acid, 37

sulphathiazole, 42
sulphur, xi, 209
  metallization, 214
sylvestrene, 225

TAG-MNT, 245
temperature measurement, 121, 122, 125, 128
terphenyl, 35, 49
tetracene, 187
tetracyanoethylene, 181, 239
thermal expansion, 77
thermoelastic constants, 47
thiophene, xii, 231, 232
TNAD, 45
topochemistry, 178, 200, 212, 219, 221, 228, 237
toroidal cells, 85, 88, 89
transition state, 159, 165, 168–172
trioxane, 191, 241
two-photon
  absorption, 148, 187, 188, 246, 249, 257
  induced fluorescence, 189, 249, 251

unsaturated hydrocarbons, xi, 211, 217, 218, 225
unsupported area, 81
urea, 36

van der Waals, 2, 36, 38, 41, 55, 56, 162, 211, 216, 230
vertical shift, 186, 249
vibrational
  frequencies, 25, 32
  pressure gauges, 137
  relaxation, 147
Voigt
  bound, 53
  notation, 27
volume contraction, ix, 1, 2, 24, 49, 162, 170

water, vii, 155
  acetylene mixture, 258
  CO mixture, 257
  compressibility, 20
  elastic constants, 35
  ethane mixture, 258
  ethylene mixture, 258
  hydrocarbons mixtures, 258
  N_2 mixture, 257
  photodissociation, 256, 258
  propene mixture, 258
  pump, 19
  two-photon absorption, 257

X-ray
  absorption, 151
  compression media, 118
  diffraction, 149, 185
  emission, 154
  gasket materials, 111
  inelastic scattering, 152, 153
  magnetic circular dichroism, 154
  pressure gauges, 130, 139
  Raman scattering, 153
  tomography, 154

Young's modulus, 30, 34, 41, 44, 49, 81